We Can Sleep Later

Alfred D. Hershey and the Origins of Molecular Biology

We Can Sleep Later

Alfred D. Hershey and the Origins of Molecular Biology

Edited by

Franklin W. Stahl
University of Oregon, Eugene

Cold Spring Harbor Laboratory Press
Cold Spring Harbor, New York

This volume was conceived by James D. Watson, President of Cold Spring Harbor Laboratory, shortly after the moving Memorial for Al Hershey at the Laboratory. My long association with Al and my respect for his achievements, as well as my continuing relationship with Cold Spring Harbor, prompted me to accept Jim's invitation to be the volume's editor.

Sincere gratitude is extended to all the contributors who so willingly shared their memories of Al. John Inglis and Elizabeth Powers deserve thanks for making this book a reality. Special mention must be made of Clare Bunce, Manager of the CSHL Library Archives, for all of her research and photocopying. In addition, the efforts of Joan Ebert, Project Coordinator, and Patricia Barker, Production Editor, for moving the project along to completion, are gratefully acknowledged.

F.W.S.

We Can Sleep Later
ALFRED D. HERSHEY AND THE
ORIGINS OF MOLECULAR BIOLOGY

Project Coordinator Joan Ebert
Production Editor Patricia Barker
Desktop Editor Daniel de Bruin
Interior Designer Denise Weiss
Cover Designer Tony Urgo

Front cover: Sketch of Alfred D. Hershey by Efraim Racker, 1949.
Back cover: Waring Blendor® used in the Hershey-Chase experiment, 1952 (see p. 137).

Library of Congress Cataloging-in-Publication Data
We can sleep later: Alfred D. Hershey and the origins of molecular biology/edited by Franklin W. Stahl.
    p. cm.
  Includes bibliographical references and index.
  ISBN 0-87969-567-6 (cloth)
  1. Molecular biology. 2. Hershey, A.D. (Alfred Day), 1908- 3. Molecular biologists--Biography. I. Stahl, Franklin W. II. Hershey A.D. (Alfred Day), 1908-

  QH506.W425 2000
  572.8--dc21   99-086195

All Cold Spring Harbor Laboratory Press publications may be ordered directly from Cold Spring Harbor Laboratory Press, 10 Skyline Drive, Plainview, New York 11803. (Phone: 1-800-843-4388 in Continental U.S. and Canada.) All other locations: (516) 349-1930. E-mail: cshpress@cshl.org. For a complete catalog of all Cold Spring Harbor Laboratory Press publications, visit our World Wide Web Site http://www.cshl.org/

Contributors

Seymour Benzer, James G. Boswell Professor of Neuroscience Emeritus (Active), Division of Biology, California Institute of Technology, Pasadena

John Cairns, Former Director of the Cold Spring Harbor Laboratory, Hollygrove House, Wilcote, Charlbury, England

Allan M. Campbell, Department of Biological Sciences, Stanford University, Stanford, California

William F. Dove, George Streisinger Professor of Experimental Biology, McArdle Laboratory, University of Wisconsin, Madison

James D. Ebert, Former President of the Carnegie Institution of Washington, Professor of Biology, Johns Hopkins University, Baltimore, Maryland

Alan Garen, Department of Molecular Biophysics and Biochemistry, Yale University, New Haven, Connecticut

Ira Herskowitz, Department of Biochemistry and Biophysics, University of California, San Francisco

Rollin D. Hotchkiss, Professor Emeritus, Rockefeller University, New York, New York

Gisela Mosig, Department of Molecular Biology, Vanderbilt University, Nashville, Tennessee

Helios Murialdo, Department of Medical Genetics and Microbiology, University of Toronto, Ontario, Canada

Mark Ptashne, Ludwig Professor of Molecular Biology, Memorial Sloan-Kettering Cancer Center, New York, New York

Anna Marie Skalka, Director of the Institute for Cancer Research, Fox Chase Cancer Center, Philadelphia, Pennsylvania

Franklin W. Stahl, Institute of Molecular Biology, University of Oregon, Eugene

Gunther S. Stent, Department of Molecular and Cell Biology, University of California, Berkeley

Neville Symonds, School of Biological Sciences, University of Sussex, Brighton, United Kingdom

Waclaw Szybalski, Department of Oncology, McArdle Laboratory, University of Wisconsin, Madison

Jun-ichi Tomizawa, National Institute of Genetics, Mishima, Japan

Elliott Volkin, Life Sciences Division, Oak Ridge National Laboratory, Oak Ridge, Tennessee

Bruce Wallace, Professor of Genetics Emeritus, Cornell University, Ithaca, New York and University Distinguished Professor Emeritus, Department of Biology, Virginia Polytechnic Institute and Virginia State University, Blacksburg, Virginia

James D. Watson, President, Cold Spring Harbor Laboratory, Cold Spring Harbor, New York

Robert A. Weisberg, Laboratory of Molecular Genetics, National Institute of Child Health and Human Development, National Institutes of Health, Bethesda, Maryland

Norton D. Zinder, John D. Rockefeller, Jr., Professor Emeritus, Rockefeller University, New York, New York

Contents

CARNEGIE INSTITUTION OF WASHINGTON
GENETICS RESEARCH UNIT
Box 200,   COLD SPRING HARBOR, NEW YORK 11724

November 2, 1970

Dear coauthors,

   Nearly all research papers, and certainly all general chapters have to be reviewed, edited, revised by the authors, and then prepared for the printers. At present, this process is taking about six weeks.

   If we dont have your manuscript, please send it. If you are reviewing, please finish. If you are revising, ditto.

   Not only must we have manuscripts now, but the filtration process must be speeded up, both here and there.

   Delay in publication will make everybody unhappy and will be especially unfair to people who had papers ready at the time of the meeting.

   We can sleep later.

                                    A. D. Hershey

(MS returned to you October 22.)

Alfred Day Hershey (1908–1997)

Alfred Hershey:
A New York Times *Tribute*

$\mathcal{M}$OST OF THE BASIC FACTS ABOUT the gene and how it functions were learned through studies of bacteriophages, the viruses of bacteria. Phages came into biological prominence through experiments done in wartime United States by the German physicist, Max Delbrück, and the Italian biologist, Salvador Luria. They believed that in studying how a single phage particle multiplies within a host bacterium to form many identical progeny phages, they were in effect studying naked genes in action. Soon they recruited the American chemist-turned-biologist Alfred Hershey to their way of thinking, and in 1943 the "Phage Group" was born. Of this famous trio, who were to receive in 1969 the Nobel Prize, Hershey was initially the least celebrated.

Al had no trace of Delbrück's almost evangelical charisma or of Luria's candid assertiveness and never welcomed the need to travel and expose his ideas to a wide audience. He framed his experiments to convince himself, not others, that he was on the right track. Then he could enjoy what he called *Hershey Heaven*, doing experiments that he understood would give the same answer upon repetition. Although both he and Luria had independently demonstrated that phages upon multiplying give rise to stable variants (mutants), it was Hershey, then in St. Louis, who in 1948 showed that their genetic determinants (genes) were linearly linked to each other like the genes along chromosomes of higher organisms.

His most famous experiment, however, occurred soon after he moved to the Department of Genetics of the Carnegie Institution of Washington in Cold Spring Harbor on Long Island. There, in 1952, with his assistant, Martha Chase, he showed that phage DNA, not its protein component, contains the phage genes. After a tadpole-shaped phage particle attaches to a bacterium, its DNA enters through a tiny hole while its protein coat remains outside. Key to Hershey's success was showing that viral infection is not affected by violent agitation in a kitchen blender, which removes the empty viral protein shells from the bacterial surface.

Although there was already good evidence from the 1944 announcement of Avery, MacLeod, and McCarty at the Rockefeller Institute that DNA could genetically alter the surface properties of bacteria, its broader significance was unknown. The Hershey-Chase experiment had a much stronger impact than most confirmatory announcements and made me ever more certain that finding the three-dimensional structure of DNA was biology's next most important objective. The finding of the double helix by Francis Crick and me came only 11 months after my receipt of a long Hershey letter describing his blender experiment results. Soon afterward, I brought it to Oxford to excitedly read aloud before a large April meeting on viral multiplication.

An abridged version of this essay appeared in the *New York Times Magazine* on January 4, 1998.

Hershey's extraordinary experimental acumen was last demonstrated through his 1965 finding that the DNA molecule (chromosome) of bacteriophage λ had 20-base-long single-stranded tails at each end. The base sequences of these tails were complementary, allowing them to find each other and form circular λ DNA molecules. This was a bombshell result, because circular molecules had two years before been hypothesized as an intermediate in the integration of λ DNA molecules into bacterial chromosomes.

Key to the esprit of the phage group was its annual late summer meeting at Cold Spring Harbor. It was not for amateurs, and the intellects then exposed had no equal in biology then or even now, in the hurly burly rush of genetic manipulation of today where perhaps a quarter-million individuals think about DNA in the course of their daily lives. Logic, never emotion or commercial consideration, set the tone in those days, and it was always with keen expectation that we awaited Al's often hour-long concluding remarks. Tightly constructed, his summaries struck those of us aware of his acute taciturnity as containing more words than he might have spoken to outsiders over the course of the past year.

By 1970, the basic features of λ DNA replication and functioning, both in its more conventional lytic phase and in its prophage stage, had become known, and a book was needed for their presentation to a broader biological audience. Because of his stature and honest impartiality, everyone wanted Al to be the editor and to see that the 52 different papers said what they should and no more. Through Al's ruthless cutting of unneeded verbiage, the book's length was kept in check and the final volume not painful to hold in one's hands. Working hard to make the changes that Al suggested, the young Harvard star, Mark Ptashne, noted with pleasure that Al had made no further changes in his revised manuscript's first ten pages. Then on page 11, Al wrote, "Start here." Only Al could be so direct and so admired.

This was his last scientific hurrah. Although he was only 63, he soon chose to retire. It bothered me that a mind so focused and inventive would willingly stop doing science, but he lived to his own standards, and the pursuit of new ideas was never easy. New moments of *Hershey Heaven* never lasted long, and his long summers sailing on Georgian Bay were out of place as the number of new scientists seeking gene secrets increased. In retrospect, he did not get out too soon. Recombinant DNA was but two years away, and soon there would have been competitors on all sides.

Retirement saw him first expanding his garden, and I could exchange words with him on my walks past his home. Later, when he became absorbed with computers, he was inside, and I never thought I had something important enough for an excuse to interrupt. I now regret my lack of courage. Al always appreciated others trying to move ahead. In his last month, now 88 and much curved over by arthritis, Al drove his wife to my house for a small gathering. In asking about our daily lives at the lab, Al likely knew that this would be the last time we saw him. The time to go was at hand, and he stopped eating. No one among his friends ever expects to see another who so pushed science to that level of human endurance.

J.D. Watson

SECTION I

Essays

Hershey

FRANKLIN W. STAHL

*A*LFRED DAY HERSHEY DIED AT HIS HOME, in the Village of Laurel Hollow, NY, on May 22, 1997 at the age of 88. Seven weeks later, a number of Al's friends met at Cold Spring Harbor to commemorate his life. The tributes paid on that occasion have informed this Perspective, and some of them are cited in what follows. Full copies are available from the Cold Spring Harbor Laboratory.

Most students of biology know of Hershey — his best known experiment is described in texts of both Biology and Genetics. This work (Hershey and Chase 1952a) provided cogent support for the hypothesis that DNA is the conveyor of genetic information.

The subject of the "Hershey-Chase experiment" was the bacteriophage T2, composed half of protein and half of DNA, a combination compatible with any of the three competing views of the chemical basis of heredity. T2, like many other phages, is a tadpole-shaped virus that initiates an attack on a bacterium by sticking to it with the tip of its tail. The Hershey-Chase experiment used DNA-specific and protein-specific radioactive labels to show that the DNA of the virus then entered the bacterium, while most of the protein could be stripped from the surface of the cell by agitation in a Waring Blendor. Such abused cells produced a normal crop of new phage particles. Previous evidence implicating DNA in heredity had shown that a property of the surface coat of the pneumococcus bacterium could be passed from one strain to another via chemically isolated DNA. The Hershey-Chase observation justified the view that the entire set of hereditary information of a creature was so encoded. This work counted heavily in making Hershey a shareholder, with Max Delbrück (1906–1981) and Salvadore E. Luria (1912–1991), of the 1969 Nobel prize for Physiology or Medicine.

In 1934, Al earned his Ph.D. in the Departments of Chemistry and Bacteriology at Michigan State College with a thesis that described separations of bacterial constituents, identified by the quaint definitions of the times. Except for its evident care and industry, the work was unremarkable, merely part of an ongoing study "... to arrive ultimately at some correlation between the chemical constitution of [Brucella species], and the various phenomena of specificity by them" (Hershey 1934).

Al then assumed an instructorship in Bacteriology and Immunology at Washington University in St. Louis, where he collaborated with Professor J. Bronfenbrenner. From 1936 to 1939, their papers reported studies on the growth of bacterial cultures. Al cer-

---

Reprinted and modified slightly, with permission, from *Genetics* (1998) **149**: 1–6 (copyright Genetics Society of America).

tainly had the background for this work — his thesis research had involved not only the preparation of liver infusion growth medium (from scratch, as was routine in those days), but also the testing of 600 better defined media, none of which supported growth of Brucella. From 1940 to 1944, his experiments dealt with the phage-antiphage immunologic reaction and with other factors that influenced phage infectivity. During both those periods, about half of the 28 papers bearing Hershey's name were sole-authored. (It was apparently here that Al learned how to handle phage. It may also have been here that Al acquired the idea that authorship belongs to those who do the experiments and should not reflect patronization, rank, title, or even redaction of the manuscript.) Some of these papers may have been important contributions to the understanding of antigen-antibody reactions. To this reviewer they appear original, thoughtful, and quantitative, especially those on the use of phage inactivation to permit the study of the antigen-antibody reaction at "infinite" dilution of antigen (e.g., Hershey 1941). But, of course, they interested an audience that did not include many geneticists or others interested in biological replication (except, perhaps, for Linus Pauling). It took Max Delbrück to move Al in that direction.

As recounted by Judson (1996), Delbrück was attracted by Al's papers. Perhaps he liked their mathematical, nonbiochemical nature. He must have liked their originality, logical precision, and economy of presentation. Max invited Al to Nashville in 1943 and recorded the following impression: "Drinks whiskey but not tea. Simple and to the point … Likes independence." Al's first "interesting" phage papers appeared soon thereafter (Hershey 1946, 1947).

A sine qua non for genetic investigation is the availability of mutants. The ease with which large numbers of phage particles can be handled facilitated the discovery and characterization of mutants that were easily scored. Al recognized that the high infectivity of phage and the proportionality of plaque count to volume of suspension assayed allowed quantification of mutation far exceeding that possible in most other viral systems. Al measured mutation rates, both forward and back, and demonstrated the mutational independence of r (rapid lysis) and h (host range). He succeeded also in showing (in parallel with Delbrück) that these mutationally independent factors could recombine when two genotypes were grown together in the same host cells (Hershey 1946, 1947). Thus "Phage Genetics" was born as a field of study, and it became conceivable that not only could the basic question of biological replication be addressed with phage but so also could phenomena embraced by the term "Morgan-Mendelism".

Al continued the formal genetic analyses of T2 with investigations of linkage. Hershey and Rotman (1948) demonstrated that linkage analysis would have to take into account the production of recombinant particles containing markers from three different infecting phage genotypes. The same authors (1949) used "mixed indicators" to enumerate all four genotypes from two-factor crosses involving h and r mutants. That trick made it feasible to analyze fully the yields from individual mixedly infected bacterial cells. The signal finding was that all four genotypes of phage could be produced by an individual cell but that the numbers of complementary recombinants, which were equal on the average, showed little correlation from cell to cell. This demonstration of apparent nonreciprocality in the exchange process leading to recombination raised the specter that "crossing over" in phages would prove to be fundamentally different from that occurring in meiosis. The desire to unify this and other apparently disparate properties of phage and eukaryotic recombination into a single theoretical framework motivated subsequent studies of recombination by other investigators.

Delbrück tried to make such a unification by algebraic legerdemain and Papal Bull (see below). He formalized phage recombination as a succession of meiosis-like, pairwise exchanges between linear linkage structures (Visconti and Delbrück 1953). The resulting algebra embraced some of the major ways in which phage linkage data differed from meiotic data. In particular, it rationalized the "negative interference" between crossovers and the appearance of progeny particles that had inherited markers from three different infecting phage types. Visconti and Delbrück assumed that the exchange process involved physical breakage of chromosomes (DNA duplexes) and reciprocal reunion of the resulting fragments. They blamed the failure to see correlated numbers of complementary types in "single bursts" on vagaries of replication and packaging (into proteinaceous "heads") of the complementary recombinant chromosome types subsequent to their formation. Delbrück stopped thinking about phage genetics after Charley Steinberg and I (1958) showed him that his final expressions were independent of both of his two assumptions, reciprocal and break-join. A fully satisfactory conclusion to these issues for T2 came only with Mosig's and Albert's elucidation of the nonreciprocal, replicative mechanism by which recombinants are formed in T-even phages (for review, see Mosig 1994). More recent advances have established that the array of recombinational mechanisms employed by bacteria and their viruses and by eukaryotes are, in fact, pleasingly similar, depending on homologous strategies and enzymes.

By most criteria, individual T2 particles are "haploid" — they contain but one set of genetic material. However, "heterozygous" particles, which contain two different alleles at a single locus, were described by Hershey and Chase (1952b) at the 1951 Cold Spring Harbor Symposium. Following the elucidation of DNA as a duplex molecule (Watson and Crick 1953), it was possible to propose heteroduplex models for those "heterozygotes." Such models played a central role in all subsequent thinking about recombination, especially that involving relationships between meiotic crossing over and gene conversion.

In 1958, Hershey, like Levinthal (1954) before him, expanded on the Visconti-Delbrück analysis in an effort to connect observations on heterozygotes, which had molecular implications, with formal concepts proposed to deal with the populational aspects of phage crosses. The effort provided few, if any, answers, but clarified the ambient questions, at least for Al.

In 1950, Al left St. Louis to join the staff of the Carnegie Institution of Washington at Cold Spring Harbor. That move put him at the geographical center of the embryonic field of microbial/molecular genetics, and he soon became the intellectual center of its phage branch. At this time, the fruits of a collaboration conducted at St. Louis were published (Hershey et al. 1951). This work showed that phage particles were "killed" by the decay of the unstable isotope ^{32}P incorporated within their DNA. After the central importance of DNA to the phage life cycle (and to genetics) had been demonstrated, this "suicide" technique was exploited in other labs in efforts to analyze the phage genetic structure and its mode of replication. Like most early experiments in "radiobiology," these analyses were fun, but not much more. From this time on, Al's studies became more down-to-earth (and successful) as he turned from mathematically based genetic analyses to serious studies of phage structure and the biochemistry of phage development. There is no doubt, however, that these studies were informed by Al's acute awareness of the genetical and radiobiological facts that had to be explained. These new studies were jump-started by the "Blender" experiment, described above.

Several subsequent papers refined the conclusions of the Blender experiment by showing, for instance, that *some* protein is injected along with the phage DNA (Hershey 1955). With Watson-Crickery well established by this time, these studies were interesting, but not threatening to the view that the genetic substance was DNA. During this period, Al's lab published works that described DNA and protein production, and relations between them, in infected cells. They provided the biochemical counterpart of the genetically defined notion of a pool of noninfective, "vegetative" phage (Visconti and Delbrück 1953; Doermann 1953). This change of emphasis allowed Hershey (1956) to write,

> "I have proposed the ideas that the nucleic acid of T2 is its hereditary substance and that all its nucleic acid is genetically potent. The evidence supporting these ideas is straightforward but inconclusive. Their principal value is pragmatic. They have given rise to the unprecedented circumstance that chemical hypotheses and the results of chemical experiments are dictating the conditions of genetic experiments. This development I regard as more important than the bare facts I have presented, which may yet prove to be of little or no genetic interest."

Biochemical studies on phage development were clouded by the lack of understanding of phage genome structure. It was not even clear how many "chromosomes" (DNA molecules) a phage particle contained. Furthermore, although Watson and Crick had specified what any short stretch of DNA should look like (plectonemically coiled complementary polynucleotide chains), they had been understandably proud of the fact that their model was structurally coherent in the absence of any specification of longitudinal differentiation. For them, it was enough to say that therein lay genetic specificity. For Al, that was not enough, and his lab pursued studies dedicated to the physical description of phage DNA. The results of these studies were succinctly reviewed by Hershey (1970a) in his Nobel Lecture. I'll briefly summarize my view of them, dividing the studies by phage type.

Al developed and applied chromatographic and centrifugal methods to the analysis of T2 chromosome structure (e.g., Mandell and Hershey 1960; Hershey and Burgi 1960). This work systematized our understanding of the breakage of DNA during lab-

oratory manipulation and had its denouement in the demonstration that a T2 particle contains just one piece of DNA (Rubenstein et al. 1961) with the length expected
of a linear double helix (Cairns 1962). That conclusion was in apparent contradiction
to genetical demonstrations that T4 chromosomes contained more-or-less randomly
located physical discontinuities (Doermann and Boehner 1963). A major insight into
the structure of T-even phage chromosomes resulted from attempts to reconcile the
apparently contradictory physical and genetical descriptions of T-even chromosomes.
The basic idea, elaborated and confirmed in a series of papers orchestrated by George
Streisinger, was that the nucleotide sequences in any clone of phage particles were circularly permuted and that the sequence at one end of a given chromosome was duplicated at the other end (the chromosomes were "terminally redundant"). The predicted circular linkage map provided an elegant frame for displaying the functional organization of the T4 chromosome, as revealed by the pioneering studies of Epstein et al.
(1963).

The terminal redundancies of the T-even phage chromosomes provided an additional physical basis for Al's heterozygotes. (See Streisinger 1966, for references and a
more detailed recounting.) These insights were exploited and elaborated upon by
Gisela Mosig, who spent the years 1962–1965 in Al's lab. There she combined her
genetical savvy of T4 with studies on the structure of the truncated, circularly permuted DNA molecules that she discovered in certain defective T4 particles. Those
studies formed the basis for an elegant demonstration of the quantitative relations
between the linkage map of T4 (as constructed from recombination frequencies) and
the underlying chromosome (Mosig 1966). Fred Frankel (1963) and Rudy Werner
(1968), in Hershey's lab, examined the intracellular state of T-even phage DNA. Their
discovery that it was a network undermined analyses of phage recombination as a
series of tidy, pairwise, meiosis-like "matings," and well aimed triparental crosses by
Jan Drake (1967) killed the pairwise mating idea once and for all.

Meselson and Weigle (1961) demonstrated that phage λ DNA, like that of *E. coli*
(Meselson and Stahl 1958), is replicated semiconservatively, in agreement with Watson
and Crick's proposal that the replication of DNA involves separation of the two, complementary strands. However, uncertainties about the structure of the semiconserved
entities identified by Meselson prevented those experiments from being taken as proof
of the Watson-Crick scheme. Careful measurements of the molecular weight of λ's
DNA (Hershey et al. 1961) demonstrated that there was just one molecule per particle. That conclusion, combined with Cairns's autoradiographic measurement of the
length of λ DNA (1962), established that λ's semiconservatively replicating structure
is, indeed, a DNA duplex, putting the issue to rest.

The chromosome of λ also provided a surprise (Hershey et al. 1963; Hershey and
Burgi 1965). Although the chromosomes in a λ clone are all identical (i.e., not permuted), each chromosome carries a terminal 12-nucleotide-long segment that is single-stranded and is complementary to a segment of the same length carried on the
other end. The complementary nature of the segments gives λ "sticky ends." These
ends anneal at the time of infection, circularizing the chromosome, which can then

replicate in both theta and sigma modes. The demonstration of a route by which the λ chromosome can circularize provided physical substance to Campbell's (1962) proposal that the attachment of λ prophage to the host chromosome involves crossing over between the host chromosome and a (hypothesized) circular form of λ. And, of course, the understanding of λ's sticky ends, whose annealing creates *cos*, is exploited by today's gene cloners everytime they work with a cosmid.

The nonpermuted character of λ's chromosome made it susceptible to analyses prohibited in T-even phage. For instance, Skalka et al. (1968) demonstrated the mosaic nature of the chromosome — major segments differed conspicuously from each other in their nucleotide composition. (That conclusion foreshadowed our current understanding of the role of "horizontal transmission" in prokaryotic evolution.) Hershey's lab demonstrated that these differing segments had distinguishable annealing ("hybridization") behavior. They exploited those differences to identify the approximate location of the origin of replication (Makover 1968) and to identify regions of the chromosome that were transcribed when λ was in the prophage state (Bear and Skalka 1969).

Al appreciated that progress in science depends on the development of new methods. Among those to which Al's lab made important contributions were fixed-angle Cs gradients, methylated albumin columns for fractionating DNA, methods of handling DNA that avoid breakage and denaturation as well as methods that would break phage chromosomes into halves and quarters, and the calibration of methods for measuring molecular weights of DNA. Al confessed that the development of a method was painful — his view of heaven was a "place" where a new method, finally mastered, could be applied over and over. Bill Dove quoted Al as saying "There is nothing more satisfying to me than developing a method. Ideas come and go, but a method lasts."

Al occasionally blessed us with his thoughts about the deeper significance of things. His papers "Bacteriophage T2: Parasite or Organelle" (1957), "Idiosyncrasies of DNA structure" (1970a), and "Genes and Hereditary Characteristics" (1970b) delighted his contemporaries and can still be read with pleasure and profit.

But how many people really knew Al? From his works, we can say he was interested in this or that, but such a contention might leave the impression that we have adequately summarized his interests. That is hardly likely. Each of Al's contributions was truly original — he never copied even himself! Consequently, each paper was a surprise to us. We can surmise, therefore, that his published works do not begin to saturate the library of ideas available to him. His papers must be but a small sampling of his scientific thoughts.

And the rest of his mind? Who knows? Al exemplified reticence. His economy of speech was greater even than his economy of writing. If we asked him a question in a social gathering, we could usually get an answer, like "yes" or "no." However, at a scientific meeting one might get no answer at all, which was probably Al's way of saying, in the fewest possible words, that he had no thoughts on that subject suitable for communication at this time.

Encounters with Al were rare, considering that he worked at Cold Spring Harbor, which hosted hundreds of visitors every summer. That's because Al spent his summers sailing in Michigan, and, except at occasional Symposia or the annual Phage Meetings, which came early and late in the season, he was not to be seen.

Thus, most of us who valued Al as a colleague and acquaintance didn't really know him. I am one of those, and I suppose that status qualifies me for this assignment — the Al about whom I write is the same Al that most other people did not really know, either.

The Phage Church, as we were sometimes called, was led by the Trinity of Delbrück, Luria, Hershey. Delbrück's status as Founder and his *ex cathedra* manner made him the Pope, of course, and Luria was the hard-working, socially sensitive priest/confessor. And Al was the saint. Why? How could we canonize Al when we hardly knew him?

Maybe some of the following considerations apply: The logic of Al's analyses was impeccable. He was original, but the relevance of his work to the interests of the rest of us was always apparent; he contributed to and borrowed from the communal store-house of understanding, casual about labeling his own contributions but scrupulous about attributing the ones he borrowed. He was industrious (compulsively so — each day he worked two shifts). He was a superb editor (*e.g.*, Hershey 1971) and critic, dev-astatingly accurate but never too harsh; he deplored that gratuitous proliferation of words which both reflects and contributes to sloppiness of thought. And his suggestions were always helpful.

Does that qualify him for sainthood? It would do if he was in all other respects perfect. And he may have been. Who could tell? Who among us knew this quiet man well enough to know if there was a dark side? Perhaps canonization was a mark of our deep respect for this quintessential scientist. Maybe, by canonizing Al we could accept the relative insignificance of our own contributions. Maybe we were just having fun.

But, in his papers, Saint Al was *there*. He talked to the reader, explaining things as he saw them, but never letting us forget that he was transmitting provisional under-standing. We got no free rides, no revealed truths, no invitation to surrender our own judgment. And we could never skim, since *every* word was important. I think this style reflected his verbal reticence, which, in turn, mirrored his modesty. Examples: "Some clarification, at least in the mind of the author, of the concepts 'reversible' and 'irre-versible' has been achieved" (Hershey 1943). "On this question we have had more opportunity in this paper to discover than to attack difficulties" (Hershey 1944). Al's modesty was dramatically documented by Jim Ebert, who recalled that Al, whose research support was guaranteed by the Carnegie Institution, argued with Carnegie Directors for the right to apply for NIH support so that he might benefit from the cri-tiques of his peers.

In science, Al appeared to be fearless. Fearlessness and modesty might seem an unlikely combination. Not so. Modesty is kin to a lack of pretense. In the absence of pretense, there is nothing to fear.

Tastes of the many flavors of Hershey's mind and the accomplishments of his lab-oratory can be best gained from the Annual Reports of the Director of the Genetics Research Unit, Carnegie Institution of Washington Yearbook. The principal investiga-tors of this Unit were he and Barbara McClintock. In 1963 Al wrote, "Our justification for existence as a Unit, however, resides in the value of our research. We like to think that much of that value is as unstatable and as durable as other human produce that cannot be sold. Some can be put on paper, however. That we offer with the usual human mixture of pride and diffidence." Those who worked with Hershey at Cold Spring Harbor include Phyllis Bear, Elizabeth Burgi, John Cairns, Connie Chadwick, Martha Chase, Carlo Cocito, Rick Davern, Gus Doermann, Ruth Ehring, Stanley Forman, Fred Frankel, Dorothy Fraser, Alan Garen, Eddie Goldberg, June Dixon Hudis, Laura Ingraham, Gebhard Koch, André Kozinsky, Nada Ledinko, Cy Levinthal, Shraga Makover, Joe Mandell, Norman Melechen, Teiichi Minagawa, Gisela Mosig, David Parma, Catherine Roesel, Irwin Rubenstein, Ed Simon, Ann Skalka, Mervyn Smith, George Streisinger, Neville Symonds, René Thomas, Jun-ichi Tomizawa, Nick Visconti, Bob Weisberg, Rudy Werner, Frances Womack, and Hideo Yamagishi.

In his retirement, Al cultivated an interest in computers and he renewed his youthful interest in music. He is survived by his wife, Harriet Davidson Hershey, and by his son, Peter.

References

Bear P.D. and Skalka A. 1969. The molecular origin of lambda prophage mRNA. *Proc. Natl. Acad. Sci.* **62:** 385–388.

Cairns J. 1961. An estimate of the length of the DNA molecule of T2 bacteriophage by autoradiography. *J. Mol. Biol.* **3:** 756–761.

Cairns J. 1962. Proof that the replication of DNA involves separation of the strands. *Nature* **194:** 1274.

Campbell A. 1962. Episomes. *Advan. Genet.* **11:** 101–145.

Doermann A. H. 1953. The vegetative state in the life cycle of bacteriophage: Evidence for its occurrence and its genetic characterization. *Cold Spring Harbor Symp. Quant. Biol.* **18:** 3–11.

Doermann A.H. and Boehner L. 1963. An experimental analysis of bacteriophage T4 heterozygotes. I. Mottled plaques from crosses involving six *r*II loci. *Virology* **21:** 551–567.

Drake J.W. 1967. The length of the homologous pairing region for genetic recombination in bacteriophage T4. *Proc. Natl. Acad. Sci.* **58:** 962–966.

Epstein R.H., Bolle A., Steinberg C.M., Kellenberger E., Edgar R.S., Susman M., Denhardt G.H. and Lielausis A. 1963. Physiological studies of conditional lethal mutants of bacteriophage T4D. *Cold Spring Harbor Symp. Quant. Biol.* **28:** 375–392.

Frankel F. 1963. An unusual DNA extracted from bacteria infected with phage T2. *Proc. Natl. Acad. Sci.* **49:** 366–372.

Hershey A.D. 1934. The chemical separation of some cellular constituents of the Brucella group of micro-organisms. Ph.D. thesis, Michigan State College. Published in co-authorship with R. C. Huston and I. F. Huddleson as *Technical Bulletin No. 137* of the Michigan Agricultural Experiment Station.

Hershey A.D. 1941. The absolute rate of the phage-antiphage reaction. *J. Immunol.* **41:** 299–319.

Hershey A.D. 1943. Specific precipitation. V. Irreversible systems. *J. Immunol.* **46:** 249–261.

Hershey A.D. 1944. Specific precipitation. VI. The restricted system bivalent antigen, bivalent antibody, as an example of reversible bifunctional polymerization. *J. Immunol.* **48:** 381–401.

Hershey A.D. 1946. Mutation of bacteriophage with respect to type of plaque. *Genetics* **31:** 620–640.

Hershey A.D. 1947. Spontaneous mutations in bacterial viruses. *Cold Spring Harbor Symp. Quant. Biol.* **11:** 67–77.

Hershey A.D. 1955. An upper limit to the protein content of the germinal substance of bacteriophage T2. *Virology* **1:** 108–127.

Hershey A.D. 1956. The organization of genetic material in bacteriophage T2. *Brookhaven Symp. Biol.* **8:** 6–14.

Hershey A.D. 1957. Bacteriophage T2: parasite or organelle. *The Harvey Lectures, Series LI*, pp. 229–239. Academic Press, New York.

Hershey A.D. 1958. The production of recombinants in phage crosses. *Cold Spring Harbor Symp. Quant. Biol.* **23:** 19–46.

Hershey A.D. 1963. Annual Report of the Director of the Genetics Research Unit, *Carnegie Institution of Washington Yearbook* **62:** 461–500.

Hershey A.D. 1970a. Idiosyncrasies of DNA structure (Nobel Lecture). *Science* **168:** 1425–1427.

Hershey A.D. 1970b. Genes and hereditary characteristics. *Nature* **226:** 697–700.

Hershey A.D. (Editor) 1971. *The Bacteriophage Lambda*. Cold Spring Harbor Laboratory Press, NY.

Hershey A.D. and Burgi E. 1960. Molecular homogeneity of the deoxyribonucleic acid of phage T2. *J. Mol. Biol.* **2:** 143–152.

Hershey A.D. and Burgi E. 1965. Complementary structure of interacting sites at the ends of lambda DNA molecules. *Proc. Natl. Acad. Sci. USA* **53:** 325–328.

Hershey A.D. and Chase M. 1952a. Independent functions of viral protein and nucleic acid in growth of bacteriophage. *J. Gen. Physiol.* **36:** 39–56.

Hershey A.D. and Chase M. 1952b. Genetic recombination and heterozygosis in bacteriophage. *Cold Spring Harbor Symp. Quant. Biol.* **16:** 471–479.

Hershey A.D., and Rotman, R. 1948. Linkage among genes controlling inhibition of lysis in a bacterial virus. *Proc. Natl. Acad. Sci.* **34:** 89–96.

Hershey A.D. and Rotman R. 1949. Genetic recombination between host-range and plaque-type mutants of bacteriophage in single bacterial cells. *Genetics* **34:** 44–71.

Hershey A.D., Kamen M.D., Kennedy J.W. and Gest H. 1951. The mortality of bacteriophage containing assimilated radioactive phosphorus. *J. Gen. Physiol.* **34:** 305–319.

Hershey A.D., Burgi E., Cairns H. J., Frankel F. and Ingraham L. 1961. Growth and inheritance in bacteriophage. *Carnegie Institution of Washington Year Book* **60:** 455–461.

Hershey A.D., Burgi E. and Ingraham L. 1963. Cohesion of DNA molecules isolated from phage lambda. *Proc. Natl. Acad. Sci.* **49:** 748–755.

Judson H.F. 1996. *The Eighth Day of Creation* (Expanded Edition) p 35. Cold Spring Harbor Laboratory Press, NY.

Levinthal C. 1954. Recombination in phage T2: Its relationship to heterozygosis and growth. *Genetics* **39:** 169–184.

Makover S. 1968. A preferred origin for the replication of lambda DNA. *Proc. Natl. Acad. Sci.* **59:** 1345–1348.

Mandell J.D. and Hershey A. D. 1960. A fractionating column for analysis of nucleic acids. *Anal. Biochem.* **1:** 66–77.

Meselson M. and Stahl F. W. 1958. The replication of DNA in *Escherichia coli*. *Proc. Natl. Acad. Sci.* **44:** 671–682.

Mosig G. 1966. Distances separating genetic markers in T4 DNA. *Proc. Natl. Acad. Sci.* **56:** 1177–1183.

Mosig G. 1994. Homologous recombination. pp 54–82. In: *Bacteriophage T4* , Edited by J. D. Karam et al. ASM Press, Washington.

Rubenstein I., Thomas C.A., Jr. and Hershey A.D. 1961. The molecular weights of T2 bacteriophage DNA and its first and second breakage products. *Proc. Natl. Acad. Sci.* **47:** 1113–1122.

Skalka A., Burgi E. and Hershey A. D. 1968. Segmental distribution of nucleotides in the DNA of bacteriophage lambda. *J. Mol. Biol.* **34:** 1–16.

Steinberg C. and Stahl F. 1958. The theory of formal phage genetics. Appendix II in Hershey (1958).

Streisinger G. 1966. Terminal redundancy, or all's well that ends well. pp 335–340. In: *Phage and the origins of Molecular Biology*, Edited by J. Cairns, G.S. Stent, and J.D. Watson. Cold Spring Harbor Laboratory, NY.

Visconti N. and Delbrück M. 1953. The mechanism of genetic recombination in phage. *Genetics* **38:** 5–33.

Watson J.D. and Crick F. H. C. 1953. A structure for deoxyribonucleic acid. *Nature* **171:** 737–738.

Werner R. 1968. Initiation and propagation of growing points in the DNA of phage T4. *Cold Spring Harbor Symp. Quant. Biol.* **33:** 501–507.

Mendelism at the Molecular Level

FRANKLIN W. STAHL

ACTERIOPHAGES ARE GREAT FOR investigating the molecular mechanisms of recombination of linked genetic factors. The vigorous exertions of T4 have facilitated the analysis of the biochemistry of recombination in that phage (for review, see Mosig 1994). Investigations of the more sedate behavior of phage lambda have provided information of an equally fundamental sort (for review, see Stahl 1998). Papers on genetic recombination in T-even phage (T2, T4, and T6) first appeared in 1947, in the Cold Spring Harbor Symposium of the previous year. And strange papers they were, when judged from today's perspective. Delbrück and Bailey (1947) called their paper "Induced Mutations in Bacterial Viruses," whereas Hershey (1947) called his "Spontaneous Mutations in Bacterial Viruses." The titles, which could not have been independently conceived, conceal reports of genetic recombination — simultaneous infection by two different mutant types of phage results in the production not only of both of the infecting types but also of particles with new combinations of the characters distinguishing those types. Neither paper made a committed effort to interpret the exchange of information in conventional genetic terms. Delbrück talked about one phage inducing mutations in the other, and Hershey wrote:

> So far the available methods of analysis have not given any indication of different degrees of structural relationship, to correspond with the alleles, crossover units, and linkage groups of conventional genetics .. [and] .. it is not likely that this difference is due solely to limitations of method. . . If it may be assumed that there is anything in common in the transformation of pneumococcal types, the fibroma-myxoma transformation, the induced mutations with respect to lysis inhibition (Delbrück, this Symposium), and the directed mutation with respect to host range described in this paper, it now appears for the first time that a biological phenomenon of general importance is involved."

It is unlikely that either Delbrück and Bailey or Hershey were *striving* for obscurity. It may be, instead, that they aimed to describe their observations in terms that did not prejudge the underlying mechanisms. They might, additionally, have been influenced by Luria's interpretation of multiplicity reactivation, the production of live phage following infection by two or more UV-inactivated phage. Luria (1947) and Luria and Dulbecco (1949) demonstrated the adequacy of a simple model for multiplicity reactivation (but see Dulbecco 1952). The only assumption that seemed to be required was that each phage particle is composed of a fixed number of different, but identically sensitive, subunits and that the ability of an infected cell to produce phage was assured if, among them, the infecting particles contained at least one unhit subunit of each type. The apparent high efficiency with which the subunits could be "recombined" to produce viable particles argued against a linkage system.

In 1948, Hershey and Rotman quantified the frequencies of wild-type phage among progeny particles emerging from cells mixedly infected with independently isolated *r* mutants. The mutants fell into two classes. Mutants from different classes gave about 15% wild-type particles when grown together ("crossed"), whereas mutants within a class gave values between 0.5% and 8%, depending on the pair. Hershey and Rotman were willing to call these data evidence for linkage, but they refrained from offering a map. Their reluctance apparently stemmed from the rather poor additivity of the recombinant frequencies, which proved later to be characteristic of phage crosses (for review, see Stahl et al. 1964). In 1949, their reluctance was overcome by additional experiments, and the first phage linkage map was presented. Despite the undeniable evidence of linkage, Hershey and Rotman did not abandon Luria's concept of independent subunits. They proposed instead that mutations in different classes were on separate Luria subunits, whereas mutations in the same class resided at different loci on the same subunit. Thus, they saw their work as elaborating Luria's views by showing that homologous radiosensitive subunits were not only freely assorted, but that they also swapped information with each other in a manner that generated linear linkage maps. The further experimental and theoretical development of such reasoning has been reviewed previously (Stahl 1959, 1995).

The 1949 paper of Hershey and Rotman was important also for its examination of recombinant yields from individual, mixedly infected bacteria. Hershey and Rotman noted that the numbers of complementary recombinants showed little correlation from one "single burst" to another. This suggested that, unlike crossing-over in meiosis, the mechanism that generated recombinants did not form the two complementary types simultaneously. This observation provoked notions of "copy choice" recombination for phage, contrasting with the apparent "break–join" mechanism of meiotic crossing-over.

Delbrück was heartened by Hershey's linkage data and by additional data collected soon thereafter by Doermann and Hill (1953). The occurrence of linkage supported the possibility that the genetic system of phages was, in fact, very much like that of eukaryotes. Such a possibility made phage a more attractive model system (in Delbrück's eyes) for examining biological replication — it gave phage genetical respectability. Delbrück published the anthropomorphic notion that a phage cross involves sequential pairwise "matings" (serial monogamy) between phage genomes (Visconti and Delbrück 1953). The participants in each mating were presumed to be drawn from a "mating pool" without regard to their genotype or prior mating experience. Visconti and Delbrück tested the notion against the limited data available and found that it could explain the following features of a phage cross: Some recombinants derive information from three different infecting particles, the recombinant frequency increases with time following infection, and coefficients of coincidence are greater than unity (negative interference = positively correlated exchanges). It made no attempt, however, to deal with phage heterozygotes, to which we now turn.

In 1951, Hershey and Chase (1952) reported heterozygous particles in T2. This was a shocker. In all respects, save that one, T2 particles appeared to be haploid, and, indeed, the observed heterozygosity was confined to short segments of the genome. Soon after Watson and Crick (1953) published their duplex structure for DNA, Levinthal (1954) offered two molecular models for these heterozygotes. One model supposed that bits of single-chained DNA were patched from one phage chromosome into another. The other supposed that heterozygosity was the consequence of phage chromosomes being built out of DNA segments spliced together by overlapping, complementary single chains. The latter model, but not the former, predicted that, in crosses with three linked factors, heterozygotes for the middle locus would be predominantly recombinant for the markers flanking that locus. Levinthal obtained linkage results supporting the latter model. Calculations based on some simple assumptions suggested that all recombinants arose by such a route and that recombination was tied to chromosome replication.

The Visconti-Delbrück theory made no effort to deal with heterozygotes and assumed that recombination of linked genes involved reciprocal breaking and rejoining of linear linkage structures. They blew off the single-burst experiments of Hershey and Rotman (1949) by asserting that post-mating randomizing factors could obscure the reciprocality assumed in their theory. The mathematics adopted by Visconti and Delbrück presumed that matings were synchronous in time. When they rewrote their equations to relax that unrealistic assumption, all traces of break–join and of reciprocality disappeared from the theory. The disappearance was pointed out several years later (Steinberg and Stahl 1959).

Hershey (1959) built upon the Steinberg-Stahl framework to better define the ambient questions in phage genetics. His analysis addressed data from both T-even and λphage. The points examined included matings in pairs versus groups, on the one hand, and copy-choice versus break–join on the other. Whereas the treatment by Steinberg and Stahl dealt with the former issue, information regarding the molecular mechanism required data other than recombinant frequencies. By making certain assumptions, data on heterozygotes and on crosses involving radiation-inactivated phages could be brought to bear. Hershey's analysis was a mathematical tour de force, and it probably did as much to kill interest in phage genetics as did the Visconti-Delbrück paper 5 years earlier.

The 1959 paper, published in, but not delivered at, a Cold Spring Harbor Symposium, was understood by few (at most) and soon became obsolete. The paper made two assumptions that were shown to be wrong within a few years: (1) the linkage map (or maps) of the T-even phages were linear, (2) except for the markers they might contain, all heterozygotes were of the same structure. Hershey was further sandbagged by artifactual data that appeared to demonstrate negative interference between closely neighboring exchanges in successive mating acts (see Steinberg and Edgar 1961) and by the removal of radiation damages from DNA by repair systems unrelated to recombination. Hershey's analysis crowned the heroic age of phage genetics.

Hershey's assumptions of linearity and of heterozygote simplicity died together when insights by Meselson and Streisinger blossomed into experiments. Their notion was that the chromosomes of T-even phages were linear but circularly permuted, and that heterozygotes were, in part, manifestations of the terminal redundancy present on each such chromosome (for review, see Streisinger 1966; Stahl 1968). With the demonstration of circular linkage, the mating theory began to unravel. The negative interference, which supported the notion of discrete matings, was attributable to that circularity and to certain trivial sources of heterogeneity encountered in a phage cross (Stahl and Steinberg 1964). An experiment by Drake (1967) supported the conclusion that the mating concept was entirely superfluous and put an end to population-genetic models of phage crosses.

However, the usefulness of phage for investigating recombination mechanisms did not end. On the contrary, it was now free to proceed with less regard for mathematical formalisms. Most importantly, in the present context, the role of heterozygotes in recombination became clarified. In T-even phage, two classes of heterozygotes arise during genetically mixed infections. One class has the terminal redundancy as its physical basis. Such heterozygotes reveal little about a genetic exchange but are (unavoidably) recombinant for markers flanking the heterozygous marker on the circular map. The other class of heterozygotes are heteroduplexes, reflecting the duplex nature of DNA. These heterozygotes do reflect the mechanism of exchange but are not usually recombinant for flanking markers (in the model of Levinthal, patches are more frequent than splices).

Despite the complexities of heterozygosity in T4, Gus Doermann and his students were able to define further their roles in recombination. The terminal heterozygotes are recombinagenic; they are not formed at the site of an exchange between two DNA duplexes, but they appear to provoke further exchanges. Their inevitable proximity to double-strand ends, which are recombinagenic, presumably accounts for this behavior, which may contribute to the clustering of exchanges over short intervals, referred to as high (or localized) negative interference (HNI). The heteroduplexes also contribute to HNI and do so in two ways: (1) since they are short and mostly parental for flanking markers (i.e., mostly patches), one of the two daughters arising by semiconservative DNA replication will look like a close double crossover. Additionally, mismatch repair operating on a heteroduplex can introduce higher orders of apparent close multiple exchanges (Doermann and Parma 1967).

Before an understanding of the heteroduplexes arising in phage crosses had been fully secured, other workers (for review, see Holliday 1990) commandeered them, along with the then visionary concept of mismatch repair, to explain associations between non-Mendelian segregation ratios ("gene conversion") and crossing-over in meiosis. Hershey's seminal observations on phage recombination promised that Mendelism could soon be addressed at the molecular level. And, indeed, it was.

References

Delbruck M. and Bailey W.T., Jr. 1947. Induced mutations in bacterial viruses. *Cold Spring Harbor Symp. Quant. Biol.* **11:** 33–37.

Doermann A.H. and Hill M. 1953. Genetic structure of bacteriophage T4 as described by recombination studies of factors influencing plaque morphology. *Genetics* **38:** 79–90.

Doermann A.H. and Parma D.H. 1967. Recombination in bacteriophage T4. *J. Cell. Physiol.* (suppl. 1) **70:** 147–164.

Drake J.W. 1967. The length of the homologous pairing region for genetic recombination in bacteriophage T4. *Proc. Natl. Acad. Sci.* **58:** 962–966.

Dulbecco R. 1952. A critical test of the recombination theory of multiplicity reactivation. *J. Bacteriol.* **63:** 199–207.

Hershey A.D. 1947. Spontaneous mutations in bacterial viruses. *Cold Spring Harbor Symp. Quant. Biol.* **11:** 67–77.

Hershey A.D. 1959. The production of recombinants in phage crosses. *Cold Spring Harbor Symp. Quant. Biol.* **23:** 19–46.

Hershey A.D. and Chase M. 1952. Genetic recombination and heterozygosis in bacteriophage. *Cold Spring Harbor Symp. Quant. Biol.* **16:** 471–479.

Hershey A.D. and Rotman R. 1948. Linkage among genes controlling inhibition of lysis in a bacterial virus. *Proc. Natl. Acad. Sci.* **34:** 89–96.

———1949. Genetic recombination between host-range and plaque-type mutants of bacteriophage in single bacterial cells. *Genetics* **34:** 44–71.

Holliday R. 1990. The history of the DNA heteroduplex. *BioEssays* **12:** 133–142.

Levinthal C. 1954. Recombination in phage T2: Its relationship to heterozygosis and growth. *Genetics* **39:** 169–184.

Luria S.E. 1947. Reactivation of irradiated bacteriophage by transfer of self-reproducing units. *Proc. Natl. Acad. Sci.* **33:** 253–264.

Luria S.E. and Dulbecco R. 1949. Genetic recombinations leading to production of active bacteriophage from ultraviolet-inactivated bacteriophage particles. *Genetics* **34:** 93–125.

Mosig G. 1994. Homologous recombination. In *Molecular biology of bacteriophage T4* (ed. J.D. Karam), pp. 54–82. ASM Press, Washington, D.C.

Stahl F.W. 1959. Radiobiology of bacteriophage. In *The viruses: Plant and bacterial viruses* (ed. F.M. Burnet and W.M. Stanley), Vol. 2. pp 353–385. Academic Press, New York.

——— 1968. Role of recombination in the life cycle of bacteriophage T4. In *Replication and recombination of genetic material* (eds. W.J. Peacock and R.D. Brock), pp 206–215. Australian Academy of Science, Canberra.

——— 1995. The amber mutants of phage T4. *Genetics* **141:** 439–442.

——— 1998. Recombination in phage λ: One geneticist's historical perspective. *Gene* **223:** 95–102.

Stahl F.W. and Steinberg C.M. 1964. The theory of formal phage genetics for circular maps. *Genetics* **50:** 531–538.

Stahl F.W., Edgar R.S., and Steinberg J. 1964. The linkage map of bacteriophage T4. *Genetics* **50:** 539–552.

Steinberg C.M. and Edgar R.S. 1961. On the absence of high negative interference in triparental crosses. *Virology* **15:** 511–512.

Steinberg C. and Stahl F. 1959. The theory of formal phage genetics. *Cold Spring Harbor Symp. Quant. Biol.* **23:** 42–46.

Streisinger G. 1966. Terminal redundancy, or all's well that ends well. In *Phage and the origins of molecular biology* (ed. J. Cairns et al.), pp. 335–340. Cold Spring Harbor Laboratory, Cold Spring Harbor, New York.

Visconti N. and Delbrück M. 1953. The mechanism of genetic recombination in phage. *Genetics* **38:** 5–33.

Watson J.D. and Crick F.H.C. 1953. Molecular structure of nucleic acids. A structure for deoxyribose nucleic acid. *Nature* **171:** 737–738.

Franklin W. Stahl *(born 1929 in Boston, Massachusetts) was infected when he took the "Phage Course" at Cold Spring Harbor in 1952, while he was a graduate student at the University of Rochester. He has been at it ever since, with emphasis on mechanisms of genetic recombination. After postdoctoral study at Caltech, and a one-year stint in Missouri, he settled in Oregon where he is now American Cancer Society Research Professor of Molecular Biology.*

In Memoriam: Alfred D. Hershey (1908–1997)

WACLAW SZYBALSKI

*T*HE WORLD OF SCIENCE IN GENERAL, and our group of phage and bacterial molecular genetics, in particular, are saddened by the loss of Alfred (Al) D. Hershey, the last one of three 1969 Nobel Prize laureates and Founding Fathers of this field. I was privileged since 1950 to know all three of the laureates, Max Delbrück, Sal Luria, and Al, and to develop lifelong friendships. Al died in his home, within walking distance of Cold Spring Harbor Laboratory, on June 22, 1997. He was 88.

Instead of any formal obituary, which I am sure will be carefully prepared in *Biographical Memoirs* by the National Academy of Sciences, and in various journals, I prefer to celebrate Al's life and to tell what I personally learned from Al, and about Al, who was a true scientist's scientist, a great pioneer in the field of bacteriophage genetics, from T2 to lambda.

I first heard about Al from the daughter of Professor J. Bronfenbrenner, with whom I shared, by chance, the same dormitory in Copenhagen in the summer of 1949, and who described him to me as an "amazingly serious" scientist. At that time Al was working in Professor Bronfenbrenner's laboratory in St. Louis. I told Al about Professor Bronfenbrenner's daughter's description of him when I first met him in the summer of 1950 in Cold Spring Harbor, where I attended the First Phage Meeting, but I do not remember Al's comments, if any. I noticed then that he was a man of very few words. At that time, I was also visiting Milislav Demerec, the Director of the Laboratories, since I was about to join the staff of the Biological Laboratory at Cold Spring Harbor.

When that happened, early in 1951, during the first staff meeting I attended, Al was presenting for the first time his results on the now famous Hershey-Chase Waring blender experiment. In his introduction, which I doubt was recorded in any way, Al said that he considered himself an immunochemist and, therefore, he believed that genetic specificity must reside in highly specific protein structures and not in the monotonous tetranucleotide "scaffold," namely DNA. That was the main reason he designed the experiment in which he hoped that ^{35}S-labeled protein would enter the host cell and be transferred to progenies, and that the ^{32}P-labeled DNA (phage particles contained only protein and DNA) would be left outside to be washed off in the Waring blender. However, Al told us that to his great dismay as to the omnipotence of proteins, the experiments showed just the opposite: Phage DNA entered the cells and proteins were left on the outside of the cells. Therefore, Al said that maybe Avery et al. (1944) were right and DNA was the true carrier of heredity. However, Al warned us that a small amount (about 1%) of ^{32}S counts did enter the cell, and maybe this special protein still could be the carrier of heredity. A lively discussion developed, with Barbara McClintock praising Al's bold and imaginative experiments.

Since it happened that I had carefully read the Avery et al. (1944) paper in Gdansk, Poland, when we received it in 1945/46 as part of the first postwar gift from UNRRA of various American post-1939 journals, including the *Journal of Experimental Medicine*, I asked Al why he ever doubted the very convincing chemical and enzymatic data on the DNA nature of the transforming principle. And Al's answer was: "Maybe you in Poland believed Avery et al. (1944) data, but nobody in America believed their data." When I asked why, Al said "because Mirsky convinced everybody that Avery et al. did not know the difference between the nucleoproteins, which must be the essence of transforming principle, and the DNA."

Against his original intention, but convinced by his own precisely designed experiments, Al turned off the anti-DNA tide and convinced America and the world that DNA is the hereditary component of the T4 coliphage. I remember it all so vividly, because this was my first staff meeting in Cold Spring Harbor, and thus I became "imprinted" for life.

In the 1950s, as I witnessed it during my stay in Cold Spring Harbor, Al worked very intensively, doing every experiment by himself and in perfect silence. Marty Chase, Al's technician, was also a very quiet person, and they seemed to communicate only by "sign language." Often I had to watch them for a long time, standing in the door and not daring to interrupt their concentrated silence, even if I had some seemingly "very pressing" questions or other business.

When he was not totally absorbed by his experiments, Al loved classical music and sailing. He organized concerts in his "Hershey's cottage" on the Laboratory grounds using his then "state-of-the-art" hi-fi equipment, and he and his wife, Harriet (Jill), invited all the interested members of our Cold Spring Harbor community. These were memorable evenings. He also invited me several times for a sail on the Long Island Sound in his brand new Thistle-class sailboat, which was a much better boat than my National-class all-wood boat, which I co-owned with Allan Garen, Al's collaborator.

Al's wife, Jill, who was the spirit behind the hi-fi concerts and sailing, was my colleague at Vernon Bryson's group in the Nichols Memorial Laboratory, whereas Al's laboratory was located first at the Animal House (present McClintock lab) and after 1953 at the Carnegie lab (presently Demerec lab). Jill told me that Al worked so intensively that he lost weight throughout the whole year, and that he would never survive if it was not for her taking him for a summer-long sail on the Great Lakes and feeding him until he gained enough weight to last for the entire next year!

I had many good friends among the collaborators and visitors in Al's lab, starting with Marty Chase (who liked dancing but bitterly complained about her very low technician's salary) and Niccolo Visconti di Modrone, an extremely brilliant geneticist and most pleasant, very aristocratic, and best-looking colleague, and followed by Allan Garen, Gus Doermann, S. S. Cohen, R. Herriot, Neal Symonds, Connie Chadwick, Norman Melechen, J.-i. Tomizawa, Dorothy Frazer, René Thomas, Ann Skalka (at present the Director of the Institute for Cancer Research in Philadelphia), and many others (list is incomplete and not in chronological order).

Al's lab had many simple and frugal "inventions" like rolls of toilet tissue at each bench instead of more expensive boxes of special tissues, and whenever some of his colleagues or students complained about some of their adversities, Al's comment was: "It is good for your soul." I also remember one of Al's official lectures, which he started by saying: "One of the tragedies of my scientific life is that I make too many discoveries." When that statement was met by general laughter, Al countered by saying: "This is no laughing matter but a real tragedy, because new discoveries interfere with completing each previous line of research."

In that vein, I remember once consulting Al on the experimental design of one of my experiments, which, I said, could not fail since the positive result would prove one hypothesis whereas the negative result would prove the opposite theory. Al thought for a moment and then commented: "I wish you luck, Waclaw, with distinguishing between two hypotheses, but the trouble is that results of most of my experiments fall just in between!" Al was so right! I no longer remember the experiment; I remember, however, that the results were not a clear "yes" or "no," but in between.

After I left Cold Spring Harbor sometime around 1955, I still met Al practically every year at the famous Cold Spring Harbor Phage Meetings. One year, I remember, he was not presenting any paper, and when I asked why, his answer was that the experiments were too specialized and nobody would be interested. He agreed, however, to summarize for me his current experiments, which turned out to be the discovery of the "sticky ends" of phage lambda DNA. This is an example of Al's modesty: At present a large part of genetic engineering and biotechnology is based on the applications of "sticky ends," the term that Al himself introduced! Another time, I mentioned to Al that we missed his presence at some "very important" Symposium or Meeting, to which I knew that he was invited. Al's comment was: "Why should I travel many miles to meetings instead of working in my lab; if anything of importance was presented, my colleagues or students will visit me in Cold Spring Harbor and tell me all about it, often more than once!"

I also remember a festive occasion to which Al was kind enough to invite me. This was Al's Harvey Lecture presentation on April 26, 1956. We all had to wear tuxedos to the dinner honoring Al. I changed to a tuxedo just before dinner and left my Harris tweed suit in my car. Al's lecture on "Bacteriophage T2; parasite and organelle" was a memorable one— short, witty, and critically discussing the fate of DNA and protein during phage replication. However, when I returned to my car parked near the Academy of Medicine in New York, I found my suit stolen. Therefore, for the next two days when walking around New York and giving seminars at Columbia University and the Rockefeller Institute, I had to wear my tuxedo, explaining that it was because of Al's Harvey Lecture.

Maybe the most moving of very memorable occasions involving Al was the dedication to him of a new building on the Cold Spring Harbor grounds. There were many toasts and speeches, followed by a short acknowledgment by Al who said: "This occasion is the nearest to what I would call the Hershey's Heaven."

One of Al's last major contributions was his editing of the very important monograph *The Bacteriophage Lambda* published by Cold Spring Harbor Laboratory in 1971.

Al retired in 1972, and I have seen him more and more rarely in his lab when visiting Cold Spring Harbor, but Jill looked to it that he attended the social hour cocktail parties during each Phage Meeting. Unfortunately in most recent years, Al did not join us anymore even at the Phage Meeting (now "Molecular Genetics of Bacteria and Phage") cocktail parties, but each year I tried to visit him and Jill in their home. He preferred to talk about his garden, trees and flowers, about music and computers, but not about the latest phage experiments. I remember the last time seeing his large portrait hanging in the living room, which Jill explained to me was done by an artist who was painting all the Nobel Prize laureates.

In 1995, I received from Al his last long-hand letter pertaining to the 50th Anniversary of the Bacteriophage Course in Cold Spring Harbor.

Dear Al, Cold Spring Harbor and the Phage World will never be the same without you. We will all miss you very much, but we will also celebrate your life as that of a truly great scientist.

Reference

Avery O.T., MacLeod C.M., and McCarty M. 1944. Studies on the chemical nature of the substance inducing transformation of pneumococcal types. Induction of transformation by a desoxyribonucleic acid fraction isolated from pneumococcus type III. *J. Exp. Med.* **79:** 137–158.

Waclaw Szybalski *(born in Lwów, Poland in 1921), came to Cold Spring Harbor in the early 1950s. He rapidly became known for his ingenuity, innovation, and imagination, and his aggressive interest in wide areas of biology. His diverse contributions, including the first genetic transformation of human cells (the foundation for gene therapy) and novel methods to facilitate the human genome project, could be loosely classified as "molecular genetics," with an early and enduring, but not monogamous, devotion to phage lambda. All geneticists know him as founder and editor of* Gene. *Since 1960, he has been at the University of Wisconsin.*

Bacteriophage λ as a Model "System"

ROBERT A. WEISBERG

Lambda is important mainly because it can recombine genetically with its host, min-
gling cellular and viral inheritance in ways that are fascinating to contemplate and,
very likely, of practical importance to humans.

HERSHEY 1967

$\mathcal{H}$ OW AND WHY DID λ become a model "system?" *System?* Perhaps I should have rejected this title from the start, but, having learned from Al Hershey to treat words and suggested titles with the respect they deserve, I decided to brand the offending term with quotes. According to Al, the word "system" is "pretentious, meaningless, and overworked" in a biological context, and should be preserved "to go with 'Copernican'." I agree, although I confess that even after long years of trying to clean up my act, I still backslide into meaningless pretension. Mea culpa! How and why did λ become a model organism? It's hard to improve on Al's succinct answer (above), but his editorial doctrine that less is more is harsh for those less gifted with words than he. Therefore, I'll elaborate.

Cells of the bacterium *E. coli* K12, *a* λ host, often survive infection by λ and give rise to descendants that inherit the phage genome in a noninfectious form called prophage. Such cells are lysogens, so called because of their capacity to lyse and to liberate λ in the absence of reinfection. Phage varieties capable of forming lysogens are called temperate, in contrast to virulent, species, which always kill upon infection. Although λ is not the first example of a temperate bacteriophage, it emerged as a paradigm of its type in the 1950s, when conjugation and generalized transduction became tools to study the *E. coli* K12 chromosome. These studies revealed the linkage of λ prophage to a unique segment of the host chromosome, and this, together with the discovery of λ-mediated specialized transduction, suggested an interesting connection between lysogeny and chromosome structure. These findings attracted new workers to the field, and by the time *The Bacteriophage* λ appeared, about 15 years later, we knew more about λ than about any other temperate bacteriophage, or, indeed, most other organisms. Since λ has a moderately complicated developmental program, complete with forks in the life cycle and numerous control points, this accumulated knowledge was (and still is) an inviting springboard for leaps into the less well-understood world of cellular organisms (see, e.g., McAdams and Shapiro 1995).

Well, Al (I assume you can read this— I still imagine you're looking over my shoulder while I write), the previous paragraph is a shortened version of the introduction to λ that I tried to slip past you as the beginning of the chapter "Prophage Insertion and Excision" in *The Bacteriophage* λ 28 years ago (Gottesman and Weisberg 1971). I

23

don't blame you for expunging it (editing is too mild a word). We used about eight pages to convey a paragraph's worth of useful information, and you and Bill Dove quite rightly wanted the subject for yourselves. Am I doing better now? If not, it's certainly not your fault.

Lambda is not only a model organism, it's also the matriarch (and patriarch [Hendrix and Duda 1992]) of a widespread bacteriophage family. Genetic diversity within the λ family is so large, and genetic exchange between family members so frequent, it is unlikely that the founder (or, more precisely, its clone) could be reisolated from nature. Given such diversity, it's difficult to define the characteristic λ family traits and limits with any precision. Hershey and Dove (1971), noting the presence and common organization of limited sequence identity among different family members, probably came closest to the essence of the matter with this characteristically cryptic statement: "These facts, together with the common structure of the molecular ends and the ability to form intervarietal hybrids, suggest the existence of a common genome that is nearly but not quite independent of specific nucleotide sequences." The idea of a common family genome, which I take to mean a collection of genes whose biological functions are analogous to those of λ genes and whose order along the chromosome corresponds to that of their λ analogs, is right on the money. The ability of an unknown phage to form viable hybrids with a known member of the λ family is an excellent sign of kinship, although we now know that the structure of the DNA ends is not (Botstein and Herskowitz 1974). However, I have always been bothered by "nearly but not quite." This is what makes the sentence memorable, but I'm not sure why. "Not quite" does not change the meaning of "nearly" (or vice versa), so the extra words add nothing but emphasis. Knowing Al's parsimony, I'm convinced there must have been a better reason that "nearly" wasn't allowed to stand alone, but I never had the nerve to ask. Perhaps reiteration simply reassures the reader that the authors have not forgotten that a genome cannot be totally independent of nucleotide sequence. This reminds me of another of Al's strictures: "The word (genome) should not be used as a synonym for DNA or chromosome. Chromosomes multiply by replication, genomes multiply by speciation." This usage implies that the entire λ family should be considered a single species. An alternative and probably more profitable view holds that the high frequency of horizontal transmission of genetic information between bacteriophage families blurs boundaries and diminishes the utility of classification by species (Hendrix et al.1999). But I digress.

Lambda's tendency to inspire models is well illustrated by my doctoral thesis project, which started in 1959 in Renato Dulbecco's laboratory at Cal Tech, and which was directed toward understanding the oncogenic transformation of cultured mouse embryo cells by the recently discovered polyoma virus. My experiments, guided by what we then knew about the interaction of temperate bacteriophages with their hosts, were designed to show that transformed mouse cells contained polyoma virus as a noninfectious provirus that resembles λ prophage. Unfortunately for me, the techniques then available were not up to the job, and it later became clear that I had taken the analogy altogether too literally. While I was vainly struggling to understand polyoma by forcing it into a lambdoid mold, understanding of λ itself was rapidly advancing: The immunity of lysogens to superinfection, whose likeness I had fruitlessly sought in polyoma-transformed mouse cells, was shown to be due to a phage-encoded repressor whose primary role is to silence genes that act early in the lytic cycle (Jacob and Campbell 1959; Jacob and Monod 1961), and the stable inheritance of the prophage results from its insertion into the host chromosome. This insertion is achieved by crossing over between the host chromosome and a circular form of the virus chromosome (Campbell 1962). It's worth mentioning that the circularity of the polyoma virus chromosome, discovered in Dulbecco's lab around this time (Dulbecco and Vogt 1963), showed that λ need not be a misleading guide to the biology of tumor viruses, and vice versa; indeed, the similarities are striking. Nevertheless, it occurred to me that I might have more fun working on the model phage than on its less tractable cousins, and, after scraping together enough experiments to satisfy a beneficent Ph.D. thesis committee, I applied for a postdoctoral fellowship to a laboratory where I could do just that. I originally intended to return to polyoma virus and mice, but I never have—there always seemed to be more amusing things to do.

I wish I could regale you with stories of Al's scientific mentoring and his (no doubt) numerous personal idiosyncrasies. Sadly for this article and also for my career, I spent only a month in his laboratory about 30 years ago, and remember little of that brief period. However, I can tell a story about one of Al's rare errors with words. The tale displays λ's talent for provoking models once again.

Al was inspired by Allan Campbell's (1962) bold proposal that λ prophages are formed by recombination of a circular form of the phage chromosome with the bacterial chromosome. Campbell's proposal (a.k.a. The Campbell Model) rested on two assumptions: first, that a mechanism exists for circularizing the linear λ chromosomes found in phage particles, and second, that prophage insertion occurs by a process akin to recombination between homologous chromosomes. Hershey and Burgi (1965) duly discovered that circularization occurs by base-pairing between the single-stranded chromosome ends, and the same principle has since been applied by genetic engineers who wish to join DNA fragments that, unlike the λ chromosome ends, are not naturally attached. However, the limited and scattered sequence similarity between the λ and *E. coli* chromosomes differs from the nearly complete identity between homologous chromosomes (Cowie and Hershey 1965) and seemed unable to account for the high efficiency, precision, and reversibility of prophage insertion. This led Al to suggest that the "region of homology" proposed by Campbell as the site of λ–*E. coli* crossing-over could be interpreted as "locus of crossover points" (Hershey 1967). My dictionary disagrees. It defines "homologous" as "corresponding in structure and evolutionary origin" (The American Heritage Dictionary, Houghton Mifflin Company, Boston, Massachusetts 1982). I suspect that this is exactly what Campbell meant, and, if so, he is almost certainly correct, although the homology is on a much smaller scale than either he or Al could appreciate at the time.

In 1977, Landy and Ross (1977) confirmed earlier genetic studies (Shulman and Gottesman 1973; Shulman et al. 1976) by demonstrating that the λ and *E. coli* chromosomes do indeed share a common sequence at which they recombine with each other, but the extent of identity amounts to only 15 base pairs. A few years later, Mizuuchi et al. (1981) and Craig and Nash (1983) showed that DNA cleavage and strand exchange during insertion occur at each edge of a 7-base pair segment, the overlap region, that is located within these 15 common base pairs. Weisberg et al. (1983) found that sequence identity between the λ and *E. coli* overlap regions is required for efficient recombination between the two chromosomes, as it is for recombination between homologous chromosomes. Finally, studies of other phages that use the same recombination mechanism as λ but insert at different host sites revealed that in some cases the overlap regions are embedded in regions of sequence similarity that are considerably larger than λ's 15 base pairs (for review, see Campbell 1992), leaving little doubt that at least these phage and host sequences have a common evolutionary origin.

Well before the work just cited was published, it was generally accepted that λ prophage insertion occurs by a special or "site-specific" recombination pathway that differs in important respects from the usual pathway of recombination between homologous chromosomes (Signer 1968; Guerrini 1969). According to this model, λ insertion is catalyzed by a phage-encoded recombinase or "integrase" that recognizes the nucleotide sequence of the λ insertion sites and specifically promotes crossing-over within them. More than 100 enzymes that are related to λ integrase and possess comparable ability to promote crossing-over within "microhomologies" or even within totally dissimilar sequences have since been discovered (for review, see Nunes-Duby et al. 1998). Of course, λ and the λ family have also been and still are models for understanding the mechanism of DNA recognition by helix-turn-helix proteins, the role of cooperative interactions between DNA-binding proteins in assuring sharp temporal transitions between gene repression and expression, the mechanisms of transcription termination and antitermination, and more. For a fuller summary of λ's versatility, I refer the reader to an excellent review with the same title as this article (except for the quotes) (Campbell 1986).

Acknowledgments

Thanks to Naomi Franklin, Roger Hendrix, and Michael Yarmolinsky for reading the manuscript and to Al Hershey for helping me to avoid meaningless pretension.

References

Botstein D. and Herskowitz I. 1974. Properties of hybrids between Salmonella phage P22 and coliphage lambda. *Nature* **251**: 584–589.

Campbell A.M. 1962. *Episomes. Adv. Genet.* **11**: 101–145.

———.1986. Bacteriophage lambda as a model system. *BioEssays* **5**: 277–280.

———.1992. Chromosomal insertion sites for phages and plasmids. *J. Bacteriol.* **174**: 7495–7499.

Cowie D.B. and Hershey A.D. 1965. Multiple sites of interaction with host-cell DNA in the DNA of phage lambda. *Proc. Natl. Acad. Sci.* **53:** 57–62.

Craig N.L. and Nash H.A. 1983. The mechanism of phage lambda site-specific recombination: Site-specific breakage of DNA by Int topoisomerase. *Cell* **35:** 795–803.

Dulbecco R. and Vogt M. 1963. Evidence for a ring structure of polyoma DNA. *Proc. Natl. Acad. Sci.* **50:** 236–243.

Gottesman M.E. and Weisberg R.A. 1971. Prophage insertion and excision. In *The bacteriophage lambda.* (ed. A.D. Hershey, pp. 113–138). Cold Spring Harbor Laboratory, Cold Spring Harbor, N.Y.

Guerrini F. 1969. On the asymmetry of lambda integration sites. *J. Mol. Biol.* **46:** 523–542.

Hendrix R.W. and Duda R.L. 1992. Bacteriophage lambda PaPa: Not the mother of all lambda phages. *Science* **258:** 1145–1148.

Hendrix R.W., Smith M.C., Burns R.N., Ford M.E., and Hatfull G.F. 1999. Evolutionary relationships among diverse bacteriophages and prophages: All the world's a phage. *Proc. Natl. Acad. Sci.* **96:** 2192–2197.

Hershey A.D. 1967. Annual report of the director of the genetics research unit. In *Carnegie Inst. Wash. Year Book* **66:** 645–664.

Hershey A.D. and Burgi E. 1965. Complementary structure of interacting sites at the ends of lambda DNA molecules. *Proc. Natl. Acad. Sci.* **53:** 325–328.

Hershey A.D. and Dove W. 1971. Introduction to lambda. In *The bacteriophage lambda* (ed. A.D. Hershey), pp. 3–11. Cold Spring Harbor Laboratory, Cold Spring Harbor, N.Y.

Jacob F. and Campbell A. 1959. Sur le système de répression assurant l'immunité chez les bactéries lysogènes. *C. R. Acad. Sci.* **248:** 3219–3222.

Jacob F. and J. Monod. 1961. Genetic regulatory mechanisms in the synthesis of proteins. *J. Mol. Biol.* **3:** 318–356.

Landy A. and Ross W. 1977. Viral integration and excision: Structure of the lambda att sites. *Science* **197:** 1147–1160.

McAdams H.H. and Shapiro L. 1995. Circuit simulation of genetic networks (see comments). *Science* **269:** 650–656.

Mizuuchi K., Weisberg R.A., Enquist L.W., Mizuuchi M., Buraczynska M., Foeller C., Hsu P.L., Ross W., and Landy A. 1981. Structure and function of the phage lambda att site: Size, int-binding sites, and location of the crossover point. *Cold Spring Harbor Symp. Quant. Biol.* **45:** 429–437.

Nunes-Duby S.E., Kwon H.J., Tirumalai R.S., Ellenberger T., and Landy A. 1998. Similarities and differences among 105 members of the Int family of site-specific recombinases. *Nucleic Acids Res.* **26:** 391–406.

Shulman M. and Gottesman M. 1973. Attachment site mutants of bacteriophage lambda. *J. Mol. Biol.* **81:** 461–482.

Shulman M.J., Mizuuchi K., and Gottesman M.M. 1976. New *att* mutants of phage lambda. *Virology* **72:** 13–22.

Signer E.R. 1968. Lysogeny: The integration problem. *Annu. Rev. Microbiol.* **22:** 451–488.

Weisberg R.A., Enquist L.W., Foeller C., and Landy A. 1983. Role for DNA homology in site-specific recombination. The isolation and characterization of a site affinity mutant of coliphage lambda. *J. Mol. Biol.* **170:** 319–342.

Robert A Weisberg *(born 1937 in Bayonne, New Jersey) abandoned the study of animal viruses early in his career to enjoy the pleasures offered by the experimentally tractable bacteriophages. Bob's postdoc work in prokaryotic genetics included one month in Hershey's lab. Following three years at the Biology Division of the Oak Ridge National Laboratory, Bob took a post as independent investigator at the National Institutes of Health. His most intense interaction with Al occurred during the writing of his contribution (with Max Gottesman) of the review, "Prophage Insertion and Excision," for* The Bacteriophage Lambda, *published by Cold Spring Harbor in 1971. Bob was one of many who thought they knew how to write until they were subjected to Al's demanding editorship. When the book needed updating in 1983 (Lambda II), Jim Watson recruited Bob to be one of four editors who, he hoped, could pool their efforts to approach Al's standards. Bob continues his research on λ at the National Institutes of Health, where he is Assistant Chief of the Laboratory of Molecular Genetics of the National Institute of Child Health and Human Development.*

My "Special Memories" of Al Hershey

JAMES D. EBERT

I AM PINCH-HITTING FOR MAXINE SINGER who unavoidably could not be here today. Recently in a note which she sent to Carnegie's trustees, Maxine wrote in part that Al Hershey was dedicated to research, deriving great joy from it. The Department of Genetics provided a comfortable setting in which he could flourish. Carnegie Institution of Washington can be extraordinarily proud of this chapter in its history. I'm speaking not just for Carnegie but for myself because I was fond of Hershey. I knew him, although I can't say I knew him intimately. Two anecdotes illustrate our interactions and perhaps reveal aspects of Hershey's personality that were rarely expressed. The first time I saw Hershey in action was in December, 1955 in Washington. Vannevar Bush was then president of Carnegie Institution and it was the Carnegie president's practice to entertain the departments' directors at dinner annually; the dinner would be followed by a lively discussion on whatever was on either the directors' or Bush's mind. This particular dinner had the then six Carnegie directors in attendance along with three visitors: Caryl Haskins who was about to become president of the institution succeeding Bush, myself (I was about to become director of the Department of Embryology succeeding George Corner), and Al Hershey. Until the discussion began to unfold I didn't know what Al was doing there. It was very rare to have a non-director present at these meetings. But it turned out that a very cordial dinner was followed by what turned out to be a very lively debate aligning on the one hand Bush and Merle Tuve, the great director of the Department of Terrestrial Magnetism, and on the other Milislav Demerec, director of the Department of Genetics. The discussion centered on the question of whether Carnegie should begin taking federal research grants. Demerec was very anxious to have research grants and Bush, although he was the architect of the National Science Foundation, did not believe the institution should take grants. This is a side of Bush that isn't generally known. Although he favored federal support for most universities, he did not want them for the "elite organizations" of which Carnegie was one. Demerec was leading the way in the battle to have them, and his arguments were based in part on the fact that he was running two organizations on the same site and he felt very uncomfortable with the Carnegie scientists not having grants. He wasn't getting anywhere with that argument. Finally I learned why Hershey was there, when he spoke up. We were well into the discussion. Hershey and I were the only two who hadn't spoken. Al very quietly put himself into the conversation. He said, "Well, Mr. President, I have to tell you that I am going to have to apply for some research grants whether you like it or not." Those were almost his exact words. And Vannevar Bush, one of the most formidable figures in American

Science, said, "Well, why? Aren't we supporting you?" Hershey replied, "You are today, but I don't know what the future will bring." And then Hershey went on, "Mostly, I want grants because I want peer review." And he said, "I must tell you, Mr. President, I know that you are enormously well informed, but I doubt that you have the capacity to understand my research." That put a damper on the conversation, but Hershey prevailed. Not with Bush but with Caryl Haskins, Bush's successor; Hershey and others in the Department of Genetics began to have grants, but only in that department. The Department of Genetics had federal support for several years while the rest of the institution remained "grantless" after the Department of Genetics closed, until I became president in 1978. It fell to me to reverse that policy both because the portfolio was slim at the end of the 1970s and because the cost of science had become so great. Carnegie had to break away from its traditional practice. That memory of Hershey is one I will never forget because on other occasions he had been such a quiet man, so unwilling to speak up in that kind of a way.

I turn now to another discussion with Al to a few months later, in June, 1956. After I became director of the Department of Embryology, I began making trips to the other departments and giving seminars. I went to Cold Spring Harbor to speak and meet the staff. I had great aspirations for the Department of Embryology, some of which involved developing molecular developmental genetics (when I could find the right people to do what I felt needed to get done). Al was at my lecture and I made the usual rounds for an hour or so to each of the staff members after the lecture. When I saw Hershey, during the first few minutes of our conversation he just didn't seem all that interested. He got around to asking me three or four questions, all of them informed questions, but not challenging. He finally came to the point. He said, "I know embryologists are finding it difficult to find the right organism to do molecular biology, but you are not going to make real headway (and I remember his phrase exactly) until you get closer to the genes." In fact, he said, "until you really get into the genes." Of course we drew upon that advice very often as the years passed.

I had invited Hershey to visit the Department of Embryology several times, but he always declined until he was coming to Washington to receive the Kimber Genetics Award in April of 1965, if I remember correctly, and he said he would stop off in Baltimore, not to speak but to visit and he said he especially wanted to meet Don Brown, about whom he had heard a great deal, because at long last Brown and others were just beginning to isolate and characterize genes. He spoke of Brown, Max Birnstiel in Scotland, and John Gurdon in England, all of whom were in the midst of exceptionally interesting work, and Al wanted to hear about it. After he had seen Brown and his colleagues, Hershey came into my office before we went out to dinner and he smiled that little smile of his and said, "Well, Jim, it seems that you took my advice in our conversation of 10 years or so ago." I said, "Yes I did, Al, yes I did," and he said, "Well, you didn't go far enough. You are going to have to get away from these frogs and get into an organism with real genetics." Those two memories, the Bush discussion and the "real genetics" discussions are my special memories of Al Hershey.

Jim Ebert, *born in 1921 in Bentleyville, Pennsylvania, contributed to chick embryology before becoming deeply, energetically, and effectively involved in the service of science. His directorships and presidencies include the Department of Embryology of the Carnegie Institution of Washington, the Carnegie Institution of Washington, Woods Hole Marine Biological Laboratory, the American Institute of Biological Sciences, the Society for the Study of Development and Growth, and the American Society of Zoologists. His extramural advisory appointments and his memberships and trusteeships in scientific and educational boards are countless.*

Growing Up into Our Long Genes

ROLLIN D. HOTCHKISS

*L*OOKING BACK AT THE LIFE AND CAREER of Alfred Hershey, I feel a rush of gratitude as I recall calm, undramatic conversations about the chemistry and biology of bacteriophage we had on some quiet occasions. A warmth rises up in me as I recall the quiet passion that sometimes came through in his no-nonsense, "no big deal!," straightforward descriptions of experiments that so often went directly to a "core" question.

I met Hershey in the summer of 1945. When Max Delbrück visited my chief, O.T. Avery, the year before, I'd talked with Max, and later he told me about his planned phage course at Cold Spring Harbor that next summer and accepted me in it. My original interest arose in the bacterial lysis that culminates phage infection. The staphylococci I had been working with in the Avery and Dubos departments at Rockefeller Institute were stimulated to incomplete bacteriolysis after being insulted with certain antibiotic and surface-active agents. Max's masterful, eye-opening, phage course and an unexpected opportunity soon drew me into participation in almost all of the early activities of the "phage group," and into a path analogous to one Hershey began to take.

In expressing my great respect for Hershey's profound contributions to phage biology in reasonable compass here, I must limit myself to a largely anecdotal retelling of a few moments in the history as I witnessed them. I hope that my specific recollections of his qualities and insights may shed light on a scholar from whom many felt a bit removed.

At Rockefeller, in early 1946, an opportunity opened for me to take an active part in bacterial "transformation" work, since Maclyn McCarty was just moving away from it. He had been largely responsible in 1944 for showing, with Avery and MacLeod, that pneumococcal capsule induction—the Griffith transformation—was a heritable change initiated by specific DNA. I was delighted at the chance, because as early as September 1938 I had asked Avery if I could work on transformation, but he had insisted that I pursue a different problem. I did so, then as we entered World War II I'd taken up work with René Dubos on the emerging antibacterial agents.

About then, Alfred Hershey at Washington University, St. Louis, was developing mutant coliphages and pretty soon reporting a genetics of heritable properties (host range and rapid lysis) that "recombined" the two phenotypes in cells infected with a mixture of the two phages (Hershey and Rotman 1949). This work implied a genetic function but with an as-yet-unknown chemical basis. Bacterial transformation presented the different challenge, that of a probable DNA agent affecting what might be construed as "gene-like" properties in organisms widely viewed as asexual. These challenges became the respective problems underlying Hershey's phage work, and my work on bacterial transformation that I had just taken up.

Chemistry of the Gene Material

Hershey spent most of the next summers at Cold Spring Harbor, and so did I. In the free atmosphere of the summer Laboratory, many conversations with young and older colleagues kept most of us thinking along broad lines. Max Delbrück continued to keep a friendly eye on both of us, although he himself was mostly concerned with various kinetic and immunologic aspects of phage populations. Chemistry was generally far in the background at that time; limited to quoting results describing coliphage as broadly composed of deoxyribonucleoprotein. Certainly Hershey and I talked about this sometimes, but we were not in the beginning working on a common angle in our respective problems. I know that neither of us considered likely what some biologists had conjectured, that pneumococcal transforming agent was slightly "contaminated" with a nucleoprotein bacteriophage. At that time and until well into the 1950s, the classic notion of "nucleoprotein" persisted, and molecular models of various uninterrupted, or modular, protein-bearing chromosomes were often postulated by the early chemically inclined biologists.

One of my tasks was therefore to furnish convincing proof that Maclyn McCarty and his coauthors were correct in their novel findings that transforming agent was simply and entirely a naked DNA. The chemical evidence was good, but essentially qualitative. I soon demonstrated that my "purest" transforming factor could have no more than 0.2% of protein as amino acid. By December 1948, I had devised convincing analyses of my hydrolyzed DNA which showed that even its tiny content of amino acid consisted entirely (100.7–102%!) of slowly released *glycine*. Old literature revealed that glycine is a regular product of the slow hydrolytic decomposition of adenine. This meant there was no protein at all, except for hypothetical error, which I generously allowed *might* be as much as 0.02% of the nitrogen. Delbrück and some other friends had been told about my findings.

---

[1] This statement may seem incorrect or gratuitous, until we recall facts from a mostly forgotten time frame. Avery et al. did report elementary analyses, of which mainly P and N/P (phosphorus, and nitrogen to phosphorus weight ratio) might seem useful. The N/P for their DNA ranged from 1.58 to 1.75, averaging close to that (1.69) of two DNAs well-defined at that time, from salmon sperm and calf thymus. However, as we well knew, proteins had very nearly the same N content as DNA (about 16%). Thus, with respect to hypothetical (activating or contaminating!) protein components, an N/P ratio was no more reliable than the P analysis (and the implied assumption that proteins do not contain P, one that could not now be supported). Therefore, according to the figures given, a genuine DNA that contained 10% of protein would show N/P of 1.89, and with 15% (or 25%) protein, 2.00 (or 2.28). Since P analysis in the 1940s amounted to an *ash* determination, and the expectation that the ash was *sodium* phosphate, there was considerable room for error in the "quantitative" analysis of that era. Individual base analyses could not be done until they were begun in the work revisited herein.

In April 1949, Alfred Hershey, back at St. Louis, wrote me, asking what I'd been telling Delbrück and Mark Adams about my evidence that "Avery's stuff is really DNA" (Fig. 1A). He wanted something definite to report at a coming roundtable. My glycine data weren't published, and I didn't get around to that until 1952 (Hotchkiss 1952), but I'd told others about them and was revealing the story in some local discussions and short lectures which abounded in this field that was about to grow into a chemistry of genetic processes. Therefore, I cooperated gladly, sending him some pages telling of the amino acid data as reported in the 1948 Paris symposium on Biological Units Endowed With Genetic Continuity (Hotchkiss 1949), together with my new glycine analyses. I quickly got a warm thank-you, indicating that he was satisfied that I had essentially a proof of the McCarty–Avery identification of the DNA nature of the transforming agent (Fig. 1B).

Although my data lay long unpublished, I take satisfaction in knowing that I'd reported them where it counted. Just think: This scholar who now agreed that my *DNA* really couldn't contain traces of such a nucleoprotein as his *virus,* was in three years to publish a classic paper (Hershey and Chase 1952), demonstrating that *his virus* indeed owed its biological activity to its phosphorus component, *DNA*, and not to its sulfur component, protein!

Recognizing that phage had a complex structure, they ingeniously sheared it away from the bacteria after infection had occurred. Much of the phosphorus but little of the sulfur remained with the bacteria! Of course, those first results with phages labeled in their phosphorus or sulfur moved in a year or two gradually up a ladder with figures like 80+, 85, 90, and 97% of DNA involved, but they had from the start something we didn't have. They were dealing with the *entire* genome of the phage, while we were accounting at that time only one by one for less than a dozen of the presumed "genes" of our bacteria. But I never heard AI Hershey overstate his case (as some tended to do); he had a clear sense of the strengths and weaknesses of the isotopic identification of DNA as phosphorus and protein as sulfur. In 1956 he summarized the accumulated knowledge of phage structure in a most clear and constructive way (Hershey 1956).

In retrospect, I see a lovely antiparallel symmetry (like the two strands of the DNA helix) in the phage and the bacterial research programs: With phage, Hershey had early established the *genetic* nature, and was moving a bit more slowly into the *chemical* entities responsible, whereas with bacteria, we had completed McCarty's strong identification of the DNA, but had to approach more slowly its genetic basis and generality. How fitting that each system contributed what it could do best! How natural that we in transformation had to face more, and earlier, the skepticism about that first revolutionary chemical identification of DNA!

The respective investigations benefited before long by participation of a few early coworkers. By the time of the 1951 Cold Spring Harbor Symposium, Harriett Taylor Ephrussi (an early disciple of McCarty and Avery) was reporting rare subtle variations in pneumococcus of a transmissible capsular trait that seemed to "recombine" to produce the normal capsule. By then I was showing that, like capsule traits, a group of laboratory-selected antibiotic resistance traits were embedded in DNA and transmitted faithfully—one mutation at a time. Other traits were being sought and found in DNA;

A

(1949)

WASHINGTON  UNIVERSITY

SCHOOL OF MEDICINE
SAINT LOUIS (10), MO.

DEPARTMENT OF BACTERIOLOGY AND IMMUNOLOGY
EUCLID AVENUE AND KINGSHIGHWAY

April 18

Dear Dr. Hotchkiss,

I hear from Adams and Dellbrück that you have convinced yourself that Avery's stuff is really DNA. I would like to be able to say something more definite about this at a round table on nucleic acids at the SAB meeting. If you are willing, I would appreciate hearing something about what you are doing.

Best regards,
A D Hershey

B

WASHINGTON UNIVERSITY

SCHOOL OF MEDICINE
SAINT LOUIS (10), MO.

DEPARTMENT OF BACTERIOLOGY AND IMMUNOLOGY
EUCLID AVENUE AND KINGSHIGHWAY

May 3, 1949.

Dr. R. D. Hotchkiss
The Rockefeller Institute for Medical Research
66th Street and York Avenue
New York 21, N. Y.

Dear Doctor Hotchkiss:

Thanks very much for the manuscript. The experiments are very beautiful. I will let you know if anything of interest comes out at the round table. My guess is nothing will.

My own feeling is that you have cleared up most of the doubts. Some people may cling to the virus theory a little longer, perhaps.

If you want the manuscript returned, please send me a post card.

Sincerely,

A. D. Hershey.

ADH/McK

Figure 1. Letters, A.D. Hershey to R.D. Hotchkiss, 1949. (*A*) April 18. (*B*) May 3, after receiving text, data, and illustrations from R.D. Hotchkiss.

Robert Austrian and very soon, Hattie Alexander, Grace Leidy, and Stephen Zamenhof, were other early colleagues participating in a similar way. By 1954 Julius Marmur and I had described a DNA that transferred a pair of unrelated, but linked, markers. Then, we could say that we too were really dealing with typical genetic markers. I will not attempt to describe the increase of enthusiastic contributions to phage biosynthesis and function (by, e.g., Seymour Cohen and Alan Garen) by the mid 1950s, but any chemical work leaned heavily on this analysis by Hershey and Chase.

Hershey and his coworkers were important contributors in the next few years to the next step in DNA chemistry—the gradual realization that in extended form, DNA was an extremely long fiber, challenging all our preconceptions of molecules, even *macromolecules*. Under the electron microscope, such material appeared long and uniform, and after a while most people were accepting that all natural DNAs presumably had the fundamental integrity indicated by our amino acid-free, or Hershey's phosphorus-rich, sulfur-deficient, specimens.

Were we interacting? cooperating? Certainly, although little of it had to do with techniques—far more about ideas and perhaps encouragement. Our first "antiparallel" reports came out in the same year, so cross-referencing did not occur then; it came later (see, e.g., Hershey 1956). The influence went in both directions, in quite a few conversations, involving also a meager correspondence. And outside, in other directions; it was an exciting time—one hurried to many short symposia and back to the laboratory, often neglecting a wider publication. The proud historians of science, studying diligently how science is documented (by dates), may miss much of how science is *made*. I suspect they can hardly ever sense the import of all the fertile, often subtle, relationships of teacher to student, and colleague to colleague, that carried the information and ideas of science forward and upward in any of its generous and productive periods and provinces!

Among the students in Max's phage course by 1946 was Mark Adams, my friend and colleague who at Rockefeller had been brought in to continue and extend my chemical work on capsular antigen of pneumococcus. By then he was a gifted teacher of microbiology at New York University Medical School. Inspired like so many by the events of phage development in bacteria, he soon moved both his research and teaching into this area. Already in the next summer, he helped Max with the course, and then Mark Adams became the official teacher of the summer phage course for six of the seven years, 1948–1954. Mark gave a popular and carefully reasoned presentation of the core material, achieving also its continued updating, and arranged supporting seminars.

Dedicated Service to the Phage Community

Although the original observations of the single burst, phage counting and cloning, use of antibody to block delayed infection, and so on, had been published, the intense productive development of the field had diverted the phage group from assembling a workbook covering the advancing methodology. Mark Adams conscientiously filled this gap with a handbook chapter (Adams 1950), properly authenticated and sanctified by a careful reading by Max Delbrück. Soon the pace of research led Adams to attempt a revision, conceived now as a volume on phage science more than methodology. Mark had this well planned and in process, when his promising career was tragically cut off at age 44 by an atypical, fatal splenic rupture during an infection of the usually mild-acting mononucleosis virus.

As both family friend and colleague, I was approached by his wife, Hazel, about possibly carrying forward Mark's book project. She gave me his preliminary copy for twenty chapters, outline plans for three or four more and for illustrations. Already too far out of the phage technology, with the backing of Alwin Pappenheimer, an NYU colleague of Adams, I asked Alfred Hershey to complete the book project.

Probably most colleagues have looked upon Hershey as a well-supported research worker; his successive sponsors, Milislav Demerec, Vannevar Bush, and Caryl Haskins (and John Cairns and James Watson of the Biological Laboratory) all recognized his genius and creativity. That and his cool reserve ("Why make things such a big deal?") have not led him to make conspicuous personal crusades. His generous trust of intelligent colleagues allowed him for a long time to escape committee or administrative posts.

But Hershey in early 1957 conscientiously agreed to proceed with Mark's phage book, despite the time and work it would demand. In a short time, he had approached a dozen colleagues and organized significant cooperation, reporting to us that he would see it through. With great good will, by four months all was arranged and in motion: new chapters and necessary revisions, new illustrations, etc. Bringing the ultimate book to press (Adams 1959) involved endless details, of course, and I know he labored much and carefully, an editor working for his community with consummate modesty.

Mark Adams as a chemist was the ultimate rationalist, but he approached biology as something of a "naturalist" to whom phage was a phenomenon. For medical students the charm of his teaching was romantic:

> Here are these bacteria growing.....what's this? —they're disappearing, clearing up!.....let's take a sample.....Well, look! They're growing again, finally!.....Now, what's all that about—what's going on?

As part of this appreciation, I want to confess what was an informed and deliberate decision we made at that time, choosing Al Hershey, who approached phage more as logic:

> So, there's this virus; nucleic acid in a protein envelope: in order to replicate, it has evolved a way to attach to suitable cells and transfer its nucleic acid.......Then the nucleic acid untwists and......

Hershey did move the book in this rational direction, in the new chapters and in some of Adams's text that he arranged to have revised. Something may have been lost, but something else was gained—for modern students, who are destined to be recipients and not performers of the research, and easily accept word pictures as if they were concepts.

But, then again, the experimental methodology had been de-emphasized. Hazel Adams had given me Mark's own personal copy of the Methods in Medical Research that bore his Phage Methods chapter (which I treasure). So, early on, I suggested that we simply reprint that chapter in the new book, after merely "dusting off" a few misleading or obsolete details. Executing this idea brought out another sidelight that tells of the good will which was general for this project.

Hershey's terseness was proverbial. Many of his missives were dashed off on the plain two-cent postal cards of the period, and often signed "Al" or, to save time, with the shorter initial, A. Even a few of the editorial communications were like that. When I proposed including the methods chapter, his *entire* response said exactly this:

> Dear Rollin,
> I think your idea about reprinting "Methods" is the best one yet
>
> A.

Since Ephraim Racker of NYU, who was friendly to all of this, and us, and Cold Spring Harbor, was at the time somehow involved with Interscience publishers, he arranged for the reprinting permission. Having seen Al's card, Racker couldn't resist an obvious playful imitation. As the quest proceeded, I received a similar card only slightly more burdened:

> E. R., P.H.R.I.
Ft.E. 15th St.
>
> Dr. R. Hotchkiss
> Rockefeller Inst. for Med. Research
> Dear R:
> Dr. P. of Interscience is going ahead with inquiries regarding best idea yet.
>
> ~~A.~~
> E.

Racker had first used Hershey's all-signifying initial, then obviously cringed at this crass "forgery," crossed it out, and supplied his own. The permission was granted and I enlisted help from Nancy Collins (early phage worker, later Nancy Bruce, in-law of the Delbrücks). We annotated, but did not update, Mark's 1950 methods chapter as a kind of parental seed in the new volume.

An Artisan Sharpens His Tools

Alfred Hershey was a scientist who certainly enjoyed designing and interpreting unambiguous experiments suited to answer a logical question. At a time in science history when one's important audience was the experts, he was a disciplined "scientist for the scientists," making reports typically compact and brilliantly clear.

In this connection, Hershey could tell a colleague about some work, adding wistfully, "I didn't suppose anybody would be interested in that!" —leaving an impression that there were more such data that he didn't think colleagues wanted to know about.

That may have been true—for a time—but I like to think that, even then, it was not diffidence about his data, but quiet confidence: He knew exactly what was significant *to him.* On the other hand, throughout his career, in publicly reporting his research to colleagues, he held back all personal feelings. Any emotional excitement, disappointment, or surprise the work gave him was rigorously excluded.

Commonly, however, I have sensed joy and satisfaction, expressed informally, in the execution of an experiment. Delbrück once reminded us that Hershey fantasized about a heaven in which one could do a successful experiment over and over—getting, I presume, new and useful data. I know that he took great pleasure simply in the *performance* of even a minor well-arranged experiment.

To illustrate Hershey's innate joy in experiment, I want to recall a simple story of a day when I "dropped in" to his lab to talk informally about something. I suppose it must have been about 1963. He welcomed me but apologized that he would from time to time take some routine readings on the spectrophotometer. I stayed with him, simply looking at a reprint or two, during three short interludes. Then, he showed me with obvious delight what he was finding out. It had to do with heat "melting" of the helical structure of phage DNA preparations. Four DNA samples were being slowly warmed to their denaturing temperature by circulating heated water around them. In a couple of hours, he had systematically moved DNA samples around, and found that the sample nearest the warmed water inlet always melted a little sooner than others. In effect, he was showing that an apparent difference between DNAs was due to an inapparent temperature gradient across the cell holder even in the rapidly circulated water flow.

This was no "big deal" result; rather, the collapse of a potentially interesting "difference" in the DNAs. But the instant pleasure of the scholar was unmistakable, and in private he could let it be revealed! Planning, performing, and interpreting experiments were joyful efforts; how fortunate that these were put, by his analytical intellect, largely at the service of asking broad and deep questions in his science.

Although somewhat reserved, Al Hershey shared with more than a few persons parts of the rich scientific life that he maintained. I also look back with satisfaction on a few quarter-hours we spent together in the general frame of tolerant amusement at the emotional excesses of other colleagues. I do not recall a time that he was critical, or more than vexed, about their behavior.

A Time for Reflection

Elsewhere I have commented that during the 1940s and 1950s not only did excitement grow in the molecular biology and genetic fields, but also during these years scientists began to address more and more of their findings to the interested, though less-than-expert, scholars outside their own fields. It is an important way for professionals to pass information into the broader public, but since those broad audiences are also less critical, it can be a tempting route to becoming a "popular" scientist.

Hershey eschewed such routes to easy "success"; for him even a great experiment could be allowed to seem "obvious." He avoided the temptation to offer pronouncements on the variety of general topics a Nobel Prize-winner is often asked about. As far as I know, he has been spared misquotations by science reporters, who can magnify the wrong aspect of a well-meant idea.

About 1962 Hershey accepted a management role, as the Carnegie Institution reduced its functions at Cold Spring Harbor to a Genetics Research Unit, basically to support Barbara McClintock and him. Their departments continued productively and developed the careers of a number of young associates. I mention it here mainly to quote bits of scientific philosophy that now appeared in the annual reports, a virtual duty of a leader at the new post.

Often the report opened with a crisp and efficient review, like one on quasi-sexual processes in bacteria which he chose to "recount in an appreciative, as opposed to an authoritative, manner" (Hershey 1967).

The scholar's goal could be compactly summed up: "Influential ideas are always simple. Since natural phenomena need not be simple, we master them, if at all, by formulating simple ideas and exploring their limitations" (Hershey 1970).

Al Hershey knew that scientists were fallible human beings, too; one preamble for me sums up at once the humility and grandeur of his vision of a research professional's problem (Hershey 1966). I quote, with some slight elisions:

> The universe presents an infinite number of phenomena. The faith of the scientist, if he has faith, is that these can be reduced to a finite number of categories. ...To speak of goals at all is... unscientific.... One cannot measure progress toward the goal of understanding.... If understanding is reached through creative acts, they are partly acts of faith. These are large questions. They are pondered by professional thinkers, who evidently believe in the power of abstract thought. If that power is efficacious, it behooves the scientist to exercise it now and then when his experiments flag. Otherwise he risks failing a personal goal: to see his work in selfless perspective.

When I made a last visit to Hershey at the time of the Phage Anniversary Celebration in August, 1995, I was admitted by his wife Jill, who cheerfully and faithfully as always supported the social/monastic balance of his lifestyle. I found him practicing with zest a sophisticated computer programming and management of musical and scientific library data files. I photographed him as he searched around to give me a couple of machine-language programs he'd written. Now I've had to darken with my computer the prohibitively bright background I impulsively accepted (Fig. 2). It was wonderful to find that this represented another in his chain of intelligently managed crafts: his science, in early days sailing, moving on to furniture making. And now, in his latest years, this genius who had described phage infection and *bursts* in the terms that have entered the very lexicon of molecular genetics, had learned to convert such word, data, or number information into—of all things!—precisely timed *electron-bursts!* In this sense, even in retirement, Al Hershey was practicing—and much enjoying—a modern reversal of what he always did so well.

Figure 2. A.D. Hershey in his home, August 27, 1995 (photo by R.D. Hotchkiss).

References

Adams M.H. 1950. Methods of study of bacterial viruses. In *Methods in medical research* (ed. J.H. Comroe, Jr.), vol. 2, pp. 1–73. Year Book Publishers, Chicago, Illinois.

———. 1959. *Bacteriophages.* Interscience, New York.

Hershey A.D. 1956. The organization of genetic material in bacteriophage T2. *Brookhaven Symp. Biol.* **8:** 6–16.

———. 1966. Annual report of the director of the genetics research unit. *Carnegie Inst. Wash. Year Book* **65:** 559.

———. 1967. Annual report of the director of the genetics research unit. *Carnegie Inst. Wash. Year Book* **66:** 3–6.

———. 1970. Genes and hereditary characteristics. *Nature* **226:** 697–700.

Hershey A.D. and Chase M. 1952. Independent functions of viral protein and nucleic acid in growth of bacteriophage. *J. Gen. Physiol.* **36:** 39–56.

Hershey A.D. and Rotman R. 1949. Genetic recombination between host-range and plaque-type mutants of bacteriophage in single bacterial cells. *Genetics* **34:** 44–71.

Hotchkiss R.D. 1949. Etudes chimiques sur le facteur transformant du pneumocoque. In *Unités biologique douées de continuité génétique* (Paris Symposium), pp. 57–65.

———. 1952. The role of desoxyribonucleates in bacterial transformations. In *Phosphorus metabolism* (ed. W.D. McElroy and B. Glass), vol. 2, pp. 426–439. Johns Hopkins University Press, Baltimore, Maryland.

Rollin D. Hotchkiss *(born 1911 in South Britain, Connecticut) came to biology through chemistry, attracted by the specificities exhibited in immunobiology and type transformations of bacteria. Rollin's detailed component analysis of purified "transforming principle," recounted in his article, removed any doubts that DNA was the responsible agent, establishing a relevance of DNA to heredity in bacteria. This relevance was further cemented by his demonstration, with Julius Marmur, of genetic linkage in bacterial transformation, whose implications inspired him to offer key concepts in the molecular modeling of genetic recombination. Rollin's latest work, in collaborations that included Magda Gabor, has revealed the provocative "silencing" of one of the two genomes, also full, sometimes reciprocal, genome recombination, in artificially produced diploid bacteria. Rollin spent his career at The Rockefeller Institute (later University) in New York City, which afforded him opportunities to participate in summer activities at Cold Spring Harbor and in Laboratory governance. As the "Oldest Living Graduate" of the Phage Course, he was known to many through his Profound Pronouncements at Graduation Ceremonies.*

Closing the Circle:
A.D. Hershey and Lambda I

WILLIAM F. DOVE

M Y STUDIES OF BIOLOGY BEGAN with phage lambda in its heady days from 1962 to the mid 1970s. Two of my senior colleagues introduced me to A.D. Hershey by reputation, but these were two *very different* reputations. Here I complete the circle by sharing with you my own experience with Al Hershey that culminated in 1971 with the publication of the Cold Spring Harbor monograph now called Lambda I (Hershey 1971).

One impression of Hershey was given to me by François Jacob.

> I have one distinct memory of Al Hershey at a Cold Spring Harbor meeting. He gave one of his particular talks spending half the time looking at the board without talking. After that there were questions. One of the guys in the audience talked for a very long time at full speed, finally asking his question. After that Al looked deeply at the floor thinking very hard for about 3 or 4 minutes... and finally said: "NO!"

The other impression of Hershey was shared by Jean Weigle, that denizen of the Delbrück laboratory and the California desert who by his enjoyment of science brought so many of us into touch with lambda in preference to the more virulent phages. As Weigle and I worked together on the change in state of the DNA of lambda after infection, he read to me from a letter of Hershey's describing the development of methods of ultracentrifugation by which he could characterize the large unbroken DNA molecules of phage. These methods led to the discovery of lambda's cohesive ends by which it circularizes after infection. Said Hershey, "There is nothing more satisfying to me than developing a method. Ideas come and go, but a method lasts!"

During the mid 1960s, I discussed with Hershey the issues surrounding lambda's DNA eclipse. I soon learned the deep reason for Jacob and Weigle's respect: Al Hershey had a passion for the truth, and even the famous success of the Hershey-Chase experiment did not close his open mind. No answer went unquestioned by Hershey. Indeed, when he had described the ^{35}S/^{32}P experiment in a lab meeting, he emphasized: note that about 1% of the ^{35}S counts *did* enter the cell, and perhaps *it* contributes to the phage heredity (Waclaw Szybalski, pers. comm.). Hershey related to my study of the lambda DNA eclipse and maturation cycle as one way to determine whether more than polynucleotide synthesis was needed for infectious DNA. Indeed, his final scientific publication (Hershey 1970) focused on phenomena such as cortical inheritance that only now have moved to stage center with prion agents. Ironically, it was Hershey's discovery of the cohesive ends of lambda DNA by ultracentrifugation analyses that afforded an explanation for the DNA eclipse/maturation cycle that Weigle and I had found.

I was not the only young lambda investigator who found Hershey's critical attitude and operationalism to be an invaluable guide. When Mark Ptashne purified lambda repressor, he wanted Hershey's critique of his manuscript. The first draft or two was returned with *everything* crossed out; an occasional phrase was marked "Good!" Finally, a draft came back that looked promising. First page ... nothing. Encouraging! Second page ... nothing. Fine! Third page ... still nothing. *How is it possible?* At the last paragraph of page 4, an arrow was found, with a message: *"Start here"* (Mark Ptashne, pers. comm.). (Postscript: Since this talk I have learned of at least one other early lambda investigator with the same story — Michael Yarmolinsky.)

Jim Watson has a talent for selecting the right person for the task. So it is not surprising that Al Hershey was asked by Watson to organize and edit the monograph on Bacteriophage Lambda (Hershey 1971)consequent to a conference in September 1970. I was in Cold Spring Harbor that summer teaching the 25th (and final) Phage Course, together with René Thomas and Ariane Toussaint from Brussels. Al asked me to work with him on this project. What Al Hershey accomplished in this final major scientific effort involved a special blend of the critical taciturnity that Jacob and Ptashne had known, the fundamental trust in sound experimental methods that Weigle had noted, and the frank but warm collegiality that I came to enjoy. He wrote notes to individuals and to groups of authors, aiming for clarity and rigor. He made sure that each research paper was critically refereed. He started with his own writing, giving us an inner view of why he was so taciturn. "Give it Hell!" he said to me, and to himself.

Here is one of his notes to everyone:

I am sending this letter to all chapter authors because it cost me days of thought and it deals with the general problem of *saying what you mean.*

I am Mr. Average Reader of your chapters. I am trying to put together a hearsay account of regulation that I have picked up from you and other authorities. In particular, I am trying to fill in things that I didn't see in your chapters, and to grasp the language and interpretations in the best way I can.

Everybody including me seems to feel a need to revise terminology at this point. ... [But] let's try *not* to introduce any more words whose meaning cannot be found in the dictionary. The end result of this sort of naming is ... to use *three* words every time you mention one thing because at most you can hope that the reader will know *one* of them. I imagine some of the people who do this (philosophers, lawyers, and a few scientists) invoke the principle that redundant messages are resistant to noise. To oppose this view I offer the proposition that three synonyms, none of which can be found in the dictionary, make pure noise.

In November of 1970 Hershey culminated his campaign for clarity with a 2-page document called WORDS. Here are a few of his comments:

map (verb). *Genes don't, geneticists do.*

lysogenic. *Means 'generating lysis', practically the same as 'lytic.' A temperate phage is not lysogenic, it is lysogengenic. 'Lysogenic' has become quite virulent.*

dilysogenic?

lysogenic excision?

dilysogenic excision??

These words have a fine ring until you ask yourself what they mean.

CHOICES OF WORDS:

Clear but nasty: replication inhibition, lambdoid, transcription initiation, and too many others to list.

Nasty but interesting: heteroduplex mapping. Here heteroduplex could be an adjectival noun, which is merely bad. It could also be an adjective and drive you crazy.

Unclear, nasty, and dull: immediate early.

I once observed to Chargaff that scientists don't have time to read each other's papers anymore. Speaking as an editor he said, "*They don't have time to read their own papers.*"

I have lots of time.

Hershey

To all authors of general chapters

November 20, 1970

After Lambda I, Al Hershey "moved on to other interests." He was in his early 60s, 25 years ago. (In the milieu of 1997, investigators of that age are "just reaching their prime!") My contact with Al and Jill Hershey came to involve opportunistic lunch or dinner meetings when I came to a Cold Spring Harbor meeting.

Al Hershey treasured his work and he treasured his friends. We can be grateful for the intensity with which he honored these treasures.

References

Hershey A.D. 1970. Genes and hereditary characteristics. *Nature* **266:** 697–700.
———. Ed. 1971. *The bacteriophage lambda. Cold Spring Harbor Laboratory, Cold Spring Harbor, New York.*

Bill Dove, *born in 1936 in Bangor, Maine, was an American pioneer of research on bacteriophage lambda, with special relevance to the mechanisms by which that phage controlled the expression of its genes and the replication of its chromosome. Bill's thoughtful expertise led Al to seek his collaboration in writing the much admired* Introduction to The Bacteriophage Lambda. *Bill's more recent interests have been in slime mold development and neoplasia in mice. All geneticists know him as co-editor of the Perspectives section of* Genetics.

The Size of the Units of Heredity

JOHN CAIRNS

The only means of strengthening one's intellect is to make up one's mind about nothing—to let the mind be a thoroughfare for all thoughts.

JOHN KEATS

AL HERSHEY ONCE ASKED ME if I knew who invented the term "negative capability," and I said it was the poet Keats. The question was not really surprising. The phrase must have attracted his attention because it so precisely described what he himself believed. For he seemed to approach every subject, scientific or not, with a uniquely uncluttered mind. Facts were viewed with the utmost suspicion; nothing was taken for granted. It was as if he had decided, early in his career, that the burden of proof rested on his shoulders alone and that it was his responsibility to protect everyone from all possible sources of confusion. He once summarized his reservations in the following way:

> ...the account of any successful phase of research in biology always contains a lesson about the communal nature of scientific effort...For biological laws (unlike physical ones, perhaps) do not emerge from the elucidation of single mechanisms, however intellectually satisfying that may be. The single mechanisms, and especially that part of it decipherable in a given species, is usually a truncated, even in a sense a pathological, variation on a theme. As such it is at best a clue to that theme, at worst a senseless diversion. To discover the theme many clues, including the false, must be run to ground (Hershey 1964).

Those who never met him can get a glimpse of his character from the Annual Reports of the Carnegie Institute of Washington. Every now and again he included in his section some unexpected words of caution. At the time, they seemed needlessly fastidious. But when we come across them, years later, we breathe a sigh of relief to find that, once again, the master's caution has paid off. For example, he was reluctant to believe that the blendor experiment ruled out any important role for the small amount of protein that came into a cell with a virus's nucleic acid; and now we wonder if he had in mind enzymes such as reverse transcriptase that are present in some viruses because they are needed to get things going. Similarly, he said in one of his later annual reports (Hershey 1970) that he did not totally accept the supremacy of nucleic acids as the sole source of information, and he waxed lyrical about cytoplasmic inheritance in *Paramecium* and *Stentor*; looking back, we wonder if he might have had prions in mind.

His name became something of a household word as the result of the famous "blendor experiment"; (in deference to him, I use his initial spelling of the word blendor). But that was just one step in his long journey to try to establish the rules of molecular biology. His first paper on the molecular biology of bacteriophages was published in 1946, and in the next 25 years he produced about 50 papers in which he wrestled with some of the major questions of the times. Not enough, these days, to satisfy an NIH study section, but it did at least get him a Nobel Prize.

Young molecular biologists, if they think about it at all, will not see why the discovery of the structure of DNA should have been followed by a time of great doubts and uncertainties. Nowadays, if you want to study the genetics of some new organism you sequence its genome, put in for a few patents and then publish the sequence on the internet. For a period of about 10 years, however, the relationship between genetic map and DNA sequence, and between chromosomes and DNA, seemed anything but clear. There was the feeling, I think, that it could not be that simple. What settled the matter was, to a large extent, the study of T2 and T4 bacteriophages and their host *E. coli*.

The Background

To explain the mindset of the times, it is necessary to backtrack a few years. In 1948, Hershey had reported that mixed infection of bacteria with marked strains of T2 resulted in genetic recombination between the infecting particles (Hershey and Rotman 1948). This allowed the construction of a genetic map containing three linkage groups. In that sense, therefore, T2 was just like any other living creature. Any doubts about the right of viruses to count as authentic organisms had been settled with the discovery, by electron microscopy in 1942, that the T-even phages have a distinct head and tail (Luria and Anderson 1942). Although their shape was the shape of a complex living creature, their chemistry was simple. T2, for example, contained roughly equal parts of protein and DNA, and that was all. Several groups therefore set about an analysis of virus multiplication by following the timing of synthesis of the DNA and protein of the phage, using ^{32}P and ^{35}S as markers. This quickly led Hershey to the "blendor experiment" (Hershey and Chase 1952), which backed up Avery's study of bacterial transforming principle and produced the doctrine that DNA was the be-all and end-all of biology.

The form of genetic maps and the structure of DNA implied that the information required to generate life in three dimensions was itself one dimensional. Yet many people found it hard to imagine that biology could be reduced to something so simple; there had to be something more than that at work. In particular, the old guard were reluctant to see proteins being demoted from their place as the supreme arbiters and managers of all biological phenomena (Allfrey and Mirsky 1957). Even if you were a believer, the way ahead was not obvious. Although the structure of the double helix showed how biological information was stored in a readily duplicatable form, X-ray crystallography could not show how molecules of DNA were arranged to form a genetic map or show what was the exact mechanism used for the duplication of double helices. Did each chromosome contain a single molecule of DNA, or was the unit of DNA the individual band in a chromosome, the individual gene, or something smaller than that? Was the functional unit composed of one double helix, or perhaps a pair of them lying side by side? Was the duplication of DNA accompanied by permanent separation of the two polynucleotide chains of the double helix (as suggested by Watson and Crick), or did the chains stay together after being copied?

Hershey, Melechen, and Tomizawa showed that when *E.coli* was infected with T2 phage, synthesis of the phage DNA that was going to end up in progeny phage particles preceded the synthesis of the progenies' protein (Hershey et al. 1956, 1957). Furthermore, the addition of chloramphenicol a few minutes after infection stopped all protein synthesis but did not stop either the replication of T2's DNA (i.e., the accumulation of DNA that would end up in the progeny when the chloramphenicol was removed) or the accumulation of recombinants; and when this intracellular DNA was subjected to UV irradiation, the resulting progeny had exactly the same properties as finished particles of T2 that had been irradiated. So, by 1956, Hershey was able to conclude that *"the chromosomes of future phage particles multiply in the form of naked molecules of DNA in the infected bacteria"* (Hershey et al. 1957). Actually, the effect of chloramphenicol had certain unexpected features that seemed to imply some form of information transfer from DNA, which depended on protein synthesis and rendered the cell radiation resistant, and this complication kept alive the notion that the DNA story might not be as simple as it seemed at first sight. (If there is a moral to be drawn from this episode in the history of phage research, it is perhaps that subtle, complicated experiments, especially experiments in radiobiology, have a tendency to produce results that are too subtle and complicated to be understood.)

If phage multiplication did indeed consist primarily in the replication of naked molecules of DNA, the next step was to identify the exact form of the molecules. The first genetic map for the T-even phages contained three linkage groups. Somewhat later all the known genetic markers were shown to be linked together (Streisinger and Bruce 1960), although the very high rate of crossing-over ensures that the linkage between distant pairs of markers does not survive very long. It remained to determine what was the relation between T2's single linkage group, or "chromosome," and its DNA. The DNA content of T2 amounted to slightly more than 100 million daltons of DNA per particle. When this DNA was extracted and purified by the routine methods of biochemistry and then analyzed by sedimentation or light scattering, it was found to be in molecules of only a few million daltons. This implied that the phage chromosome was made up of many molecules of DNA somehow joined together to generate a single genetic linkage group.

An entirely novel way of looking at this DNA was invented by Levinthal (Levinthal 1956). If the phage was labeled with ^{32}P and broken open by osmotic shock, and the contents then embedded in an autoradiographic emulsion, the size of the DNA could be determined by measuring under a microscope the number and phosphorus content of the sources of β-particles. This technique gave a completely different answer. The average yield of DNA per phage particle was found to be one large piece of DNA of about 50 million daltons plus several much smaller pieces. This encouraged the horrible thought that there might be two kinds of DNA; the large piece (perhaps made up of many molecules joined end to end) was the phage's "chromosome" and contained all the genes, whereas the small pieces by default had to have some other, non-genetic function.

Further experiments tended, if anything, to support the idea that there were two kinds of DNA. Earlier studies had shown that almost half of the DNA in an infecting T2 particle is transferred to its progeny (the exact proportion depending on how exactly the experiment was done). A similar study of the fate of the large and small pieces of T2 DNA showed that roughly half of the big piece was, on average, transferred to one of the progeny, and that that half was then transferred intact during the next cycle of infection (Levinthal and Thomas 1957). This, of course, was exactly the behavior expected of a double helix replicating according to Watson and Crick, and it supported the idea that the big piece was orthodox "genetic" DNA and the small pieces were something else.

One other problem contributed to the confusion. If a single T2 particle can produce roughly 2^8 progeny within about 20 minutes, each round of replication cannot last more than 150 seconds. If all the DNA in the big piece were simply a single molecule of 50,000 daltons which had to separate its strands in order to replicate, it would have to be unwinding at a rate of several thousand revs per minute. Although physically possible, such a process was thought by some to be implausible.

In 1956, these and other early problems of molecular biology were discussed at a meeting entitled "The Chemical Basis of Heredity," organized at Johns Hopkins University along the lines of the Cold Spring Harbor Symposia, and it was at that meeting that Delbrück and Stent invented the terms "conservative," "semi-conservative," and "dispersive" for the possible ways in which the DNA double helix might be duplicated (Delbrück and Stent 1957). Anyone who wants an idea of the confusion of the times should read the published volume, because it represented a turning point in the history of molecular biology. Interestingly (and this may be a general rule), the shortest articles were often those that gave a foretaste of what was to come, and some of the longest articles were the protests of the old guard on their way out.

In the same year, confidence in the essential simplicity of molecular biology received a great boost with the discovery of the first DNA polymerase (Kornberg 1957); this, in a way, was the experimental justification for believing that DNA really did have the innate property of being copyable without the intervention of yet another, overriding source of information.

The dispersive model of DNA replication did not survive very long. In 1957, the chromosomes of the broad bean, *Vicia faba*, were shown by labeling with [³H]thymidine to replicate semiconservatively, as if each contained two units of DNA that segregate during replication but do not disperse further during subsequent divisions (Taylor et al. 1957). Although there was no proof that the two units were the two polynucleotide chains of a double helix, they were like the chains of DNA because they behaved as if they were of opposite polarity during sister chromatid exchange.

A year later, the ^{15}N/^{14}N density-transfer experiment showed that the DNA of *E. coli* replicates semiconservatively (Meselson and Stahl 1958). Although each bacterium apparently contained several hundred molecules of DNA each of a few million daltons, the replication of these molecules was plainly under some kind of rigid control because every molecule was replicated once during each round of replication. In other words, no molecule was replicated twice until every molecule had been replicated once.

These results, using radioactive and density labels, were exactly what would be expected if each chromosome consists of a single molecule of DNA with two polynucleotide chains that segregate when the chromosome is duplicated. The experiments could, however, be interpreted in a more complicated way. It seemed almost inconceivable that the entire DNA in the largest chromosome of *Vicia faba* (almost 10^{10} base pairs) could be in a double helix that unwinds from one end to the other each time it is replicated; it was more far reasonable to imagine that the chromosome consisted of a linear array of short molecules joined by some special form of linker. As for the separation of the chains of the duplex, there was apparently something peculiar about *E.coli* DNA because the independently segregating units could be separated by heating whereas salmon sperm DNA did not appear to contain separable units (Meselson and Stahl 1958); thus it seemed conceivable that *E. coli* DNA consisted of two double helices lying side by side (and separable by heating), whereas eukaryotic DNA did not. In short, if you really wanted to be difficult you could find support for many very different models for the relation between the structure of the double helix and the structure of chromosomes.

In 1957, I think that Hershey found himself facing a difficult choice. The easiest course would have been to continue his studies on the biochemistry of phage infection; after all, his study of what he called the phosphorus economy of the phage-infected cell had yet to explore what happens to RNA. His way of reaching a decision was to write reviews of the two aspects of phage that had interested him in the past. The first review was a comprehensive account of what was known about "Bacteriophages as genetic and biochemical systems" (including a brief discussion of the possible importance of the minor class of RNA found to turn over rapidly after phage infection!) (Hershey 1957); the second was an exposition of the mathematical complexities in analyzing phage recombination (Hershey 1959). Having in that way, as it were, cleared his desk, Hershey turned his attention to what he considered the main source of confusion, namely the lack of any reliable method for determining the molecular weight of DNA.

Column Chromatography of DNA

In 1956, several methods were available for determining the molecular weight of large molecules. The most widely used was the measurement of sedimentation rate. Unfortunately this, and all the other methods, were dependent on theory and therefore were no more precise than the underlying theory. In particular, application of the theory to an exceptionally long thin molecule such as DNA could easily contain an error of a factor of two, and that would make it impossible to distinguish between, for example, simple double helices and pairs of double helices lying side by side. Obviously, a study of bacteriophage DNA offered less chance of being misled by theory, because the total amount of DNA in each particle was known very precisely and it was known to comprise, at the very most, a rather small number of molecules.

Various forms of column chromatography had, in the past, been used to separate different species of DNA. For example, DNA would bind to kieselguhr coated with basic proteins such as histone, and the salt concentration required to elute it depended on its molecular weight and base composition (Brown and Watson 1953; Brown and Martin 1955). Hershey, Mandell, and Burgi chose to study the behavior of columns that had been developed using kieselguhr coated with methylated serum albumen (Lerman 1955; Hershey and Burgi 1960; Mandell and Hershey 1960).

There followed what must have been a period of intense frustration. At first, it seemed that T2 DNA came in three distinct sizes, the smallest of which was the kind that is transferred from one generation to the next. As Hershey wrote in his annual report, *"This result is incomprehensible in terms of any of the existing ideas about the mechanism of transfer"* (Hershey et al. 1957). After a full year's work, however, it turned out that the column's separation of the DNA into three fractions was an artifact due to channeling in the column. So it was back to the beginning again, and a study of how to make and load the columns.

Meanwhile, two other groups had been measuring the molecular weight of T2 DNA by sedimentation and had arrived at answers that, in terms of molecular weight, were different by a factor of roughly four (Fleischman 1959; Thomas and Knight 1959). Their methods of preparation were, however, somewhat different, and this suggested that very large molecules of DNA might be extremely fragile. It had long been known that DNA could be broken into small pieces by sonic vibration, but no one had thought to wonder about the fragility of molecules the size of the big piece of T2 DNA. When Davison reported in 1959 that simple pipetting was enough to break T2 DNA (Davison 1959), everything changed, as it were, overnight.

When due precautions were taken in the way it was extracted, handled, and loaded onto properly constructed columns, T2 DNA gave a single peak in Hershey's columns and behaved like a single species on sedimentation. Unfortunately, since there was no absolutely reliable theory about the sedimentation of such large molecules, this result could have meant that T2 contains one double helix of DNA, or two double helices side by side, or two separate double helices of exactly equal size. Column chromatography did, however, allow Hershey to investigate the relative sizes of the fragments produced by different levels of turbulence. T2 DNA was shown to have no preferred breakage points; on stirring at increasing speed, the molecules first broke in half and then into quarters. Each of these species had its own characteristic fragility and characteristic point of elution from the column (Hershey and Burgi 1960).

The question of their absolute size was resolved in a collaboration with Davison, Levinthal, Rubenstein, and Thomas (Davison et al. 1961; Rubenstein et al. 1961). T2 was labeled with high-specific-activity ^{32}P, and its DNA was extracted intact (as judged by column chromatography). When this unbroken DNA was embedded in autoradiographic emulsion, the number of radioactive centers ("stars") and their ^{32}P content was the same as that of the intact phage particles from which they had been derived. This meant, therefore, that each phage particle contained a single piece of DNA. And once that was known, it became possible to determine, once and for all, the correct relation between DNA sedimentation rate and DNA molecular weight.

The next question was whether T2's chromosome was simply a double helix or, alternatively, was made up of two double helices somehow linked together, side by side. To resolve this it was necessary to measure the length of the chromosome. This was first done by electron microscopy of quarter-length fragments, which showed that they had the expected length and width for double helices (Beer and Zobel 1961). Shortly afterward, intact T2 chromosomes were measured by autoradiography after labeling with [^{3}H]thymidine and shown to have the right length for double helices with a molecular weight of 130 million (Cairns 1961).

What now remained to be established was that the semiconservative duplication of chromosomes, demonstrated in the broad bean and in *E. coli*, really did represent separation of the strands of the double helix, rather than the separation of pairs of double helices lying side-by-side. T2 was not ideal for this exercise because its DNA was subject to so much fragmentation by recombination and therefore could not be easily followed from one infectious cycle to the next. Fortunately, the DNA in lambda phage indulges in less recombination, and it could be shown to undergo semiconservative replication (Meselson and Weigle 1961) and to have the length expected if each particle (like the particles of T2) contained a single molecule (Cairns 1962; Ris and Chandler 1964). While these studies were going on, a quite independent proof emerged that the strands of the double helix do separate during replication. This was the demonstration that when *E. coli* was forced to use 5-bromouracil instead of thymine it made hybrid BU-T DNA that had a melting point halfway between BU-BU and T-T DNA, implying that the units that separate during replication are indeed the chains that separate when DNA is heated in vitro (Baldwin and Shooter 1963).

Once the nature of the T2 chromosome had been settled, Hershey moved on to study other phages, such as T5 and lambda. It quickly became clear that their chromosomes were somewhat different. The behavior of lambda DNA seemed particularly mysterious. Although density-labeling experiments seemed to show that each virus particle contained a single molecule that replicated semiconservatively, its DNA did not give a clean band on sedimentation and would not pass through columns that would nicely separate other species of DNA. The resolution of this mystery forms part of the subject of the next chapter.

Suffice it, by about 1962 the initial doubts had been laid to rest. The essential ingredients of Watson and Crick's vision had been established. Enzymes exist that can make new polynucleotide chains which are complementary to the chains of a DNA double helix, and during this process the two parental chains come apart and are transferred to the two daughter molecules.

Hershey's Place in the History of Biology

Hershey's main accomplishment was to identify the units that are the basis for molecular biology—first by showing (in the blendor experiment) that the genes of a virus are contained in its DNA, and second by showing that the "chromosome" in phages such as T2 and lambda consists of one double-stranded molecule of DNA. Put that way, the exercise sounds like the kind of story that is turned into a best-seller and a TV documentary. But Al's style of research seemed purposely to avoid being glamorous, as if it were solely up to him to ensure that truth would never be sacrificed in the excitement of the moment. I cannot imagine anyone else spending two or three years in developing yet another way of measuring molecular weight. But that was what he did because he knew it had to be done.

For those who were lucky enough to have known him, the memory is of a gaunt figure watching over us all to make sure that we would get it right. This role of his as the guardian of others was brought home to me most vividly in a letter he wrote in response to a manuscript I had sent him. His list of cautionary comments culminated with the words: "Not that I object to your specific interpretations: on the contrary. But you might draw a clearer line between them and what the reader is forced to accept. To define the latter is in my opinion the deepest obligation and the hardest job of the scientific writer...." And then, as if to remove some of the sting, he added the parenthesis "(if you will excuse the short homily)".

Throughout his career he never once offered any opening for the social constructivists' view of science. Never was there any sign that his conclusions were designed to give him peace of mind; indeed, it almost seemed that he sought the very opposite. For example, shortly after doing the blendor experiment, he listed all the features of DNA that made it so exciting and then added:

> None of these, nor all together, forms a sufficient basis for scientific judgment concerning the genetic function of DNA...My own guess is that DNA will not prove to be a unique determiner of genetic specificity, but that contributions to the question will be made in the near future only by persons willing to entertain the contrary view (Hershey 1953).

It was part of his demanding character that his role of intellectual ascetic had to be accompanied by a certain level of physical asceticism. Each year he allowed himself one brief respite from the rigors of his life — a holiday sailing on the Great Lakes — and I remember, after one of those holidays, that he ended a letter with the words "I am home again, too healthy to think very hard about experiments."

References

Allfrey V. and Mirsky A.E. 1957. The nucleus and protein synthesis. In *The chemical basis of heredity* (ed. W.D. McElroy and B. Glass), pp. 200–231. Johns Hopkins Press, Baltimore, Maryland.

Baldwin R.L. and Shooter E.M. 1963. The alkaline transition of BU-containing DNA and its bearing on the replication of DNA. *J. Mol. Biol.* **7:** 511–526.

Beer M. and Zobel C.R. 1961. Electron stains. II. Electron microscopic studies on the visibility of stained DNA molecules. *J. Mol. Biol.* **3:** 717–726.

Brown G.L. and Martin A.V. 1955. Fractionation of the deoxyribonucleic acid of T2r bacteriophage. *Nature* **176:** 971–972.

Brown G.L. and Watson M. 1953. Heterogeneity of deoxyribonueclic acids. *Nature* **172:** 339–342.

Cairns J. 1961. An estimate of the length of the DNA molecule of T2 bacteriophage by autoradiography. *J. Mol. Biol.* **3:** 756–761.

———. 1962. A proof that the replication of DNA involves separation of the strands. *Nature* **194:** 1274.

Davison, P.F. 1959. The effect of hydrodynamic shear on the deoxyribonucleic acid from T2 and T4 bacteriophages. *Proc. Natl. Acad. Sci.* **45:** 1560–1568.

Davison P.F., Freifelder D., Hede R., and Levinthal C. 1961. The structural unity of the DNA of T2 bacteriophage. *Proc. Natl. Acad. Sci.* **47:** 1123–1129.

Delbrück M. and Stent G.S. 1957. On the mechanism of DNA replication. In *The chemical basis of heredity* (ed. W.D. McElroy and B. Glass), pp. 699–736. Johns Hopkins University Press, Baltimore, Maryland.

Fleischman J.B. 1959. "A physical study of phage DNA." Ph.D. Thesis, Harvard University, Cambridge, Massachusetts.

Hershey A.D. 1953. Functional differentiation within particles of bacteriophage T2. *Cold Spring Harbor Symp. Quant. Biol.* **18:** 135–139.

———. 1957. Bacteriophages as genetic and biochemical systems. *Adv. Virus Res.* **4:** 25–61.

———. 1959. The production of recombinants in phage crosses. *Cold Spring Harbor Symp. Quant. Biol.* **23:** 19–46.

———. 1964. Some idiosyncrasies of phage DNA structure. *Carnegie Inst. Wash. Year Book* **63:** 580–592.

———. 1970. Genes and hereditary characteristics. *Carnegie Inst. Wash. Year Book* **68:** 655–668.

Hershey A.D. and Burgi E. 1960. Molecular homogeneity of the deoxyribonucleic acid of phage T2. *J. Mol. Biol.* **2:** 143–152.

Hershey A.D. and Chase M. 1952. Independent functions of viral protein and nucleic acid in growth of bacteriophage. *J. Gen. Physiol.* **36:** 39–56.

Hershey A.D. and Rotman R. 1948. Linkage among genes controlling inhibition of lysis in a bacterial virus. *Proc. Natl. Acad. Sci.* **34:** 89–96.

Hershey A.D., Burgi E., Mandell J.D., and Melechen N.E. 1956. Growth and inheritance in bacteriophage. *Carnegie Inst. Wash. Year Book* **55:** 297–301.

Hershey A.D., Burgi E., Mandell J.D., and Tomizawa J. 1957. Growth and inheritance in bacteriophage. *Carnegie Inst. Wash. Year Book* **56:** 362–364.

Kornberg A. 1957. Pathways of enzymatic synthesis of nucleotides and polynucleotides. In *The chemical basis of heredity* (ed. W.D. McElroy and B. Glass), pp. 579–608. Johns Hopkins University Press, Baltimore, Maryland.

Lerman L.S. 1955. Chromatographic fractionation of the transforming principle of pneumococcus. *Biochim. Biophys. Acta* **18:** 132–134.

Levinthal C. 1956. The mechanism of DNA replication and genetic recombination in phage. *Proc. Natl. Acad. Sci.* **42:** 394–494.

Levinthal C. and Thomas C.A. 1957. Molecular autoradiography: The β-ray counting from single virus particles and DNA molecules in nuclear emulsions. *Biochim. Biophys. Acta.* **23:** 453–465.

Luria S.E. and Anderson T.F. 1942. The identification and characterization of bacteriophages with the electron microscope. *Proc. Natl. Acad. Sci.* **28:** 127–130.

Mandell J.D. and Hershey A.D. 1960. A fractionating column for analysis of nucleic acids. *Anal. Biochem.* **1:** 66–77.

Meselson M. and Stahl F.W. 1958. The replication of DNA in *Escherichia coli. Proc. Natl. Acad. Sci.* **44:** 671–682.

Meselson M. and Weigle J.J. 1961. Chromosome breakage accompanying genetic recombination in bacteriophage. *Proc. Natl. Acad. Sci.* **47:** 857–868.

Ris H. and Chandler B.L. 1964. The ultrastructure of genetic systems in prokaryotes and eukaryotes. *Cold Spring Harbor Symp. Quant. Biol.* **28:** 1–8.

Rubenstein I., Thomas C.A., and Hershey A.D. 1961. The molecular weights of T2 bacteriophage DNA and its first and second breakage products. *Proc. Natl. Acad. Sci.* **47:** 1113–1122.

Streisinger G. and Bruce V. 1960. Linkage of genetic markers in phages T2 and T4. *Genetics* **45:** 1289–1296.

Taylor J.H., Woods P.S., and Hughes W.L. 1957. The organization and duplication of chromosomes as revealed by autoradiographic studies using tritium-labeled thymidine. *Proc. Natl. Acad. Sci.* **43** : 122–128.

Thomas C.A. and Knight J.H. 1959. The sedimentation properties of the large subunit of DNA released from labeled T4 bacteriophage. *Proc. Natl. Acad. Sci.* **45:** 332–334.

John Cairns *(born 1922 in Oxford, England) is an Oxford-educated M.D. who pursued postdoctoral studies in animal virology in Australia and Uganda. He got hooked on phage and bacteria during a sabbatical visit to Caltech in 1957. An early product of that interest is the widely published autoradiogram of a replicating* E. coli *chromosome. John toiled at Cold Spring Harbor from 1963 to 1973, serving as Director for the first five years. In 1973 he accepted the Headship of the Mill Hill Laboratories of Imperial Cancer Institute in London.* E. coli *continued to serve as a tractable model for his studies of cell replication and mutation as they may apply to tumorigenesis. During this period, John wrote a book about cancer that may have been the first to include discussion of the biology, the molecular biology, the epidemiology, and the results of treatment. In 1980, John took a position at the Harvard School of Public Health, where he remained until his retirement in 1991.*

The T2 Phage Chromosome

Jun-ichi Tomizawa

*I*N 1947, I GRADUATED FROM THE UNIVERSITY OF TOKYO and enrolled in the National Institute of Health of Japan. From reading the scientific literature, I got a strong sense of the emerging new biology centered around Max Delbrück, Salvadore E. Luria, and Alfred Day Hershey. Thus, in 1954, when I had a chance to start my own research work, I decided to study phage DNA replication. The concept of the genetic nature of DNA was well established by that time (Avery et al. 1944; Hershey and Chase 1952).

The news of the discovery of the structure of DNA (Watson and Crick 1953) reached me when I was thinking about starting phage experiments. The accompanying mechanistic model of DNA replication, although attractive, had no experimental support. I thought that it would be fun to test such an attractive hypothesis, on the possibility that it could be incorrect. In fact, there was a proposal that, in the process of DNA duplication, genetic information in DNA is transferred to protein and then transferred back to progeny DNA. If this proposal was correct, the protein that receives the genetic information should be species-specific, and therefore synthesis of phage DNA after infection would have to depend on a phage-specific intermediary protein. Considering this possibility, I wanted to examine the effect of inhibition of protein synthesis on phage DNA synthesis. First, I needed authentic materials to work with. Impressed by the beauty of his work on phage genetics, I wrote a letter to Al Hershey asking for strains of phage T2 and host bacteria.

The Phage Chromosome as DNA

Soon, we found that inhibition of protein synthesis by chloramphenicol before or immediately after infection of phage T2 prevented phage DNA synthesis. When some protein synthesis was allowed after phage infection prior to chloramphenicol addition, phage DNA was then synthesized in the presence of chloramphenicol, in an amount proportional to the amount of protein that had been made. Furthermore, when phage-infected bacteria were irradiated with ultraviolet (UV) light at different times after infection, the average target size of the infective centers was gradually decreased soon after infection, prior to any increase in the target number due to the increase of the number of DNA copies. I concluded that phage genetic information is transferred to a material which is insensitive to irradiation, and that the transfer is prevented if protein synthesis is inhibited. These results were consistent with the interpretation that protein might be an intermediate in the duplication of DNA molecules. Once the information had been transferred to protein, the corresponding portions of the phage chromosome would no longer be part of the UV-sensitive target (Tomizawa and Sunakawa 1956). The DNA synthesized in the presence of chloramphenicol was considered to be authentic phage DNA, because the DNA that had been irradiated with UV light was incorporated into inactive phage particles which were reactivated by irradiation of infected bacteria with visible light.

We had most of these results in several months. I told Al of our results and they surprised him. In the summer of 1955, Al was kind enough to present our results at the Phage Meeting, where they were accepted with interest. Later, I learned that Al, together with Norman Melechen (Hershey and Melechen 1957), also found requirement of new protein synthesis for initiation of T2 phage DNA synthesis, and with Betty Burgi (Hershey and Burgi 1957) found that UV-irradiated parental phage DNA appeared in both infective and noninfective progeny particles. Thus, our approaches were somehow overlapping. In the spring of 1956, Al proposed a collaboration with me and contacted Francis Ryan, who was at that time in Japan, and arranged a Fulbright Fellowship for me. Al and Haruo Ozeki picked me up at LaGuardia airport. On our way to Cold Spring Harbor, we stopped at a cafeteria for lunch where I asked Al, "What is this?" indicating the remote control unit of a music box on the table. Al answered, "Noise". This was typical of our conversations, which rarely continued more than a few words. I joined Hershey's group in February 1957.

A few years after our discovery that prior protein synthesis is required for initiation of phage DNA synthesis, it was found that phage DNA synthesis requires phage-specific enzymes, such as T2 DNA polymerase. Therefore, the requirement of protein synthesis does not necessarily mean the presence of proteins that act as intermediates in the duplication of DNA. Thus, my original aim to learn the mechanism of duplication of DNA by inhibition of protein synthesis was not fulfilled. Rather, the work was the first indication of the requirement of synthesis of specific protein for viral DNA synthesis.

Meanwhile, Al was using the system mainly to know the nature of the precursor pool of phage DNA. Thus, during my stay in Hershey's laboratory in 1957 to 1958, I worked along his line and further characterized the precursor pool made in the presence of chloramphenicol. Extensive characterization of phage particles that had incorporated the UV irradiated pool DNA showed unambiguously that the replicating phage chromosome consists almost exclusively of DNA (Tomizawa 1958). Furthermore, the fact that individual phage particles contained DNA synthesized at different times and enhancement of recombination frequency by irradiating the pool DNA showed that replication and recombination occurred during a single biochemical phase of phage growth (Tomizawa 1958; Hershey et al. 1958). These results provided a firm basis for further experimental analyses of the precursor pool and gave strong support for considering phage genetics as a manifestation of the behavior of DNA molecules.

Chromosomes in Phage Particles

In 1950s, one of the most important questions for understanding the genetics of phage T2 and its close relative, phage T4, was the number of chromosome(s) in a single phage particle. It could have been a single DNA molecule or a bundle of not more than 10 molecules. However, the exact number could not be determined by physical analysis, because there was no reliable method for determining the size of large DNA molecules. As a result, the size and kind of chromosomes had only been inferred indirectly by combinations of physical and biological approaches (summarized in Hershey and Burgi 1957). If a phage particle contains two or more DNA molecules, they are likely to be transferred separately to different progeny particles. Therefore, the pattern of distribution of ^{32}P-labeled parental phage DNA among progeny particles may tell composition of chromosomes of a phage particle. One view prevailing in the mid to late 1950s was that a T2 phage particle carries two kinds of DNA: a large piece containing about 40% of the DNA and several small pieces (Levinthal and Thomas 1957). The DNA in the large piece, which was thought to be the phage chromosome, was supposed to be transferred after replication to one or two progeny particles, whereas the DNA in the small pieces became dispersed among many progeny particles or lost. However, there was some contradictory evidence, such as the absence of any special class of molecules that could maintain the high efficiency of transfer in successive cycles of growth. However, no comprehensive alternative view was presented, except for continuous fragmentation and association of the fragmented material, possibly by recombination, for which the molecular mechanism was totally unknown. Thus, in the 1950s, our understanding of the composition and behavior of phage chromosomes was very confused. Nonetheless, that decade was a very enjoyable time because we could invent many ingenious experiments that satisfied our own imagination. However, the studies that proved to be mainly responsible for the advancement of our knowledge of molecular genetics were not such fancy experiments.

I have described the general climate of thought in the 1950s among people studying the material bases of genetic continuity. Although Al shared in creating this climate, he was at the same time very critical of the experimental basis of arguments about the physical structure of DNA. He felt that because the methods used in all transfer experiments were highly technical, critical evaluation of the results was essential (Hershey and Burgi 1957). This dissatisfaction led him to study the nature of T2 DNA by more straightforward approaches. Using chromatographic (Mandel and Hershey 1960) and sedimentation (Hershey and Burgi 1960) analyses, Al and his associates found that DNA extracted mildly from phage particles had little sign of heterogeneity. However, because of the absence of a reliable method for measuring the size of large DNA, the size of these apparently homogeneous molecules could not be determined. At the turn of the decade, with the advent of milder methods of handling of DNA, combined with reliable quantitation of DNA length by autoradiography of single ^{32}P-labeled DNA molecules, conclusive evidence was obtained by Al and associates and others that a T2 phage particle has a single large DNA molecule (Davison et al. 1961; Rubenstein et al. 1961).

Chromosomes in the Replication Pool

Although phage chromosomes in the replication pool of infected cells were shown to be composed almost exclusively of DNA, it had not been possible to study their physical structure because of technical limitations. When methods were developed to determine the size of DNA as large as the intact DNA from T2 phage particles, they were applied by Fred Frankel of Hershey's laboratory to determine the size of DNA in the precursor pool. The structure of the phage DNA accumulated in the presence of chloramphenicol was first examined, because it can be analyzed without interference from DNA in completed phage particles. It was found that the pool DNA contained molecules whose sizes were estimated to be more than twice as large as those from phage particles (Frankel 1963). Later, similar large molecules were found in the nascent phage DNA in the precursor pool formed in the absence of chloramphenicol (Frankel 1968). Although the exact structure of these large molecules was unknown, they were likely to contain DNA longer than that found in the mature phage particle. Such molecules could have arisen by replication and/or recombination. In fact, by recombination of T4 DNA, a linear molecule in which two molecules were joined together at their ends was made (Tomizawa and Anraku 1964; Tomizawa 1967). In addition, the circularly permuted genetic map of phage T4 (Séchaud et al. 1965) and the terminally redundant chromosome from phage particles (Streisinger et al. 1964) implied that DNA in the replication pool must contain molecules longer than unit length, from which unit-length pieces were cut out at the time of formation of the phage particles.

Logical Inquiry and the Invention of New Methods

During most of the time I was in Hershey's laboratory, Al was trying to formulate a comprehensive view of phage genetics. He presented the contemporary concept of a phage cross as a system of population genetics (Hershey [1959] and its Appendix by Steinberg and Stahl [1959]). He succeeded to present the view that, even though phage genetics is somewhat different from the genetics of other systems, it is no less important, because it reveals some relationships between genetic problems and nucleic acid chemistry. In addition, the study provided the bases for understanding of the general outcome of a phage cross. However, the approach was necessarily remote from providing an explanation of the elementary acts of the genetic material. As his concern shifted back to molecular understanding of mechanisms of elementary acts, Al did not return to further study of formal phage genetics.

In the late years of his career, Al and associates invented a number of useful methods for analysis of large DNA molecules and extended the subject of research to yet another important phage. His group found that a λ phage particle contains a single molecule of DNA that can form a circular structure, in which it interacts with the bacterial chromosome. The functional and structural diversities of T2 and λ DNAs are summarized beautifully in the paper on idiosyncrasies of DNA structures (Hershey 1970). Thus, throughout his research career, Al continued to invent new methods that played critical roles for his discoveries, all of which were landmarks of progress of our knowledge of the phage genetic materials.

An extremely logical style of presentation characterizes both his papers and his formal talks. At the same time, Al knew well the limits of logical inquiry unrestrained by direct experimental analysis and avoided uncertainty by continually inventing new methods most suitable for specific experimental purposes.

For the Memory

Al Hershey had a sense for grasping the central problem and concentrating his efforts on solving it. He was truly unique for his ability to judge the importance of a subject through impeccably accurate logical reasoning. Although not necessarily ignorant about other problems, he had a clear sense of relative levels of importance and did not pay much attention to subjects he considered less significant. He did not rush to conclusions and examined all possibilities before presenting his views. Sometimes his words appeared to be abrupt and illogical, but this may have been due to the long process he needed to reach a conclusion and to select a proper way to express it. Thus, I believe that the most characteristic aspects of his style of scientific research were to grasp the most fundamental problem, to focus on it continuously, to search for the best way to solve it, and to express the problem and its solution in the most logical form.

In the laboratory, Al was very kind and instructive, but seemed to prefer to help us behind the scenes. Thus, we could work efficiently and comfortably. The place was a little too quiet, but this just compensated for the noisy summer of Cold Spring Harbor. I feel pride and satisfaction on my belonging to his laboratory.

Last, I should mention his fair and unselfish consideration of the contribution of others. In my case, without his consideration and care, I would have been buried in complete obscurity, a helpless scientist of an ignored country.

When I met Al last time after the Symposium in 1978, he said that he would like to write a book on Darwinism. We have forever lost the chance to see it.

References

Avery O.T., MacLeod C.M., and McCarty M. 1944. Studies on the chemical nature of the substance inducing transformation of pneumococcal types. Induction of transformation by a desoxyribonucleic acid fraction isolated from pneumococcus type III. *J. Exp. Med.* **79:** 137–158.

Davison P.F., Freifelder D., Hede R., and Levinthal C. 1961. The structural unity of the DNA of T2 bacteriophage. *Proc. Natl. Acad. Sci.* **34:** 89–96.

Frankel F. 1963. An unusual DNA extracted from bacteria infected with phage T2. *Proc. Natl. Acad. Sci.* **49:** 366–372.

———.1968. Evidence for long DNA strands in the replicating pool after T4 infection. *Proc. Natl. Acad. Sci.* **59:** 134–138.

Hershey A.D. 1959. The production of recombinants in phage crosses. *Cold Spring Harbor Symp. Quant. Biol.* **23:** 19–46.

———.1970. Idiosyncrasies of DNA structure. *Science* **168:** 1425–1427.

Hershey A.D. and Burgi E. 1957. Genetic significance of the transfer of nucleic acid from parental to offspring phage. *Cold Spring Harbor Symp. Quant. Biol.* **21:** 91–101.

———.1960. Molecular homogeneity of the deoxyribonucleic acid of phage T2. *J. Mol. Biol.* **2:** 143–152.

Hershey A.D. and Chase M. 1952. Independent functions of viral protein and nucleic acid in growth of bacteriophage. *J. Gen. Physiol.* **36:** 39–56.

Hershey A.D. and Melechen N.E. 1957. Synthesis of phage-precursor nucleic acid in the presence of chloramphenicol. *Virology* **3:** 207–232.

Hershey A.D., Burgi E., and Streisinger G. 1958. Genetic recombination between phages in the presence of chloramphenicol. *Virology* **6:** 287–288.

Levinthal C. and Thomas C.A., Jr. 1957. Molecular autoradiography. The β-ray counting from single virus particles and DNA molecules in nuclear emulsions. *Biochim. Biophys. Acta* **23:** 453–465.

Mandel J.D. and Hershey A.D. 1960. A fractionating column for analysis of nucleic acids. *Anal. Biochem.* **1:** 66–77.

Rubenstein I., Thomas C.A., Jr., and Hershey A.D. 1961. The molecular weights of T2 bacteriophage. DNA and its first and second breakage products. *Proc. Natl. Acad. Sci.* **47:** 1113–1122.

Séchaud J., Streisinger G., Emrich J., Newton J., Lamford H., Reinhold H., and Stahl M.M. 1965. Chromosome structure in phage T4. II. Terminal redundancy and heterozygosis. *Proc. Natl. Acad. Sci.* **54:** 1333–1339.

Steinberg C. and Stahl F. 1959. The theory of formal phage genetics (Appendix II of Hershey 1959). *Cold Spring Harbor Symp. Quant. Biol.* **23:** 42–46.

Streisinger G., Edgar R.S., and Denhardt H. 1964. Chromosome structure in phage T4. I. Circularity of the linkage map. *Proc. Natl. Acad. Sci.* **51:** 775–779.

Tomizawa J. 1958. Sensitivity of phage precursor nucleic acid, synthesized in the presence of chloramphenicol, to ultraviolet irradiation. *Virology* **6:** 55–80.

———.1967. Molecular mechanisms of genetic recombination in bacteriophage: Joint molecules and their conversion to recombinant molecules. *J. Cell. Physiol.* **70:** 201–214.

Tomizawa J. and Anraku N. 1964. Molecular mechanisms of genetic recombination in bacteriophage. II. Joining of parental DNA molecules of phage T4. *J. Mol. Biol.* **8:** 516–540.

Tomizawa J. and Sunakawa S. 1956. The effect of chloramphenicol on deoxyribonucleic acid synthesis and development of resistance to ultraviolet irradiation in *E. coli* infected with bacteriophage T2. *J. Gen. Physiol.* **39:** 553–565.

Watson J.D. and Crick F.H.C. 1953. A structure for deoxyribonucleic acid. *Nature* **171:** 737–738.

Jun-ichi Tomizawa *(born 1924 in Tokyo, Japan) entered the phage field in 1954, as soon as circumstances allowed. He was from the beginning oriented to testing hypotheses regarding basic molecular mechanisms, which led him inevitably to Hershey's lab in 1957. Tomi's scientific and personal styles earned him the local title of "Japan's Hershey." He spent brief periods at MIT (Cambridge, Massachusetts) and at the Institute of Molecular Biology in Eugene before taking up posts at the National Institute of Health in Tokyo and then at Osaka University. Unsettled conditions in Japan's universities promoted a return to America, where Tomi worked at the Laboratory of Molecular Biology, NIDDK of the NIH before returning to Japan as Director of the National Institute of Genetics. He is now adjunct professor of the Institute and Editor in Chief of the journal* Genes to Cells.

Incomplete Genomes in Small T4 Particles

GISELA MOSIG

*T*HE HYPOTHESIS-DRIVEN APPROACH to biological questions, implemented and advocated by Max Delbrück, has been a hallmark of phage genetics. This was a natural way of investigation for Al Hershey, who was attracted by Max for this reason, and outdid him with an uncanny ability to put hypotheses to the most direct unbiased experimental tests and to confirm, modify, or abandon any hypothesis accordingly.

Max Delbrück's accounting of Hershey's blendor experiment, which I heard in 1955 when I was a graduate student in botany, and Waclaw Szybalski's recounting of that experiment at the 1997 phage meeting in Madison, suggested that Hershey was as skeptical as Delbrück that nucleic acids would be sufficiently versatile to be carriers of genetic information. But Hershey's experiments convinced himself and the world otherwise. Hershey went on to develop the most direct methods of that time to analyze how that information in DNA is expressed, transmitted, and exchanged.

I was fortunate to experience first-hand Al Hershey's critical open-mindedness. The circumstances that led to the discovery of small T4 particles with incomplete genomes and their relationship to concatemeric intracellular phage T4 DNA are an example of this characteristic. By gentle guidance and with few words, Al Hershey conveyed his message that critical experiments to test hypotheses and an open mind to interpret results can lead to new insights, even when results have to overcome dogmas derived from false leads.

Hershey had studied recombination in T2, a member of the T-even phages, to see whether inheritance of their genes follows rules similar to those of higher organisms. The fact that phage genes recombine in patterns that allow the construction of linkage maps, as in higher organisms, convinced him that phages could reveal fundamental aspects of inheritance (Hershey and Rotman 1948,1949). Throughout the 1950s, the prevailing hypothesis was that the T-even phages contained several "chromosomes" (three, on average), that these "chromosomes," sometimes called partial replicas (of the genome), would mate and recombine repeatedly during growth in an infected bacterium and that the partial replicas would assort independently into progeny phage particles (Luria 1947; Visconti and Delbrück 1953). This hypothesis was apparently supported by several independent lines of evidence:

1. Recombination between the few mutant sites that were known at that time suggested three, however ill-defined, linkage groups.

2. DNA extracted from phage particles seemed to be of variable lengths.

3. After Hershey's demonstration that ^{32}P would label the phage DNA selectively, several labs had reported that approximately 50% of this isotope was transmitted from parental to progeny particles in large segments, and the other half was distributed in very small segments. It appeared that the large transmitted fragments corresponded to approximately one-third of the total isotope. (Many years later, Tamar Ben-Porat, who participated in one of these studies but was not a co-author, told me she was convinced that the results were an artifactual consequence of radiation damage from the high levels of ^{32}P that had been used. In retrospect, they were probably due to the stimulation of recombination by double-strand breaks.)

In a review at the 1958 Cold Spring Harbor Symposium, Hershey pointed out several inconsistencies of the prevailing phage recombination hypotheses (Hershey 1959), and Steinberg and Stahl (1959) provided the theoretical underpinning to these arguments.

Realizing that solid data based on physical evidence were needed, Hershey developed reliable methods to extract unbroken DNA molecules, to measure their sizes, and to compare these sizes to the amount of DNA packaged in virions. He showed that DNA is broken by stirring, or pipetting, or by decay of ^{32}P, and he developed methods to avoid or minimize such damages. These experiments culminated in the demonstration that there is but one unbroken DNA molecule, of defined length, in each phage particle (Rubenstein et al. 1961). At about the same time, extensive crosses revealed that all known mutations can be assigned to a single linkage group (Streisinger et al. 1964). However, studies with heterozygotes (Doermann and Boehner 1963; Womack 1963) had suggested that there are partial overlaps between chromosomes (detectable as heterozygotes) at different map positions. These were interpreted as overlaps between the partial replicas. As mentioned before, independent assortment of these partial replicas was thought to imply that different amounts of DNA might be packaged in different particles, depending on the numbers and sizes of partial replicas that had been packaged.

After working in plant genetics in Cologne, where Doermann had initiated his studies on heterozygous overlaps during a sabbatical, I joined Doermann's lab to study phage recombination. As one test of the partial replica hypothesis, I centrifuged two genetically marked phage stocks in a CsCl gradient and subsequently performed crosses with the different phages of different densities. Recombinant frequencies increased with decreasing density of the parental phage particles (Mosig 1963) and were particularly high when phage from some very light fractions ("light" particles) were crossed. These results appeared to support certain aspects of the partial replica hypothesis. Shorter fragments might generate shorter overlaps, which might be more prone to exchanges. But how would these results fit with the single chromosomes (Rubenstein et al. 1961) reported at the annual phage meeting where I presented my results?

This is when I bundled my courage to ask Dr. Hershey whether I could come to his lab to analyze the DNA of the "light" phage by his methods. I was thoroughly indoctrinated with the partial replica hypothesis and not ready to accept the idea of one chromosome per phage particle. The longer I think about it, the more remarkable it is to me that Hershey did not tell me how ridiculous my idea was. Instead, when I arrived in his lab in June 1962, he simply suggested that, since he was going on his annual sailing vacation, I should do the experiments that I had planned using the reagents in his refrigerator. Fred Frankel (Frankel 1966) would show me how to extract DNA without breaking it, and how to use the "Hershey column" (Mandell and Hershey 1960). Betty Burgi (Burgi and Hershey 1961) would show me how to use sucrose gradients, and Ed Goldberg, who was perfecting T4 transformation (Goldberg 1966), would introduce me to other tricks of the lab and help me find my way around.

At that time, Fred Frankel was looking at the fate of T-even chromosomes after infection. Were they of the same size as the parental chromosomes? The results appeared, at first, confusing. In some experiments, the DNA appeared to be of the same size, or smaller, and in other experiments it appeared much larger than the DNA from virions. Fred used chloramphenicol to inhibit late protein synthesis and to prevent packaging of this vegetative DNA. After he found that timing of addititon of chloramphenicol was crucial, large vegetative DNA was isolated reproducibly. Eventually most intracellular phage DNA was found to be very large and complex (Frankel 1966; Hubermann 1969), especially when mutations were used instead of chloramphenicol to prevent packaging (Frankel 1969), and this DNA could be chased into particles, which again contained chromosomes of a unique size.

In retrospect, timing of chloramphenicol addition was critical because many middle genes are important for recombination and for recombination-dependent DNA replication. Because expression of T4 middle genes requires phage-encoded activators, both their transcription and synthesis of their proteins are inhibited by chloramphenicol. If chloramphenicol is added too early, only origin-dependent initiation of DNA replication occurs, which generates DNA molecules of smaller or similar size as the infecting DNA. In contrast, later addition of chloramphenicol allows formation of branched, highly complex structures due to recombination and replication. This DNA is not cut because neither the cutting enzymes, endonuclease VII and terminase, nor the proteins from which heads are assembled are made when late protein synthesis is inhibited. If chloramphenicol is added too late, the intracellular DNA is cut and packaged. Eventually, mutants were required to unravel these complexities, and because more mutants were isolated and characterized in T4 (Epstein et al. 1964), this phage became the favorite member of the T-even family.

George Streisinger's (Streisinger 1966) ingenious model, that T4 chromosomes are cut processively and with a redundancy at the termini from these large DNA structures (called "concatemers") to fill the heads (the "headful" hypothesis), gave a simple explanation for many otherwise conflicting results. Most importantly, it reconciled the circularity of a single linkage group, the single, linear chromosome, the terminal redundancies that could carry different alleles (be heterozygous), and the fact that overlap heterozygous regions could be found at many random positions of the genome. Later experiments (Werner 1968,1969; Kemper and Brown 1976) showed that the concatemers were not simple linear head-to-tail structures, but were highly branched. Conceptually, this is only an extension of the general headful packaging hypothesis, since the branches are trimmed prior to or during packaging.

Soon, Merv Smith and Ann Skalka joined the Hershey lab and showed that similar large concatemers were made in bacteria infected with phages T5 and lambda. Nevertheless, their packaging mechanisms do not generate circular permutations, nor headful sizes of chromosomes. Because lambda DNA, but not T4 DNA, could be broken by shearing into fragments of different GC content, containing different genes, and these could be fractionated on the "Hershey column," most subsequent investigations in the Hershey lab concentrated on phage lambda.

My very first experiments in Hershey's lab convinced me that I could extract, from T4 particles of average density, unbroken DNA molecules of the size that he had shown. In contrast, the "light" T4 particles contained smaller DNA molecules, which were eluted from the column as a single peak. Were these the "partial replicas" that I had been looking for? By the time I could repeat the entire series of experiments, Al was back in Cold Spring Harbor, and I do remember one very, very early morning when he watched the UV absorption of the column eluant, apparently driven by curiosity whether the DNA from the light particles would be eluted as I had anticipated from the first set of experiments. He joked then that he had automated the column to save time watching it. A few days later we discussed how one could reconcile all the results if these short DNA molecules were packaged into separate particles, and how to test this idea.

The experiments which followed showed that most of the smaller DNA molecules measured approximately two-thirds of the full-length chromosomes. Any of these particles alone did not initiate a full infection, but two or more would, with the probability expected if their DNA molecules were segments derived from a circularly permuted chromosome. The intracellular DNA established after multiple infection was concatemeric and was packaged into viable particles, most of which had normal-sized chromosomes (Mosig 1964). The results also suggested that the smaller DNA molecules reside in smaller heads. Electron micrographs of mutant T4 particles shown at the same symposium (Epstein et al. 1964) appear to contain one such small particle, which was labeled, however, as "contracted."

All these results suggested that breakage of DNA due to packaging could be separated from the recombinational joining that had to occur in subsequent, cooperative infection by several particles. To analyze recombination in more detail, I labeled the small chromosomes with genetic markers. I was most intrigued by the observation that recombination was especially high near the ends of the "incomplete" chromosomes. Moreover, the pattern of recombination suggested a "copy-choice" mechanism, a hypothesis proposed by Belling (1933). Of course, the copy-choice notion was rather unpopular, because the Meselson-Stahl experiment (Meselson and Stahl 1958) had clearly shown that DNA replication is semiconservative, apparently not easily reconciled with the copy-choice idea. After discussion of several alternative models, Hershey told me that my explanation was as plausible as the alternatives and that I should be entitled to explain the results, since I had done the experiments. Much later, I discovered, in reading a recollection of the Meselson-Stahl experiments (Meselson and Stahl 1966), that density shift experiments to study T-even DNA replication had been attempted but that they had given less clearly interpretable results than those with *E. coli* DNA. (More recently, the copy-choice idea has been resurrected [Skalka 1974; Formosa and Alberts 1986; d'Alençon et al. 1994; Tang 1994; Haber 1995]).

More extensive subsequent single-burst experiments with the small particles, using 19 genetic markers distributed over the entire map (Mosig et al. 1971), confirmed and extended the evidence that any chromosomal end at any map position enhanced recombination and that either patch-type or splice-type recombination could occur. Each of these two modes had different consequences for allele frequencies of the terminal markers among the progeny phage. These patterns were best explained by recombination-dependent DNA replication, a kind of copy-choice mechanism in which hybridizing homologous DNA guides the replisome to different templates. This recombination-dependent mechanism for initiation of DNA replication was eventually demonstrated by Luder and Mosig (1982).

After I had moved to Nashville, Rudolf Werner, in Hershey's lab, measured the numbers of and distances between replication forks in replicating T4 DNA. He showed that forks could follow each other at distances approximately 1/10 of the T4 genome. He interpreted these results in terms of the most popular hypothesis of that time: Replicons had to be circular and replication was initiated either in "Theta- or Cairns-type" bubbles, or from a nick in the double-stranded template, generating "rolling circles." To explain his results, Rudy proposed that the T4 chromosome would circularize after infection and that there was repeated rolling circle initiation from a unique origin to generate a "firewheel" from the circular template (Werner 1968, 1969). The incomplete T4 chromosomes appeared to be excellent subjects to test these ideas, because they lack the terminal redundancies and cannot circularize. Dr. Hershey arranged two working visits for me to Cold Spring Harbor, to find out whether the incomplete chromosomes could replicate. First I used his T4 DNA agarose columns, which suggested rather limited T4 DNA synthesis (Mosig et al. 1969). Then Rudy and I collaborated in measuring density shifts of the incomplete chromosomes. The results were clear. Those incomplete chromosomes expected to contain the DNA polymerase and accessory genes did replicate. However, the results posed new problems: Replication of the incomplete chromosomes, unlike that of the complete chromosomes, was restricted to one round (Mosig and Werner 1969). Efforts to uncover why that should be revealed unexpected complexities and led to many controversies. Eventually, these experiments showed that various branched structures, including firewheels, were formed by recombination-dependent DNA replication (Dannenberg and Mosig 1981, 1983; Luder and Mosig 1982), that there are several origins (Kozinski 1983), and that origin-dependent DNA replication is programmed by transcriptional controls to cease after one or at most a few rounds. Additionally, these studies revealed that there is more than one class of small particles (Mosig et al. 1972) whose chromosome lengths fit precisely the predictions of Streisinger's headful hypothesis. The geometry of the different heads and effects of certain mutations on proportions of different particles contributed to understanding architectural principles of T4 head assembly (Walker et al. 1972; Haynes and Eiserling 1996). A map constructed from the probabilities of cutting ends between genetic markers is congruent (in most intervals) with the recombination map and with distances based on DNA sequence (Mosig 1966, 1968).

What made the experiences in Al Hershey's lab so unusual is what he conveyed by example, without words, and without dwelling on it. He was driven by the desire to understand, without regard of who "was right" and with no room for pretenses. (Once he put a copy of a paper that Jean Weigle had sent to him on our desks, with the comment "The perfect paper.") When Dr. Hershey's experiments or interpretations did not agree with someone else's, the discrepancy never led to a controversy, but instead to an effort to understand the reasons behind the discrepancy. Mentioning the analogy to particle physics, he kept reminding us that experimental conditions can prejudice results and that we must be alert to avoid bending nature to fit our preconceived notions.

Dr. Hershey demanded rigor of himself and others and tolerated no compromises. Writing was probably the hardest experience for any of us, his postdocs, but was softened by the realization that he never demanded more of others than of himself. He was considerate, more than generous with his time and thoughts, and never insulting. Despite his contributions, insights, facilities, and important editing, Hershey's name appeared only on papers to which he had contributed experiments, done by himself. This may have reflected his generosity, or it may have been the consequence of his strong belief that "one thinks with one's hands," and that he had not thought enough about experiments that he had not done himself.

I doubt that I was the only one awed by "Dr. Hershey" upon entering his lab. At a later occasion, when some ice had melted, he remarked on how he would have liked to be more at ease in social contacts, and how he envied people to whom this comes naturally. Perhaps, sharing the music that he loved or a book that he enjoyed was for him the best way to share his thoughts.

Al was likened to a saint. Yes, Al was a saint, and he made his lab the "Hershey heaven" for many mortals. Thank you, Dr. Hershey.

References

Belling J. 1933. Crossing over and gene rearrangement in flowering plants. *Genetics* **18:** 388–413.

Burgi E. and Hershey A.D. 1961. A relative molecular weight series derived from the nucleic acid of bacteriophage T2. *J. Mol. Biol.* **3:** 458–472.

d'Alençon E., Petranovic M., Michel B., Noirot P., Aucouturier A., Uzest M., and Ehrlich S.D. 1994 Copy-choice illegitimate DNA recombination revisited. *EMBO J.* **13:** 2725–2734.

Dannenberg R. and Mosig G. 1981. Semiconservative DNA replication is initiated at a single site in recombination-deficient gene *32* mutants of bacteriophage T4. *J. Virol.* **40:** 890–900.

————.1983. Early intermediates in bacteriophage T4 DNA replication and recombination. *J. Virol.* **45:** 813–831.

Doermann A.H. and Boehner L. 1963. An experimental analysis of bacteriophage T4 heterozygotes. I. Mottled plaques from crosses involving six *rII* loci. *Virology* **21:** 551–567.

Epstein R.H., Bolle A., Steinberg C.M., Kellenberger E., Boy de la Tour E., Chevalley R., Edgar R.S., Susman M., Denhardt G.H., and Lielausis A. 1964. Physiological studies of conditional lethal mutants of bacteriophage T4D. *Cold Spring Harbor Symp. Quant.* Biol. **28:** 375–392.

Formosa T. and Alberts B.M. 1986. DNA synthesis dependent on genetic recombination: Characterization of a reaction catalyzed by purified T4 proteins. *Cell* **47:** 793–806.

Frankel F.R. 1966. Studies on the nature of replicating DNA in T4-infected *Escherichia coli. J. Mol. Biol.* **18:** 127–143.

————. 1969. DNA replication after T4 infection. *Cold Spring Harbor Symp. Quant. Biol.* **33:** 485–493.

Goldberg E.B. 1966. The amount of DNA between genetic markers in phage T4. *Proc. Natl. Acad. Sci.* **56:** 1457–1463.

Haber J.E. 1995. In vivo biochemistry: Physical monitoring of recombination induced by site-specific endonucleases. *BioEssays* **17:** 609–620.

Haynes J.A. and Eiserling F.A. 1996. Modulation of bacteriophage T4 capsid size. *Virology* **221:** 67–77.

Hershey A.D. 1959. The production of recombinants in phage crosses. *Cold Spring Harbor Symp. Quant. Biol.* **23:** 19–46.

Hershey A.D. and Rotman R. 1948. Linkage among genes controlling inhibition of lysis in a bacterial virus. *Proc. Natl. Acad. Sci.* **34:** 89–96.

————. 1949. Genetic recombination between host range and plaque-type mutants of bacteria in single bacterial cells. *Genetics* **34:** 44–71.

Hubermann J.A. 1969. Visualization of replicating mammalian and T4 bacteriophage DNA. *Cold Spring Harbor Symp. Quant. Biol.* **33:** 509–524.

Kemper B. and Brown D.T. 1976. Function of gene *49* of bacteriophage T4. II. Analysis of intracellular development and the structure of very fast-sedimenting DNA. *J. Virol.* **18:** 1000–1015.

Kozinski A.W. 1983. Origins of DNA replication. In *Bacteriophage T4* (ed. I.K. Mathews, E.M. Kutter, G. Mosig, and P.B. Berget), pp. 111–119. American Society for Microbiology, Washington D.C.

Luder A. and Mosig G. 1982. Two alternative mechanisms for initiation of DNA replication forks in bacteriophage T4: Priming by RNA polymerase and by recombination. *Proc. Natl. Acad. Sci.* **79:** 1101–1105.

Luria S. E. 1947. Reactivation of irradiated bacteriophage by transfer of self-reproducing units. *Proc. Natl. Acad. Sci.* **33:** 253–264.

Mandell J.D. and Hershey A.D. 1960. A fractionating column for analysis of nucleic acids. *Anal. Biochem.* **1:** 66–77.

Meselson M. and Stahl F.W. 1958. The replication of DNA in *Escherichia coli. Proc. Natl. Acad. Sci.* **44:** 671–682.

————. 1966. Demonstration of the semiconservative mode of DNA duplication. In *Phage and the origins of molecular biology* (ed. J. Cairns et al.), pp. 246–251. Cold Spring Harbor Laboratory, Cold Spring Harbor New York.

Mosig G. 1963. Coordinate variation in density and recombination potential in T4 phage particles produced at different times after infection. *Genetics* **48:** 1195–1200.

————. 1964. Genetic recombination in bacteriophage T4 during replication of DNA fragments. *Cold Spring Harbor Symp. Quant. Biol.* **28:** 35–42.

————. 1966. Distances separating genetic markers in T4 DNA. *Proc. Natl. Acad. Sci.* **56:** 1177–1183.

————. 1968. A map of distances along the DNA molecule of phage T4. *Genetics* **59:** 137–151.

Mosig G. and Werner R. 1969. On the replication of incomplete chromosomes of phage T4. *Proc. Natl. Acad. Sci.* **64:** 747–754.

Mosig G., Ehring R., and Duerr E.O. 1969. Replication and recombination of DNA fragments in bacteriophage T4. *Cold Spring Harbor Symp. Quant. Biol.* **33:** 361–369.

Mosig G., Ehring R., Schliewen W., and Bock S. 1971. The patterns of recombination and segregation in terminal regions of T4 DNA molecules. *Mol. Gen. Genet.* **113:** 51–91.

Mosig G., Carnighan J.R., Bibring J.B., Cole R., Bock H.-G.O., and Bock S. 1972. Coordinate variation in lengths of deoxyribonucleic acid molecules and head lengths in morphological variants in bacteriophage T4. *J. Virol.* **9:** 857–871.

Rubenstein I., Thomas C.A., Jr., and Hershey A.D. 1961. The molecular weights of T2 bacteriophage DNA and its first and second breakage product. *Proc. Natl. Acad. Sci.* **47:** 1113–1122.

Skalka A. 1974. A replicator's view of recombination (and repair). In *Mechanisms in recombination* (ed. R.F. Grell), pp. 421–432. Plenum Press, New York.

Steinberg C.M. and Stahl F.W. 1959. The theory of formal phage genetics. *Cold Spring Harbor Symp. Quant. Biol.* **23:** 42–46.

Streisinger G. 1966. Terminal redundancy, or all's well that ends well. In *Phage and the origins of molecular biology* (ed. J. Cairns et al.), pp. 335–340. Cold Spring Harbor Laboratory, Cold Spring Harbor, New York.

Streisinger G., Edgar R.S., and Denhardt G.H. 1964. Chromosome structure in phage T4. I. Circularity of the linkage map. *Proc. Natl. Acad. Sci.* **51:** 775–779.

Tang R.S. 1994. The return of copy-choice in DNA recombination. *BioEssays* **16:** 785–788.

Visconti N. and Delbrück M. 1953. The mechanism of genetic recombination in phage. *Genetics* **38:** 5–33.

Walker D.H., Mosig G., and Bayer M.E. 1972. Bacteriophage T4 head models based on icosahedral symmetry. *J. Virol.* **9:** 872–876.

Werner R. 1968. Distribution of growing points in DNA of bacteriophage T4. *J. Mol. Biol.* **33:** 679–692.

————.1969. Initiation and propagation of growing points on the DNA of phage T4. *Cold Spring Harbor Symp. Quant. Biol.* **33:** 501–507.

Womack F.C. 1963. An analysis of single-burst progeny of bacteria singly infected with a bacteriophage heterozygote. *Virology* **21:** 232–241.

Gisela Mosig *(born 1930 in Schmorkau [Kreis Oschatz], Germany) took her degree in botany in Cologne before entering the phage field as a postdoctoral student with A. H. Doermann at Vanderbilt University. Gisela then worked in Al's lab for three years, following which she returned as a faculty member to Vanderbilt, where she is now Professor. Gisela's work has illuminated the molecular mechanisms of replication and recombination as they occur within the bizarre life cycle of the phage T4.*

The Peculiarities of Lambda DNA

ANN SKALKA

Phage research... is beset with the following difficulty. One notices a phenomenon that looks interesting. After suitable thought and labor, one performs an experiment that ought to be instructive. What it actually does is to turn up a second phenomenon not related in a simple way to the first. This sequence of events makes for lively research. It does not provide explanations.

A.D. HERSHEY, 1969

I FELL IN LOVE WITH THE BACTERIOPHAGE LAMBDA in 1962 when I took the CSH Phage Course, taught that year by Frank Stahl and George Streisinger. Thus, it was with great enthusiasm that I joined the Hershey lab to work on lambda as a postdoctoral fellow in the autumn of 1964. My colleagues in the lab that year included a diverse, creative, and wonderfully congenial group of scientists: Ruth Ehring, Gisela Mosig, and Eddie Goldberg were studying various aspects of T4 DNA structure and genetics; my benchmate Merv Smith had brought with him a project on T5 DNA, which Hershey's long-time assistant, Laura Ingraham, had also studied; and Betty Burgi and Hershey had been analyzing lambda DNA. All of us were, in one way or other, seeking to understand the functional significance of the structure and various forms taken by these different phage chromosomes: T4 with its terminal redundancy, T5 with its single-strand interruptions, and lambda with peculiarities that seemed most exotic of all.

Lambda DNA

While analyzing lambda phage DNA in the early 1960s, Hershey and colleagues ran into a strange phenomenon: the DNA seemed to aggregate in very peculiar ways. In a typically short, but beautifully reasoned paper, Hershey, Burgi, and Ingraham (Hershey et al. 1963a) reported the reason for this peculiar behavior. They showed that the DNA could exist in a set of characteristic and interconvertible stable complexes: linear multimers ↔ linear monomers ↔ rings, the linear monomer representing the individual phage chromosomes as isolated from particles. Using careful sedimentation measurements and sucrose gradient analysis (via methods recently developed in the lab [Burgi and Hershey 1961; Hershey et al. 1963b]), they deduced that this phenomenon was due to the presence of "cohesive sites" close to, or at the ends of, the linear monomer. The three forms were later visualized in the electron microscope by Kaiser and Inman (1965).

In a similarly brief and elegant follow-up study, Hershey and Burgi (1965) deduced that lambda DNA cohesion could be explained by the presence of short, unpaired, complementary ends. Some of the experiments presented in this 1965 paper rested upon the authors' knowledge of another peculiarity of lambda DNA, the fact that molecular halves (produced by mechanical shear) differed considerably in base composition and, thus, could be resolved in CsCl equilibrium density gradients formed in the analytical model E centrifuge (Hershey 1964). Following the terminology suggested by Hogness and Simmons (1964), Hershey and Burgi called the denser half the "left" half, and the less dense the "right" and showed that left halves only joined with right. Although there was no comment on the possible significance of the variation in GC content along the lambda chromosome in this paper, concerning the cohesive ends they noted (prophetically): *The structure may be thought of as a means for achieving the initial step in formation of the stable molecular rings found in bacteria infected with lambda (Young and Sinsheimer 1964). More generally, structures of the postulated type may play a role in the synaptic phase of molecular-genetic recombination.*

A third peculiarity of lambda DNA lay in the different genetic linkages observed in the phage (vegetative) map and the prophage (integrated) map, the integrated being a cyclic permutation of the vegetative map. This was explained by the Campbell model of lambda DNA integration, which had postulated an intracellular circular intermediate for the lambda chromosome, followed by recombination with host *E. coli* DNA at a site approximately opposite to the joined phage DNA ends. The demonstration that lambda DNA could form circles via its cohesive ends lent substance to this model. Hershey hypothesized that the integrative recombination event might depend on homology between lambda and *E. coli* DNAs. He and Laura Ingraham spent a good deal of time trying to pin this down, lured off track, as it turns out, by the small amount of cross-hybridization that could be detected between lambda and *E. coli* DNAs, which, bewilderingly, seemed to map across the entire lambda chromosome. We now know that although it is a site-specific reaction, lambda integrative recombination does depend on a very short region of homology with *E. coli*, unfortunately too short to have been detected by the tools of the time.

The task that I undertook was to try to understand the significance of the first two peculiarities of lambda DNA: its asymmetric distribution of nucleotides and the existence of the cohesive ends. Because it seemed that the differences in base composition might have some relevance to transcription or replication, I began projects to study both.

The Segmental Distribution of Nucleotides in Lambda DNA

The tools available to study transcription at the time were quite crude by today's standards, but sufficient to answer the simple questions we posed. The scheme I employed, initiated by previous postdoctoral fellow Nada Ledinko, was to label phage mRNA at various times after infection and to ask how much was made, what was its base composition, and in which molecular half did it originate. The latter experiments depended on molecular hybridization with denatured DNA immobilized in agar, a method developed by Bolton and McCarthy (1962) that would seem quaint to today's experimentalists.

Just as I was beginning to prepare for these experiments, Davidson and coworkers (Wang et al. 1965; Nandi et al. 1965) reported that the density difference between molecular halves could be increased by addition of mercury ions, which combine preferentially with AT-rich regions. This trick, together with an angle rotor (Hershey et al. 1965) to establish a density gradient, provided even better resolution than we had hoped for. Analyses of a series of fragments of different length and molecular origin eventually yielded a physical map of surprising diversity, in which adjacent regions differed in composition as much as DNAs from entirely different organisms (Skalka et al. 1968). The regions corresponded, generally, with blocks of related function (Fig. 1), and the differences in the regions seemed to be preserved down to gene-sized fragments. What could this mean?

One simple possibility was that the several segments originated independently of one another and came together too recently to have the common species character of a uniform base composition. This was considered to be a "failed theory" by Hershey because (like Darwin's evolution hypothesis) it assumed unique events that cannot be reproduced, and also lacked predictive value. Other possibilities, such as some direct influence of nucleotide composition on DNA function, seemed much more interesting and certainly worth testing. As noted above, one possibility that I had prepared to test was the notion that base composition might somehow influence transcription. These experiments did, indeed, yield interesting results. In collaboration with Hatch Echols, we found that the program of transcription of lambda DNA was controlled by specific lambda genes (Skalka 1966, 1967; Skalka et al. 1967). Production of early transcripts, which originated from the right half of the chromosome, depended on gene N, and production of late transcripts, which came from both halves, depended on gene Q. Both N and Q mapped in the right half. Furthermore, the increased rate of mRNA synthesis at late times could be ascribed to increased numbers of DNA templates, as blocking DNA replication (by mutation in genes O or P) did not change the tran-

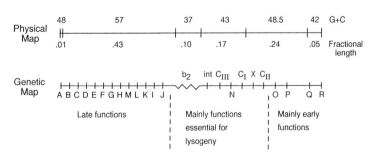

Figure 1. Physical and Genetic Maps of Phage Lambda DNA, circa 1968.

scription program, only the amount of late mRNA produced. Another Hershey fellow, Phyllis Bear, and I also showed that prophage mRNA, corresponding to the product of the *cI* and *rex* genes, originated in the 43% GC region of the right half of the DNA (Bear and Skalka 1969). These results generated considerable interest, as they were the first demonstration of temporal and spatial control of transcription. However, none of this work revealed any obvious correlation with GC content.

In a later analysis of the distribution of nucleotides in several other phages, I found that they could be relegated to two general classes: DNAs from the lambdoid phages, as well as P2 and P22, are composed of a few long segments of differing base composition, whereas DNAs from T5, T7, and P1 have relatively uniform composition (see Hershey 1969). Because all of these phages must perform the same replication functions, and all grow in the same *E. coli* host, these results suggested that we may be left, after all, with the "failed theory," in which the unique distribution of nucleotides in some phage chromosomes reflects (fairly) recent recombinations of genes that are transmitted in blocks because they encode related functions.

Cohesive Ends and DNA Concatemers

My benchmate Merv Smith had been interested in analyzing the properties of newly replicated T5 DNA isolated from infected cells, and found that much of it behaved like molecules longer than phage DNA length. Our subsequent studies of replicating lambda DNA (Smith and Skalka 1966) provided additional interesting results. First, they revealed a supercoiled circular form, as previously described (Young and Sinsheimer 1964; Bode and Kaiser 1965), which we showed was produced mainly from incoming phage DNA. In addition, we found that, like T5, replicating lambda DNA behaved as very long linear, covalent molecules that were later chased into linear monomer phage DNA. These novel forms of long phage DNA precursors were later named concatemers. In thinking about how such lambda molecules might be produced, I had proposed to Hershey that an obvious and simple model would have the circles function as templates for the long linear molecules. As replication proceeded, these linear copies would be pushed out, as from a rolling "printing press," later to be sectioned into phage DNA-sized pieces. Hershey thought it was an interesting possibility, but neither of us thought to write about it. Thus, we were somewhat surprised later on at the great fanfare that surrounded Gilbert and Dressler's (1969) proposal of the "rolling circle" model of replication — which was essentially their independently derived version of my printing press model, based on data not very different from our own. But then, science marketing was not something one would learn in the Hershey lab.

Study of lambda DNA replication was the project that I took with me when I left to set up my own laboratory in 1969. I and my coworkers ultimately showed that there are two stages of replication for lambda DNA, an early stage of circle replication and a later stage of rolling circle replication. We also showed that concatemers are obligatory intermediates in the formation on progeny DNA and they can be formed either by replication or by recombination. Cohesive ends are essential for formation of the early circular forms, but then are created once again as the concatemer precursor is fed into phage heads. As far as I know, there has never been a satisfactory explanation for why lambda doesn't simply package circular genomes. A rationale for the need to form a circle, or in the case of T4 to produce terminal redundancies, became apparent later on when details of the mechanism of DNA replication were revealed. As one DNA strand is copied continuously and the other discontinuously, a region at the 3′ end of a linear duplex template is always left uncopied. If a chromosome is to avoid losing genetic information, it must either have no end, as in a circle, or have terminal redundancies, part of which can be sacrificed.

Hershey as a Mentor

I have always felt extremely fortunate to have had the experience of working under the tutelage of Al Hershey. I was lucky, as it turned out, because this period was close to the end of his career in science. However, I don't believe that our experience was very different from that of postdoctoral fellows from earlier times. I and my laboratory colleagues enjoyed a wonderful scientific environment; we were totally engrossed in our work and shared our results and enthusiasm with each other and with many others in the Demerec building — among them John Cairns, Ric Davern, Joe Speyer, and Barbara McClintock. All of us in the Hershey lab referred to our mentor as "Dr. Hershey," and a sort of genteel formality pervaded the lab, brought about, I presumed, by Hershey's unusually reserved nature and by his treatment of each of us as colleagues rather than apprentices.

Hershey maintained a "hands-off" policy with his postdoctoral colleagues. We were free to do what experiments we wished and he never pestered us for results. At first I thought it odd that discussions about our scientific projects were almost always initiated by us, although he was invariably most helpful when approached, and very willing to think about any problem presented to him. Often, a day or two after exchanging a few words in the lab, we would receive written notes offering thoughts and ideas (see example on next page). Hershey later explained his philosophy to me in his usual economical fashion: By the time we were postdocs he felt we had "earned the right to make our own mistakes."

Besides his work on the lambda-*coli* homology problem, Hershey's personal preoccupation in the years that I was in the lab seemed to be mainly in perfecting methods of analyzing DNA molecules, and in squeezing all the theory and quantitation possible from the methods we commonly used. The results of his thoughts, calculations, and experiments were routinely distributed, again in written form, to all of us. I filled a thick notebook with the fruit of these labors that included such treasures as: a 13-page treatise on "Mercury Complexes of DNA," the theory of "RNA-DNA interactions" during annealing, and a 10-page document "On the hotness of radioactive DNA," among others. Generation of these documents was Hershey's way of organiz-

June 3, 1966

Ann,

If λ is like T2, "late" genes actually have to function **early**, because head protein at least can be detected just about as early as DNA. Than you must be right that early left messenger is the same as late except for rate of synthesis.

Recall ideas we talked about a long time ago. You and Mervyn showed that parental DNA is converted into supercoiled rings. Suppose early transcription occurs in these, mostly AT-rich sections but also other sections. For purposes of argument, the preference for AT sections can be ascribed to the strain in the rings.

When long molecules appear, the strain is gone and the preference for AT sections is lost. *Perhaps single strand cuts appear in the new DNA.*

The rise in rate of transcription from right halves at late times is due simply to more DNA. The much greater rise in rate of transcription from left halves at late times is due both to more DNA and to different DNA.

This is of course similar to Jerry Hurwitz's ideas but I don't think the invitro studies need to bother us much because they are so far nonsensical in themselves. Your gimmick is the ring structures.

This idea is important in two respects.

If it is correct, you and Echols are on the wrong track, because any mutation that stops late messenger synthesis will stop DNA synthesis too, and almost any mutation in an early gene will stop late transcription.

If you believe it may be correct, you should do some experiments that, if they couldn't prove it, would at least make it worth talking about.

Questions:

1) Does rise in rate of left and right transcription occur simultaneously? And simultaneously with appearance of newly synthesized DNA?

2) Does blocking DNA synthesis prevent conversion of parental DNA to rings? It shouldn't and probably doesn't (Bode and Kaiser). Naono and Gros found I guess that blocking DNA synthesis directly blocks late but not early transcription. In Bode and Kaiser's case, transcription is presumably blocked by repressor, which probably has nothing to do with the shape of the DNA.

3) Is all or most of the parental DNA that gets into the cells converted to rings? It was under Bode and Kaiser's conditions.

None of this is very wonderful. Maybe you will think of something better.

Al

P.S. The strongest single fact in favor of a model of this sort is the finding of Naono and Gros that blocking DNA synthesis at a peripheral level prevents late transcription.

The model would be clearly disproved by one good mutation that blocks late transcription but not DNA synthesis. This may be the only kind of mutant worth looking for.

(Editor's note: Ann found that mutants in the Q gene had that phenotype.)

ing his thoughts. But the products were object lessons in the importance of logical exposition and accurate quantitation.

Hershey's hands-off policy extended to publications. He did not put his name on a paper unless he had actually contributed an experiment. This was nice for our careers as young investigators, but should not be misconstrued, as he was truly a guiding force in all that we did. There was never an important thought or result that wasn't run by him for his ideas and comments (we were no fools!), and all of our papers were thoroughly edited (some mostly rewritten) by him. Even in the editing there was so much to learn from this lean, quiet man of piercing intellect and uncompromising integrity. Many of these lessons have been passed on to my trainees throughout the ensuing years, and remain my guiding principles to this day.

References

Bear P.D. and Skalka A. 1969. The molecular origin of lambda prophage mRNA. *Proc. Natl. Acad. Sci.* **62:** 385–388.

Bode V.C. and Kaiser A.D. 1965. Changes in the structure and activity of lambda DNA in a superinfected immune bacterium. *J. Mol. Biol.* **14:** 399–417.

Bolton E.T. and McCarthy B.J. 1962. A general method for the isolation of RNA complementary to DNA. *Proc. Natl. Acad. Sci.* **48:** 1390–1397.

Burgi E. and Hershey A.D. 1961. A relative molecular weight series derived from the nucleic acid of bacteriophage T2. *J. Mol. Biol.* **3:** 458–472.

Gilbert W. and Dressler D. 1969. DNA replication: The rolling circle model. *Cold Spring Harbor Symp. Quant. Biol.* **33:** 473–484.

Hershey A.D. 1964. Some idiosyncrasies of phage DNA structure. *Carnegie Inst. Wash. Year Book.* **63:** 580–592.

———.1969. Introduction. *Carnegie Inst. Wash. Year Book* **67:** 555–556.

Hershey A.D. and Burgi E. 1965. Complementary structure of interacting sites at the ends of lambda DNA molecules. *Proc. Natl. Acad. Sci.* **53:** 325–328.

Hershey A.D., Burgi E., and Davern, C.I. 1965. Preparative density-gradient centrifugation of the molecular halves of lambda DNA. *Biochem. Biophys. Res. Commun.* **18:** 675–678.

Hershey A.D., Burgi E., and Ingraham L. 1963. Cohesion of DNA molecules isolated from phage lambda. *Proc. Natl. Acad. Sci.* **49:** 748–755.

Hershey A.D., Goldberg E., Burgi E., and Ingraham L. 1963. Local denaturation of DNA by shearing forces and by heat. *J. Mol. Biol.* **6:** 230–243.

Hogness D.S. and Simmons J.R. 1964. Breakage of λdg DNA: Chemical and genetic characterization of each isolated half-molecule. *J. Mol. Biol.* **9:** 411–438.

Kaiser A.D. and Inman R.B. 1965. Cohesion and the biological activity of bacteriophage lambda DNA. *J. Mol. Biol.* **13:** 78–91.

Nandi U.S., Wang J.C., and Davidson N. 1965. Separation of deoxyribonucleic acids by Hg(II) binding and Cs$_2$SO$_4$ density-gradient centrifugation. *Biochemistry* **4:** 1687–1696.

Skalka A. 1966. Regional and temporal control of genetic transcription in phage lambda. *Proc. Natl. Acad. Sci.* **55:** 1190–1195.

———. 1967. Multiple units of transcription in phage lambda. *Cold Spring Harbor Symp. Quant. Biol.* **31:** 377–379.

———. 1969. Nucleotide distribution and functional orientation in the deoxyribonucleic acid of phage ϕ80. *J. Virol.* **3:** 150–156.

Skalka A., Burgi E., and Hershey A.D. 1968. Segmental distribution of nucleotides in the DNA of bacteriophage lambda. *J. Mol. Biol.* **34:** 1–16.

Skalka A., Butler B., and Echols H. 1967. Genetic control of transcription during development of phage λ. *Proc. Natl. Acad. Sci.* **58:** 576–583.

Smith M.D. and Skalka A. 1966. Some properties of DNA from phage-infected bacteria. *J. Gen. Physiol.* **49:** 127–142.

Wang J.C., Nandi U.S., Hogness D.S., and Davidson N. 1965. Isolation of λdg deoxyribonucleic acid halves by Hg(II) binding and Cs$_2$SO$_4$ density-gradient centrifugation. *Biochemistry* **4:** 1697–1702.

Young E.T., II and Sinsheimer R.L. 1964. Novel intra-cellular forms of lambda DNA. *J. Mol. Biol.* **10:** 562–564.

Anna Marie Skalka *(born 1938 in New York) applied her formal training in nucleic acid biochemistry to studies of the structures and the modes of replication of viral chromosomes. Her investigations of phage λ, initiated as a postdoctoral colleague of Hershey's (1964–1969), were continued at the Roche Institute of Molecular Biology in New Jersey. Her studies on replication led her to a prescient understanding of the role of genetic recombination in the replication and maturation of the phage λ chromosome. In the 1970s she left the comforts of λ for the tougher challenges of animal viruses. Ann is well known for her studies of molecular aspects of retroviral replication. Over the last several years she has made major contributions to our understanding of retroviral DNA integration. Since 1987, Ann has been Director of the Institute for Cancer Research and Senior Vice President for Basic Science at the Fox Chase Cancer Center.*

Section II

Reminiscences

Sketch of Alfred D. Hershey rendered by Neville
Symonds in Dallas, Texas in 1964

*A*LTHOUGH I FIRST MET AL HERSHEY in 1951, we did not become seriously acquainted until 1957–58, when I spent a year in the Carnegie Institution Department of Genetics at Cold Spring Harbor. The Cold Spring Harbor labs were much smaller and more isolated than they are today. There was already an influx of summer visitors attracted by meetings, summer courses, or the prospect of reviewing their research results in a relaxed atmosphere (Barbara McClintock once called it "the intellectuals' Coney Island"), but after Labor Day they departed, leaving the resident scientists (Hershey, McClintock, Berwind Kaufman, and Milislav Demerec) to interact mainly with their colleagues and their lab groups. That year Al's group included several talented young scientists who had been attracted by his reputation in phage genetics and molecular biology—René Thomas, Jun-ichi Tomizawa, and Gebhart Koch, to name a few. I was not formally a member of Al's group but rather a one-year replacement for George Streisinger, who was on leave; however, Al cordially invited me to attend their group meetings and was almost always accessible for discussion of experiments and ideas. Although he could be harshly critical, I found him a very patient mentor. As long as he considered another scientist competent and willing to be ruthlessly honest, he was tolerant both of inconclusive results and incompletely formulated ideas.

I arrived in Cold Spring Harbor several years after the Hershey-Chase blendor experiment had made Al's name familiar in circles beyond the Phage Group. Before the blendor experiment, his most important contributions had been in purely genetic studies—the discovery of recombination and heterozygosis, analysis of phage crosses, etc. In the 1950s he carried out some important work on isotope transfer from parent to progeny phage, but he also continued with pure genetics, including crosses among irradiated phages designed to elucidate some properties of the intracellular mating pool. By 1958, most of the world was happy to accept the blendor experiment as proof that the

Jill, Al, and Peter Hershey, 1971

phage genome was pure DNA, but Al characteristically still endeavored to eliminate any residual doubts so as to make the proof even more solid.

That year also, Al authored a critical review of formal phage genetics (Hershey 1959). Although some readers found it so complex as to be virtually impenetrable, it was in fact an impressively thorough critique of the logical structure of the models that had been proposed up to that time. He noted (correctly) that one major experimental result on which Visconti and Delbrück had based their picture of a mating pool of intracellular phage (namely, the observation of 50% recombination in 3-factor crosses among those progeny shown to be recombinant in another interval) was not really a new finding but was deducible from earlier two-factor crosses; that the number of rounds of mating calculated from their theory could not be extricated from the amount of interference in the individual mating cycle; and that most previous discussion of high negative interference blithely ignored the relationship between subadditivity of recombination frequencies and interference coefficients that had been explicitly stated by Haldane for classic eukaryotic crosses many years previously. He did all this in the gentlest, least confrontational manner possible.

The paper was not addressed to the intellectually lazy. Al thought long and hard about every aspect of this and other subjects before putting his conclusions to paper, and he was not anxious to spare the casual reader from making similar effort. Indeed, what permeated all his work was the extremely conscientious manner in which he addressed the details of both theory and experiment. Being around that year, I was able to participate in some of the discussions that accompanied the development of the ideas he was trying to formalize.

My year at Cold Spring Harbor was followed by a year in Paris. When I stopped to see Al on my return, it was clear that the review had marked the close of that phase of his career. He had decided that it should be possible and profitable to bite the bullet and analyze phage DNA molecules directly by physical and chemical means. To put things in historical perspective, experiments of the 1950s, many of them with phage, had laid the foundations of modern molecular genetics; however, the successful ones almost always finessed the issue of the structure of large DNA molecules the size of a phage genome or of a bacterial genome freed from possible protein connectors. During the next decade, working increasingly with λ DNA rather than T4, Al helped to put the study of such large molecules on a rigorous basis.

Al took an active interest in all facets of λ research in the 1960s, culminating in the Cold Spring Harbor Lambda Meeting of 1970 and the editing of *The Bacteriophage Lambda* (Hershey 1971), where his intense attention to every aspect of the product set a standard that has seldom been surpassed.

As happened with many of Al's friends and admirers, some of his remarks, both spontaneous and prepared, have stuck in my mind permanently. Realizing that I respond to things that may seem completely flat to others, I nevertheless provide a sampling here (not exact quotes, but pretty close, I hope). To me, they recall several aspects of Al's approach to science and to life.

On his discovery of (I think) genetic recombination in phage: "Knowing two phenomena, A and B, we design experiments where we can use B to investigate A. Frequently this fails because we find a new phenomenon C, that we were unaware of. This has been the history of phage work since the early 1940s. No one has been able to do experiments without being bothered by discoveries."

Shortly after the blendor experiment: "I predict that DNA will not turn out to be the only genetic substance, and also that the most important progress in the next few years will be made by those willing to assume that it is."

On why physicists went into biology: "Cy Levinthal gave the only honest answer to that question. To work in a mature field (as physics had become) is just too hard."

In a letter to the authors of the Lambda Book, suggesting we all tear up our first drafts and start over: "Try reading any manuscript other than your own and you'll see what I mean."

Waclaw Szybalski and Al

A few years ago, in a biographical chapter on Barbara McClintock (Campbell 1993), I commented on the Cold Spring Harbor book *The Dynamic Genome* (Fedoroff and Botstein 1992): "Each author saw McClintock from his or her own perspective and the chapters may ... reveal [as much] about their authors as [shed light] on McClintock's life and career." I'll be surprised if a similar statement does not apply to this volume as well.

In concluding, I return to Al's research direction in 1957–58. After Al's death, a friend of his was quoted as saying that Al had switched to the DNA studies because phage genetics had become too complex for him. That may well be true. And 40 years later, it's hard to interest any large audience in the subtleties of genetic mapping when the major end result can be read from a DNA sequence. However, phage crosses remain a unique source of information on the dynamics of DNA transactions within the infected cell, and both Al's pioneering experimental contributions and his critical analysis of later developments deserve full appreciation.

Allan M. Campbell

𝓕OR MY FIRST POSTDOCTORAL POSITION STARTING IN 1952, I joined Al Hershey's laboratory in the Genetics Section of the Carnegie Institution of Washington at Cold Spring Harbor, which then had three laboratories in addition to Hershey's (Demerec's, McClintock's, and Kaufman's). At my first meeting with Al to discuss a postdoctoral arrangement, I found him staring at a fresh graph containing two curves, one marked ^{32}P and the other ^{35}S, which I failed to appreciate and about which Al, with characteristic taciturnity, did not presume to enlighten me. Only later did I realize I was privy to a momentous scientific event.

The members of Al's laboratory when I arrived were Martha Chase and June Dixon, who worked closely with him, and Nicolo Visconti. Although my research during the three years I was at Cold Spring Harbor did not, to my regret, directly involve Al, his influence on me was deep and lasting. His total dedication to his science, combined with his unshakable integrity and honesty and his remarkable ability to assess critically and succinctly the contributions of others, set a standard that was both intimidating and inspiring.

Because my laboratory was adjacent to Al's, I had the pleasure of spending long working hours near him, usually continuing into the late night. He did not encourage interruptions, but he was willing to share ideas and information, which did not require much of his time because he could convey so much in so few words. I also remember other sides of Al's personality that were not always recognized. When his advice and help were needed he responded generously. When his impressive knowledge of subjects outside of science had the chance to surface it was fascinating to hear.

Phage Group 1952. (Left to right) Niccolo Visconti di Modrone, Martha Chase, Al Hershey, Constance Chadwick, Neville Symonds, June Dixon, and Alan Garen

Al Hershey's seminal and central role as a pioneer of molecular genetics should be evident to all who delve into this volume. His more subtle role as a paragon of a vanishing world of science, where an individual scientist working alone or with few colleagues could have a lasting impact, should be cherished.

Alan Garen

I FIRST ENCOUNTERED AL HERSHEY, as I recall, by reading one of his papers as a graduate student at Harvard in 1963. Reading a paper that conveys a personal and compelling style — that is an experience to be cherished. I can recall a handful of papers that affected me that way back then, by François Jacob, Arthur Kornberg, and Meselson-Stahl among others; Gunther Stent's textbook on viruses made a similar bracing impression.

The Hershey paper (Hershey et al. 1963) was his description, or perhaps I should say discovery, published with Burgi and Ingraham, of the sticky ends of the phage lambda chromosome. The conclusions seem straightforward: Each lambda chromosome bears a cohesive site near (or as we now know at) each of its ends; intramolecular interaction of those sites causes formation of folded molecules (now called Hershey circles), and, as would be expected, intermolecular interaction causes formation of dimers, trimers, etc. What a powerful and still inspiring feat of imagination this analysis and its recounting are. The introduction comprises three sentences:

> Aggregation of DNA is often suspected but seldom studied. In phage lambda we found a DNA that can form characteristic and stable complexes. A first account of them is given here.

There follow some 17 lines of materials and methods — no need for more, because the analysis was performed simply by heating, cooling, and/or shearing (by stirring with a "thin steel blade") at various DNA and salt concentrations and examining the products by sedimentation.

The text then, step by step, leads us through the construction of a beautiful picture — beautiful because sense is made of findings that would strike most of us as uninterpretable and because each bit, on its own of marginal importance, assumes, in light of the conclusion, a powerful position in the overall argument. Along the way the authors invent a set of terms and justify their use. For example, they noticed that stirring the aggregates, formed when lambda DNA is stored at high concentration, reduces the average size in a fashion that depends on the speed of stirring: At 1300–1700 rpm a ^{32}S species is formed, whereas at 200 rpm a smaller fragment is formed. "We call the former process disaggregation by stirring" and "We call [the latter] phenomenon breakage." They then show that disaggregation, but not breakage, is reversible — when warmed to 45 degrees and cooled slowly, the disaggregated DNA re-forms large aggregates, whereas the broken DNA does not. And so: "The reversibility of disaggregation, and the dependence of rate of aggregation on concentration of DNA, justify our language" (i.e., the use of the contrasting terms aggregation and breakage). Thus far the paper, including title, introduction, materials and methods, a substantial figure, and text, has taken up about one and one half pages of PNAS.

To take another example: DNA, at low concentration, heated to 75 degrees and cooled slowly forms "folded" molecules, i.e., molecules that sediment uniformly (37S) and somewhat faster than do linear molecules

Betty Burgi

(which are produced by cooling the same sample quickly). Folded molecules, unlike linear ones, do not aggregate when incubated at 45 degrees, even at high concentration, hence the term "folded." There follows this remarkable paragraph:

> The similarity between the conditions, other than DNA concentration, controlling formation and destruction of folded molecules, and formation and destruction of aggregates, suggests that similar cohesive forces are involved in both phenomena. The folding implies that each molecule carries at least two mutually interacting cohesive sites, which join to form a closed structure. The uniformity of structure of folded molecules, indicated by the narrow zone in which they sediment, suggests that there are not more than two cohesive sites, and that these are identically situated on each molecule.

I invite the reader to consult the paper to see how this economy and depth of analysis is maintained throughout the rest of the paper.

I recall particularly vividly two personal interactions with Al. One occurred at a symposium in Oregon in the early 1980s. I was invited to speak and he was invited to do the various introductions. Al misunderstood his invitation (as explained recently to me by Frank Stahl) and arrived with a prepared text. But by that stage in his life he spent his days mostly in his study where he searched for deep meanings in life by listening to music. (In typical Al fashion the music was recorded from the radio on a small tape recorder. As I recall he was particularly absorbed in Elgar's *Dream of Gerontius.*) It seems that he had come to some rather gloomy conclusions about the ultimate meaning, or lack thereof, of life (as gleaned from his studies of music) and the organizers convinced him to act as master of ceremonies (their original intent)

Anna Marie Skalka and Al

instead of delivering his rather uncheerful message. He introduced each speaker, after scratching his head in the characteristic Hershey manner, with one or two sentences. Ham Smith: "a very tall fella"; Linus Pauling: "winner of two Nobel prizes...seems to be overdoing it a bit"; Gunther Stent: "I am a simple man; I think maybe A, maybe B, and then try to decide which is wrong. This next speaker thinks maybe A, maybe B, maybe C, maybe D, maybe E, and then tries to decide which is correct." (Gunther, with perfect aplomb, walked to the platform, paused, and then said: "You know, I think Al has figured out my problem.") I was one of the last speakers: "Here we have the smartest guy of the lot— this I know because he tells me he likes my papers."

The other personal encounter I will recall has been widely retold, but it is too good to pass up here. In the early 1970s Al agreed to edit, for Cold Spring Harbor, a book on phage lambda (Hershey 1971). I was thrilled to be writing for The Master, the "novel-ist manqué" as Sydney Brenner once described him to me, and so I worked especially hard on my contribution. It came back with more or less every line crossed out, with the occasional "good phrase" encircled. So I fasted, prayed, and rewrote to perfection. This time I was thrilled to see no marks on the first page and none on the second — but on the third page was the command "START HERE," followed by lots of cross-outs. I doubt that any of us reached the standards Al was setting for us, but finally a book of model clarity and consistency emerged. I said to him that, having gone through that process with this book, he must know more about lambda than anybody. He replied no, he actually took in very little of what he read as he edited, instead rely-ing on a sentence-by-sentence check of clarity and logic to produce a coherent text. Hmm, another lesson from Al.

Mark Ptashne

*A*L HERSHEY HAD BEEN NO MORE than a name to me when he turned up, along with almost all of the dozen or so communicants of the "American Phage Church" (as André Lwoff called it), at a Phage Meeting run by Salva Luria at Indiana University in April, 1949. They included Salva's resident crew — his postdoc Renato Dulbecco and his graduate student Jim Watson — as well as the Church's Pope, Max Delbrück, and his first two Caltech post-docs, Élie Wollman and me.

Al hadn't been at Cold Spring Harbor in the previous summer of 1948, when I took the Phage Course and learned that "Hershey agar," the optimal solid *E. coli* growth substrate for scoring plaque-type phage mutants, derived its name not, as I thought at first, from the chocolate bar, but from its inventor. He came to Bloomington from St. Louis, with his graduate student, Raquel Rotman (the only woman at the meeting).

Élie created a sensation when he summarized Lwoff's recent, definitive demon-stration of the reality of the lysogeny phenomenon, because Max had decreed long ago that belief in lysogeny was heresy. But now that Lwoff had shown that a covert, non-infective prophage is responsible for the long-term coexistence of phage and host cell, it seemed obvious that a new age was dawning for understanding the long-term sur-vival of viruses in nature.

Then Al gave what I thought was the best talk of the meeting, a wonderfully lucid account of his and Rotman's studies of genetic recombination in phage. They had found that there occurs an exchange of genetic markers between two or more phage genomes infecting the same host cell. Hence it transpired that viruses can engage in genetic recombination, which, ever since the rediscovery of Gregor Mendel's paper at the turn of the 20th century, had been regarded as being reserved for sexually repro-ducing organisms. Ironically, the recombination mechanism in phage turned out to be formally analogous to the crossing-over of chromosomes first elucidated in fruit flies by T.H. Morgan's group, which Max had found too arcane for his taste when he came to Morgan's Caltech lab as a postdoc in 1937 and chose to work on phage instead. Hershey and Rotman showed that phage genes, like fruit fly genes, are deployed according to a linear map, on which the relative distances of three genes, A, B, and C, are provided by the frequencies of crossing-over between A and B, B and C, and A and C. This finding opened the way for the molecular biological reform of genetics, which began with the refinement of Hershey and Rotman's (1948, 1949) recombination methods by the 1948 Phage Course alumnus, Seymour Benzer, and his empirical reduction in 1955 of the classic, abstract bead-on-a-string model of the gene to the concrete DNA-nucleotides level.

Also at the meeting were Leo Szilard and his erstwhile Los Alamos adjutant, Aaron Novick, who had set up a Biophysics Laboratory at the University of Chicago, in an abandoned synagogue just south of the Midway. Szilard and Novick presented their recent discovery that the progeny of a mixed infection of one bacterium with two parent phages differing genetically in their host range phenotype include some seemingly schizoid phage particles: They have the genotype of one parent and the host-range phenotype of the other parent. Max and the rest of us considered this phenomenon—termed phenotypic mixing—as a mere curiosity, perhaps a trivial epiphenomenon. Had we foreseen the results of the 1952 Hershey-Chase experiment, our quest to fathom the mechanism of intracellular phage multiplication would have been shortened by

George Stent and Al

three years. For we would have certainly inferred from the phenotypic mixing story that intracellular self-replication of the injected parental phage DNA proceeds independently of the DNA-controlled synthesis of the phage tail proteins and that the maturation of intact progeny particles proceeds by a random encapsulation of progeny phage DNA by progeny phage proteins.

I first learned of the Hershey-Chase experiment at a Symposium on the Nature of Virus Multiplication held by the British Society for Microbiology in Oxford in the spring of 1952. Salva Luria had been invited to present a general overview of the present knowledge of the mechanism of phage multiplication. But the early 1950s were the heyday of anti-Red hysteria, and, virtually at the last moment, Salva (who had been an avid supporter of leftist causes) was refused a passport for travel to England. So the Symposium's organizers asked Salva's student, Jim Watson — by then John Kendrew's postdoc at Cambridge— to stand in for his teacher at the Oxford meeting. Jim was to summarize the manuscript that Salva had submitted for advance distribution to all registered Symposium attendees in compliance with the Society's requirement.

Most unfortunately, in his manuscript Salva proposed that the genes of the phage reside in its protein rather than in its DNA. He grounded this proposal on electron-micrographic examinations (carried out in collaboration with Cyrus Levinthal) of phage-infected bacteria at different stages of the latent period. The only phage-specific intracellular structures Luria and Levinthal could discern prior to the appearance of intact progeny phage particles were empty shells of phage proteins—termed "doughnuts." They inferred on this basis that the doughnuts are responsible for the transgenerational genetic continuity of the virus.

This placed Jim in an awkward position, because just a few days before the Symposium, the news of the Hershey-Chase experiment had reached him in Cambridge (as they had not yet reached me in Paris, whence I had come for the meeting). So in his summary of Salva's intended presentation, Jim informed the audience that in view of very recent, as-yet-unpublished experiments done at Cold Spring Harbor, a slight emendation of the Luria manuscript was called for: The genes of the phage reside in its DNA, rather than in its protein.

The tidings of the Hershey-Chase experiment made a tremendous impression on me. On the trip back to Paris from Oxford, I decided that when I would set up shop at the Berkeley Virus Laboratory in the following fall, I was going to follow the intracellular fate of the injected phage DNA in the course of its replication and genetic recombination in the infected cell. I also decided that the method I was going to use for this purpose was one that Al had discovered in the previous year, in collaboration with his St. Louis colleagues Martin Kamen, J.W. Kennedy, and Howard Gest (Hershey et al. 1951). This was the "suicide" of ^{32}P-labeled phages, which ensues upon the cut of both polynucleotide strands of the phage DNA double helix by one out of ten decays of the incorporated radioactive ^{32}P atoms.

Martha Chase and Al, 1953

These studies of the replication and recombination of phage DNA by means of the ^{32}P-decay suicide method occupied me and my first graduate students and postdocs until the early 1960s. Thus, I am indebted to Al Hershey's brilliant work for critically influencing and inspiring my research for the first decade of my career as an independent investigator. Moreover, I am also indebted to Al as a role model of a deep thinker and scientist of unimpeachable integrity.

Gunther S. Stent

T HE FIRST TIME I MET HERSHEY WAS IN THE SPRING of 1951. I was a young theoretical physicist with thoughts of becoming a biologist and had been sent down to Cold Spring Harbor by Warren Weaver of the Rockefeller Foundation to learn something about phage research. When I arrived, Hershey, Luria, and Delbrück were chatting in Hershey's office. During the course of the day they proceeded to talk to me in turn. Hershey sat me down in front of his desk in the office, Luria took me outside where we sat on some steps in the sun, and Delbrück walked with me down to the beach. I understood hardly a word, but nonetheless in the fall found myself at Caltech, where I took the phage course and started a new life.

In the following spring, Visconti, who was working in Hershey's laboratory, came out to California to write the seminal paper with Delbrück on "The mechanism of recombination in phage" (Visconti and Delbrück 1953). I got involved in some theoretical aspects of the paper, had a brief note added to it as an appendix, and became friendly with Visconti, who asked me to spend the fall in Cold Spring Harbor collaborating with him on some genetic experiments. I duly arrived there at the end of September.

In the fall of 1952 Hershey's laboratory was located in the old Animal House. The phage group consisted of Martha Chase, June Dixon, and Constance Chadwick, who were assisting Hershey in experiments designed to follow the differential synthesis of bacterial and phage DNA in T2-infected cells; Alan Garen and Nick Visconti, who were carrying out some ingenious superinfection experiments on the same sys-

Eddie Goldberg and John Cairns

tem to characterize aspects of the vegetative phage pool; and I, who also with Visconti, began to study the properties of a recently isolated T2 mutant which produced "star" plaques that threw off large numbers of **r** mutants, suggesting the parental particle had unstable characteristics.

Hershey lived in a house close to the laboratory and was always first there, often last to leave, and treated weekends as any other day. Sometimes if the weather was fine and the tides were right, he and Jill used to take us sailing in a small yacht they had moored in the bay. Occasionally, too, Hershey invited Alan Garen and me to his house in the evenings. He would bring out a bottle of Irish whisky and soon Alan and I would be avidly listening to his candid views on science and scientists. Because of the notoriety of the blendor experiment, which had recently been published, numerous visitors used to appear in the laboratory wanting to talk to Hershey. Some he liked and respected, but with others he resented giving up time, and after 10 to 15 minutes they would emerge from his office commenting that after a couple of monosyllabic replies to their questions there would be a long silence, so they thought it time to leave. Soon after, Hershey would appear with a slight smile on his face and get back to work. Visconti, or to give him his full name, Niccolo Visconti di Modrone, belonged to an old aristocratic Italian family and was a huge presence in the laboratory. He and Hershey were good friends and held each other in high regard. He was a man of immense energy and charm. Most evenings he would drive up to New York, where rumor had it he met up with the local mafia. On a couple of occasions he appeared in the laboratory with a glamorous young girl on his arm; one was an exotic Indian princess, the other a film star. He introduced them to Hershey, who chatted benignly with them for a time, and then they disappeared in a puff of perfume.

Those three months were as stimulating and enjoyable as any I can remember. I did return to Cold Spring Harbor the following summer, but things were different. Hershey was away much of the time, sailing with Jill on the Great Lakes. Visconti had left America and returned to Italy. The laboratory had moved to a new building which initially had no air-conditioning. And of course there were all the visitors, the conferences, and the courses.

After 1953, I continued to see Hershey on the intermittent occasions I crossed the Atlantic and visited New York. Initially this was in his laboratory and later, after he had retired, at his home on the hill above Cold Spring Harbor. On one of these latter occasions, when he was well into his seventies, the BBC recorded an interview I made with him in which he was reminiscing about his career. Listening to a tape of this the other day brought back memories of some of the mannerisms of his conversation. A question to him would usually be followed by a pause, sometimes quite lengthy, then would come a measured reply, and more often than not at the end there would be a chuckle when something in the reply caught his fancy. For example, at one point in the interview I asked whether he wasn't surprised how quickly the conclusions of the blendor experiment had been accepted, as 20% of the protein from the infecting phage had stayed with the infected cells after the blending. First came the pause, then the reply, "I wasn't too impressed by the results myself. But of course the reason for its final acceptance was the beautiful structure of DNA that came up soon

Al Hershey and Jim Watson, 1966

after this time, in 1953, which made DNA so irresistible intellectually, whatever the facts might be." Then the chuckle.

Perhaps the most illuminating passage on the tape concerning Hershey's attitude to research came in response to a query as to what aspects of his work had given him most satisfaction. He thought for a while and then replied, "I can't distinguish too much. Of course I was very excited by the blendor experiment because I realized that other people were going to be excited by it. But as far as my own private interests were concerned I always liked what I was doing. Of course there are depressing periods when nothing appears to be happening. But whenever anything was happening, and even when nothing was happening, it was fun just to do phage experiments. Later when I started playing around at being a physical chemist, I enjoyed very much doing work on the structure of DNA molecules, something which I would never have dreamed of doing before I started. But that's the nice thing about doing research. Whatever you do is novel, so you always have this sense of novelty even if you are only using a new gadget. That's nothing to be proud of, but it's fun. And if you get some results it's doubly fun."

As a corollary to this reply I commented that it was intriguing how often his research was in fields in which he had no experience or training, and mentioned a few examples including the theoretical studies he had published on the population biology of recombining phage particles. He laughed, and then added, "At my present stage, I'm least proud of my theoretical contributions. Yes, I once had a dream in which I wrote a book called *The Public Education of A.D. Hershey*. The implication was that, as a dilettante, I was never afraid to expose my ignorance." And to me that's a nice way to remember him: as one of the last scientists who, by a combination of their ability, the time in which they lived, and enlightened patronage, were able to carry their research into whatever avenue they felt inclined, and to succeed in and enjoy their endeavors.

Neville Symonds

*W*E SAT THERE, JUST HERSHEY AND I, in what appeared to be an office consisting of a table and a few chairs thrown into the middle of a working laboratory. At the request of Mark Adams, I had earlier talked about our findings of an RNA that mimicked the composition of the infecting phage's DNA to my classmates taking the Phage Course in Cold Spring Harbor, but it was much more intimidating describing our experiments to this man. I was strictly a biochemist who had only recently entered the area of the bacteriophage, and I found myself often lost in the maze of phage information, mostly oriented toward genetics. But I think I had read every paper written by this luminary sitting next to me. He knew every aspect of the bacteriophage field, genetic as well as biochemical.

At that time, most of us biochemists were convinced that heredity was a property directed by those wonderfully complex and sophisticated molecules, the proteins. Then, in 1952, Hershey and Chase set the record straight forever with that experiment where they physically separated the infecting phage DNA from the phage protein— an experiment that in anybody else's hands would have been considered a technically goofy approach. An experiment that elevated the ordinary Waring® blender to a most prominent place in scientific history. Here was the ultimate proof (quite disturbing to us biochemists) that the genetic material was not in a protein with an intricately arranged set of twenty-some amino acids, but, instead, was contained in a package of not more than four nucleotides! Perhaps it was because Hershey was universally held in such high regard by geneticists, microbiologists, and biochemists alike that this experiment became the defining moment for DNA, rather than the much earlier work, albeit quite thorough, of Avery, MacLeod, and McCarty (1944) with pneumococcal DNA.

1954 Phage Meeting

We talked at length about this RNA that was labeled with a T2 phage-like high AU/GC ratio (T2 DNA has an AT/GhmC ratio of about 2/1). But, as I remember it, the topic was only about every technical detail that went into our experiments. I don't think Al asked me one question about what this finding meant. For this I was grateful; I didn't feel obligated to sputter some ridiculous nonsense. I recall that afterward, when one of Hershey's colleagues asked how the session went, I related that we just discussed the experimental details and not much more. He assured me that it went well, for if Hershey was not interested in what he heard, he would have simply stood up and walked away.

When I decided to do what I knew to be an experimental approach that would unambiguously determine whether RNA was synthesized after lytic phage infection, there were two observations that influenced me. It was known that upon infection with these phages, material RNA production immediately stopped. But in all other biological systems where DNA and protein were actively synthesized, so too was RNA. The other motivating factor came from Hershey's experiments involving the uptake of $^{32}PO_4$ into the phage-infected host. I believe his main purpose was to follow the isotope's route into phage DNA, but with the meticulous detail of accounting for every molecule labeled with ^{32}P, this complete biochemist found a small amount of label in a fraction that had the properties of RNA. If Hershey found it, it must be there!

Sal Luria and Al Hershey were the two strongest supporters of our work. Sal, who was then editor-in-chief of a relatively young journal called *Virology*, urged me to publish in that journal (to this day, my biochemistry peers do not understand the choice of this "obscure" journal). Sal, in turn, had Hershey review the papers. If I had thought that I was to get any favorable treatment from Hershey, that notion came to an abrupt halt. Every detail in these manuscripts was scrutinized critically and had to be accounted for. At the same time, however, I was pleased to be given such serious consideration by the man I respected most in the phage business. For in reading his works, the painstaking detail of his experimentation made me think of him as somebody I could call a biochemist's biochemist.

Over the next few years, we determined that this previously unrecognized kind of RNA, which we called DNA-like RNA, was a quantitatively minor species of RNA that underwent rapid metabolic turnover. Its nucleotide composition appeared to alter with time after infection, and it could be found upon infection with the unrelated T7 bacteriophage, as well as in uninfected *E. coli*. It was Hershey, then, who suggested that we see how this system responded to the protein synthesis inhibitor, chloramphenicol. Indeed, as a result of these experiments, we showed that the synthesis of the DNA-like RNA was intimately associated with the synthesis of protein—but we didn't know how. This connection was resolved about three years later with the realization that RNA with these properties was the missing part necessary to formulate the elegant concept that this RNA was a cellular messenger (mRNA) and an essential component for the translation of the genetic information in DNA.

And now, Al, I think it is time for you to get back to the laboratory. Forty years ago, we left a phage problem hanging. We observed that after T2 infection, ^{14}C formate not only entered the purine rings, as expected, but also somehow became part of the RNA pyrimidines. It is not supposed to. This is not the case with uninfected *E. coli*, T7-infected *E. coli*, or as we demonstrated years before, in chick or rat tissues. What new enzyme information is brought into the host with this pattern of four deoxyribonucleotides unique to T2? This is a job for a biochemist's biochemist, like Al Hershey.

Elliott Volkin

*A*LTHOUGH THE TOTAL AMOUNT OF TIME I spent with Al Hershey was not great, his spirit has been with me, through the years, because of two of his unique ideas.

One was the now widely known concept of Hershey Heaven. When asked (by Alan Garen, I believe) what his idea of happiness would be, Al replied, "To have an experiment that works, and do it over and over again." Al really pinpointed the essence of joy in science.

Less well-known, and an important contribution, I call Hershey's Razor. Everyone is familiar with Occam's razor: "If anything can be cut out of a manuscript without significant loss, cut it out." Hershey's correlate is: "When one has a word, phrase, or sentence in a manuscript that somehow doesn't fit, regardless of twisting it every which way you can, just cut it out."

Al and Seymour Benzer

Hershey's Razor has worked for me like a charm, innumerable times. It is brought to mind with every experiment and manuscript, including the writing of this note. So, Al, I can never forget you.

Seymour Benzer

*E*VER SINCE I SPENT MY FIRST SUMMER IN graduate school at the Laboratory over 30 years ago (it was 1968), Cold Spring Harbor has been woven into the fabric of my scientific career. It is thus no surprise that I had the great pleasure of interacting with Al Hershey. A few memories stand out in connection with the exciting lambda meeting in 1970 and the epic monograph, *The Bacteriophage Lambda*, that followed in 1971.

Ira Herskowitz and Al

Somehow Al allowed himself to be convinced to be editor of this book, a monumental undertaking. It was probably for this reason that he attended the sessions. In one truly classic moment that I am sure is remembered by many, Al stood up after a talk and said something like the following to the speaker: "In 19—, I published a paper along with . . . I've always wondered why nobody ever cites this paper. Is it that our work was wrong, or has no one read the paper?" I don't remember the response of the speaker, but it was wonderful to hear Al ask the question that many of us have wondered so many times about our own work.

An unforgettable memory from the lambda book was fearfully awaiting Al's response after sending him the manuscript that I'd written with Costa Georgopoulos. All of the authors knew that Al went over each paper with a fine-tooth comb, editing the papers with the kind of care that people rarely experience and necessitating a lot of dreaded rewriting. There were many tales of manuscripts coming back covered with red pencil. I felt like I was receiving a final examination as I opened the package from Cold Spring Harbor with our edited manuscript. Fortunately, the results were much better than my college examination performances: Al made not one single mark on the MS, accompanying it with a note (now misplaced, but you can be sure I'll find it) saying something about how his red pencil muscles never twitched while editing our paper. A true thrill.

Another lambda book memory is Al's preparation of a memo called "Words," in which he clarified certain genetic terms. I sure would like to get my hands on a copy of that memo. (*Editor's note:* See page 347 of this volume.)

A particularly wonderful memory of that period was working with Jill on preparing the index for the lambda book. I give her my condolences and warmest regards.

Ira Herskowitz

W HEN I WAS A GRADUATE STUDENT, I read the reports that Al Hershey wrote for the Carnegie Institution Year Book. I am sure others are going to comment on the many qualities of Al Hershey; all I want to say is that those scientific reports, because of their insight and their concise language, became part of my scientific education.

Al, 1971

On the occasion of the Tribute to Max Delbrück (CSHL, August 29, 1981), Al Hershey spoke about the beginnings and the history of the Phage Group. In front of an audience of about 200 phage molecular biologists, summarizing the work of Delbrück and others (like himself) of the Phage Group, he ended by saying "something must have been done right; the ship is still seaworthy."

And the ship is still navigating the sea of knowledge.

Helios Murialdo

I MIGHT BE EXCUSED FOR REVIEWING THE list of potential contributors to this remembrance of Al Hershey. In a sense, I am an outsider, an intruder. On the other hand, I feel rather close to at least half of the other contributors, most of whom were colleagues at Cold Spring Harbor. One contributor I knew while he was an undergraduate at Columbia College; still another as a fellow graduate student's child. Consequently, if I am an outsider, I am a familiar one. Much, in fact, like the tomcat that discovered Al and Jill Hershey's complex, tunnel-like window ingress–exit that they made for their cat (one of my cat's former kittens). That tom entered the Hershey's house seemingly at will, and marked chairs, sofas, and other objects as tomcats will. Indeed, there was a rumor that Al did his best one night to mark the tom in turn, but failed. Only in a figurative sense can Al mark me for what follows.

Before I laid eyes on Al, I knew that he was coming to Cold Spring Harbor. I was in charge of seminars that year—probably 1949—when Dr. Demerec informed me of a special seminar to be given by one Al Hershey. Demerec made it clear that Al might join the Carnegie staff, but he also implied that Al was in the neighborhood by accident. Now, I am not articulate; introducing seminar speakers is a task more nerve-wracking for me than presenting a research seminar. What to say regarding Hershey, a person I did not know? The answer lay in one of the many songs that the young persons—technicians, postdocs, and summer visitors—often sang during evening get-togethers. Along with the sinking of the Titanic, the foggy, foggy dew, and the height at which an apron was worn, we also sang of waylaying wayfaring strangers. That was it! Demerec had said that Hershey just happened to be in the neighborhood.

Unfortunately, as I arose to introduce Hershey at his special seminar, I realized that I would never be able to utter "waylaying a wayfaring stranger" without horrible stammering, so I switched to a poor man's version: ". . . in line with our policy of obtaining seminars from those who happen to be passing by . . . ." Al studied me carefully before rising to speak. Indeed, when I recounted this horrible moment to him many years later, I was startled to learn that he also still remembered it. There was a sense of relief—of closure—in his voice as he said, "I have never understood before how you came to give that introduction." Oh, to be articulate like Dobzhansky, who once introduced Al at Columbia by saying that those persons who have difficulty grasping the concept of the Holy Trinity might contemplate Al Hershey. Dobzhansky then detailed the three areas of Hershey's scientific expertise: serology, microbial genetics, and virology.

Professionally, I had little reason to visit Al in his laboratory. On the few occasions that I did go there, I stood in respectful silence outside the hallway door if he was busy at the lab bench. I stood there until he looked up from his work as if he had completed some task; a slight movement on my part was then sufficient to catch his eye— only then did we talk. Something that he said to me much later implied that he treated his technician/colleague, Martha Chase, with the same respectful deference. Only Martha, he once confided, had the concentration needed to carry out the protocol that led to the Hershey-Chase experiment. He also told how he once began explaining to his glassware washer why she would have to rinse his petri dishes nine rather than six times; residual radiation had to be greatly reduced if the experiment was to succeed. "Dr. Hershey," the young lady interrupted, "if you want me to rinse your glassware nine times, just say so but, please, don't tell me *why* I have to rinse them nine times."

Al was not garrulous; he chose his words carefully. They were always apt and to the point. He detested four o'clock seminars; he preferred a pre-dinner nap. He once confided that he often composed comments that would devastate the speaker, but never uttered them. For this restraint, he was grateful because, as he told me, most of his devastating comments were based on his own misunderstanding of the speaker's message.

I do remember, however, going with Miriam, my wife, to Al's home (Urey Cottage) for dinner one evening. We were early, so Jill was still toiling in the kitchen. Miriam, Al, and I sat silently in the living room. After some moments, overcome by the silence, Miriam announced that she would help Jill in the kitchen, and left. Al and I sat in silence. Referring to a seminar we had just attended, Al suddenly asked quietly, "I wonder if he is as stupid as he sounded?" More silence, and then he added as an afterthought, "I guess not; it's impossible." The seminar in question was on the nature of the gene. It dealt with the mutational outcome (*E. coli*) of UV exposures that varied both in intensity and duration and exposure to visible light ("photoreactivation") that also varied in intensity and time duration. The outcome was a vast number of slides illustrating curve after curve, none of which shed light on the nature of the gene. "If one does not understand A," Al continued, "and one does not understand B, how can one interpret any result obtained by combining A and B?"

Although terse and insightful, Al's comments were not unkind. I once reported to the staff at Cold Spring Harbor (staff meetings consisted of members of both the Biological Laboratory and the geneticists at the Carnegie Institution) on an observation that I had made. In that era, population geneticists were often limited to the determination of allelic ("gene") frequencies in populations. Often these studies were limited to two dissimilar alleles. Having determined the zygotic frequencies, the investigator proceeded to ask, "Do these fit Hardy-Weinberg (i.e., binomial) expectations?" As a rule at that time (and occasionally even now), the investigator also asked (if the distribution of genotypes did not fit the expected one), "What genotype occurs in higher than expected frequency?" Common sense dictated that natural selection favored that genotype. That common sense conclusion is correct, however, only if no change has occurred in gene frequencies! This was the point of my report. As I talked, Al reclined in a corner chair looking much like a relaxing integral sign, eyes seemingly closed. When I finished my account, Al had a summarizing comment: "Bruce, if I understand correctly, you have said that if you know where you are, but you don't know where you started, you can't say how you got there." Succinct, and to the point!

A remarkable incident occurred once during Al's presentation—probably of the Hershey-Chase experiment. Having explained a point to the apparent satisfaction of all present, Al was interrupted by Barbara McClintock; she had not followed Al's argument—could he be clearer? Al faced the notes he had made on the blackboard; he remained silent for several long moments. Turning to Barbara, he said that he had explained matters as clearly as he could; all that he could do would be to repeat what he said, but louder. At that point Barbara prepared to leave the room. "Others," she said, "seem to have understood, and I do not want to hold up your talk. We can talk later, in private." With that, she was gone.

An incident similar to the above occurred when Al gave a graduate seminar to the genetics, biochemistry, and plant breeding students at Cornell. Circular DNA was known, but staggered cuts and "sticky ends" were less familiar. Following his seminar, a student wanted more information about the reannealing of the "sticky ends" created by restriction enzymes. Again, Al thoughtfully studied the sketches that he had made while lecturing. Turning to the student, he said, "No, I won't explain. If I did, you would see how it works. If I don't and if you think about it, you will see how it works, and you will have done it yourself. I think that's best." In the following days, the grad students were evenly split: About half thought Al's response was condescending and rude; the other half thought it showed faith in the questioner's ability, and in the end would be best for the student's personal satisfaction.

Years after my departure from Cold Spring Harbor, I wrote a fable ("The Weaver and Her Marvelous Fabric") that appeared in *Psychology Today* (January 1984). It was inspired by Barbara McClintock, her work, and the reaction of others to that work.

Jim Watson, Frank Stahl, Jim Ebert, Al Hershey, Max Delbruck, and Sal Luria

When the fable appeared in print, I sent a copy of the magazine to Jill and Al for their amusement; in doing so, I neglected to put a return address on the mailing label. Later, I received a long letter from Jill. The magazine came on the very day she had gone to the local Social Security office to arrange for her retirement. Not knowing the source of the magazine, she admitted being irritated by noting articles on Alzheimer's disease, therapy after sixty, and sex never ages. Eventually, however, she saw Barbara's photograph and my name; with those clues she realized that she was not the target of someone's tasteless joke. Al added the following footnote to Jill's letter: "As I see it, you found your way out of a dilemma. Best wishes to you both. Al." And, as I see it, any attention one received from Al was, and is, to be treasured.

Bruce Wallace

OVER THE YEARS, WHEN I WAS AT Cold Spring Harbor, Al Hershey wasn't there. In the early years while I taught the bacterial genetics course in the summer, Al was off on his boat sailing on Lake Michigan. In fact, for the summers of 1954–1956 I and my family house- or rather cat-sat for him and Jill. During the later years when I was a trustee at CSH, Al had retired. Still we managed to see much of each other over the years. However, we tended to exchange words only as needed.

In the mid 1950s, I was collaborating with Alan Garen on some radiobiological experiments with phage P22. Alan was in Hershey's laboratory and, since some of the experiments required large amounts of ^{32}P, we did these experiments at CSH, as I did not have access to a hot laboratory at Rockefeller. We did most of the experiments in the middle of the day when Al went home for his nap. Still, we awaited his return so that we could discuss with him what we were doing. P22 plaques could be counted in four hours, so, with proper planning, several experiments could be done in one day. I don't know how Alan felt, but it was my impression that Al considered us beyond the pale and politely didn't dispute our ideas. How wrong one can be in the impression of others. Many years later apropos of nothing in particular, just an apparent impulse, I received a note from Al saying how much he missed those days in the laboratory when Garen and I were doing experiments.

Jill and Al Hershey, 1969

Different people do the different parts of science in different ways. I was trained by Lederberg, who was and still is an omnivorous reader of the literature. He is always searching out articles in many areas of science. Luria, whom I met at CSH in the summers, was of the opinion that the less one knew the better off one was. He seemed to rely mostly on what people told him. Al, on the other hand, seemed to know just what to read in order to be up on the developments that were of concern to him—no more, no less. Since few can be like Lederberg and know everything, and I was too neurotic to rely on circumstance as Luria did, I tried to be like Al. (In retrospect, I ended up more like Josh—afraid I would miss something.)

While the work proceeded, Alan and I were continually discussing, debating, and interpreting results. Whether things were going well or not for us, Al always had a few words of encouragement. Al and his own experiments always seemed to have come to an accommodation. I cannot imagine him allowing an experiment to harass him, in that he would come and would leave at precise times regardless of how things seemed to be going. Typical of Al's serenity (or not) was a story that I believe Garen told me.

At that time we would add KCN to synchronize phage infection. Generally, we had a one-molar solution which we would dilute into the culture to 1/100. Somehow, Al forgot and mouth-pipetted rather than letting the pipet fill by capillary action; no pipettors in those days. Al got a mouthful and, as recounted, just spat it out and went back to work saying, "What will be will be."

Through the rest of the 1950s I would see Al at the annual winter phage meetings, which he organized but always found others to run. Although Max Delbrück was rarely there, his rules were kept. There was no agenda, a few chairmen were chosen, and people volunteered to speak at the various sessions. No slides were allowed unless they were pictures of things that one really had to see. Al rarely intervened, but when he did it was always in a positive and clarifying way. He either wanted some point to be further developed or was trying to make sure that everyone was aware of something that he felt to be important. This was unlike Max, who terrified one and all, as he did not hesitate to question the data as well as the competence of the speaker.

My next significant encounter with Al was after Demerec retired. After a number of scientists had turned down the position as his replacement as Director of the lab, it was offered to me. The details of this happening are another story, but Al and I talked a great deal about it, going over budgets and the lab's needs. In part because Al was concerned that he might have to take over, he really wanted me to come, and I probably would have had Carnegie been interested in other than a maintenance operation. In an era when genetics was flowering, and CSH should have been expanding, they refused to supply, or allow me to obtain, the necessary money. Al was the only person at the lab allowed to have a government grant. In our discussions, two matters were of prime importance: allowing all scientists to apply for grants and making available a reasonable capital budget. Many of the laboratory buildings were in need of considerable repair. Carnegie also wanted to cease all relationships with LIBA, thereby dividing the laboratory. Carnegie believed that in the absence of its support, the LIBA Lab would default to it.

Al helped me to obtain support from other scientists, such as Luria, Szilard, and Tatum, but all to no avail. Carnegie would not budge. When I turned down the position, Carnegie withdrew completely. Ultimately, given major effort by F. Ryan and the cooperation of some 14 local universities, the independent CSH Laboratories were incorporated and struggled along until, in 1968, James Watson took on the Directorship. Even though I continued to teach a bacterial genetics course in the summers, I rarely saw Al, even after becoming a trustee of the laboratory, since trustees met in New York City in those days. Moreover, as my career developed, I became more and more involved with the public side of science, while Al stayed with the lab. After Al retired, I saw him only on rare occasions.

My last encounter with Al was in many ways typical of our interactions over the years. In 1993 I was asked by Rockefeller to prepare some events to celebrate the fiftieth anniversary of the Avery, MacLeod, and McCarty (1944) paper; the first results which indicated that DNA was the genetic material. I planned a roundtable with a number of scientists from the area who were still alive. They included M. McCarty, E.

Chargaff, S. Cohen, R. Hotchkiss, J. Lederberg, and Al. The moderator was to be R. Olby, the historian of *Science*. I met Al soon thereafter at the dedication of the McClintock building. I told him what I had planned, offering a car to pick him and Jill up and return them after the celebratory meeting and dinner. Al was his usual non-committal self. He thought it was an interesting idea but was perhaps too much for him. Others joined us in conversation and the matter was not settled. Some weeks later I had another occasion to see Al. I was now able to tell him that most of the other invitees had accepted. He thought that was fine but was still to me noncommittal. As the time approached when I had to prepare the program for the printer, being still uncertain about Al, I called on Rollin Hotchkiss to help. He called Al, who told him he didn't understand why I was concerned since he had told me that he was certainly coming. Quickly calling Al, I arranged the time and place for his pickup.

Nobel Prize Ceremony, 1969

The meeting itself was a surprise in that our 400-seat auditorium was filled with postdocs and students. We had feared that no one would come to see and hear these scientists from the past. With Olby's guidance, they were each to tell us what they had been doing at the time of the DNA paper and how it influenced their further work. Most revealed their different but well-known personalities; a bit verbose with their egos hanging out. However, not Al, who was the hit of the meeting. His terse and sharp comments on the relevance of any chemical genetic experiments after the Watson-Crick model of DNA and his reason for shifting from studies of phage recombination to phage biochemistry drew the most applause and laughter from the audience. Al was so different from the others. He said what he meant and in only a few clear words. He seemed to enjoy meeting the many who thronged about him afterward. Shortly thereafter he sent me an E-mail saying how much he had enjoyed the occasion.

People do science for different reasons. Al loved to solve finite problems. Thus, his works are each things in themselves. I doubt that he ever was concerned about what others thought. I also liked to solve problems, but I craved the interest of others. The fact that whenever I saw Al over the years he'd always mention something that I'd done scientifically, indicated as I noted above, that he found it worth thinking about—a great compliment.

Norton Zinder

*T*HE STATEMENTS BELOW ARE from a transcript of a symposium titled "Historical Roundtable," a discussion with key scientists active between the publication of the Avery, MacLeod, and McCarty paper in 1944 and the discovery of DNA's double-helical structure in 1953.

This is an excerpt from a letter that Dr. Norton Zinder sent with the transcript...

"I enclose statements of Al's from the transcript of a symposium that was held on February 3, 1994 (at Rockefeller University) to celebrate the 50th anniversary of the Avery, MacLeod and McCarty paper on DNA mediated transformation. I was the organizer, Robert Olby the moderator of a panel including E. Chargaff, Seymour Cohen, Al Hershey, R. Hotchkiss, M. McCarty and J. Lederberg. Al's statements catch his terseness, intelligence and honesty..."

Excerpts from the transcript...

Olby: *Thank you Dr. Hotchkiss, very much. (Applause) Dr. Hershey, you came from a background in chemistry through to bacteriology, and then on to work on the phage system. Would you like to tell us about that move, what brought you finally to work on phage.*

Hershey: *The answer is very simple. When the work of Delbrük and Luria appeared, I saw a reason to study phage. (Laughter and Applause) Ah...may I talk now?*

Olby: *Please!*

Hershey: *Two things that struck me about the work of Avery/MacLeod/McCarty. First of all it's wonderful. It was wonderful and still is wonderful. But second, it had so little influence, and that's been brought out here. Now why is that? Well, first of all, as long as you're thinking about inheritance, who gives a damn what the substance is—it's irrelevant. Now, once you know that the genetic material is DNA, there's only one inference: you should study DNA! And Dr. Chargaff did, but few people did seriously, until of course Watson and Crick. Furthermore, another reason for the work of Avery and his colleagues not attracting quite as much attention as* *they might was that they were just too modest. They refused to advertise. A third thing, and this is really not an explanation, but a curiosity. This was an extremely awkward system to work with, the pneumococcus, and a close homolog of course was the tobacco mosaic virus which was the first vehicle for demonstrating the role of nucleic acid in viral infection. These are the last systems you would have chosen, if you'd been looking for material to study from this point of view.*

Well, so, some 10 years had elapsed before the structure of DNA was elucidated. The first fortunate thing about that discovery was that it did have some meaning. As Jim Watson once said, the structure of DNA could have been completely uninteresting. But it was not. But the elucidation of the structure had another effect. It made the work of Avery/MacLeod/McCarty, Hershey, and many other people completely unneeded, superfluous. Because once you have that structure, it has to be genetic material!

That's really all I wanted to say. (Applause)

Later in the transcripts...

Zinder: *I'd like to ask Al Hershey a question. In 1951 Al Hershey was writing a paper in Cold Spring Harbor Symposium on phage heterozygotes. In it the recombination of bacteriophage was performed. Yet, in 1952, you published the Hershey-Chase experiment, which was a biochemical discovery. Can you tell us how—what made you give up the phage genetics and go against the grain to biochemistry?*

Hershey: *I mentioned this afternoon, talking to Dr. Montgomery (but that's irrelevant).... I'd gone to considerable trouble to set up the equipment necessary for working with isotopes, following the work of Seymour Cohen, who first used isotopes to study bacteriophage. Having done this all by myself, buying the equipment and learning how to use it, more or less, I had to—now, to ask what shall I do with it? Well, that answers your question.*

References

Avery O.T., MacLeod C.M., and McCarty M. 1944. Studies on the chemical nature of the substance inducing transformation of pneumococcal types. Induction of transformation by a desoxyribonucleic acid fraction isolated from pneumococcus type III. *J. Exp. Med.* **79:** 137–158.

Campbell A. 1993. Barbara McClintock. *Annu. Rev. Genet.* **27:** 1–6.

Fedoroff N. and Botstein D., Eds. 1992. *The dynamic genome: Barbara McClintock's ideas in the century of genetics.* Cold Spring Harbor Laboratory Press, Cold Spring Harbor, New York.

Hershey A.D. 1959. The production of recombinants in phage crosses. *Cold Spring Harbor Symp. Quant. Biol.* **23:** 19–46.

———, Ed. 1971. *The bacteriophage lambda.* Cold Spring Harbor Laboratory, Cold Spring Harbor, New York.

Hershey A.D. and Chase M. 1952. Independent functions of viral protein and nucleic acid in growth of bacteriophage. *J. Gen. Physiol.* **36:** 39–56.

Hershey A.D. and Rotman R. 1948. Linkage among genes controlling inhibition of lysis in a bacterial virus. *Proc. Natl. Acad. Sci.* **34:** 89–96.

———. 1949. Genetic recombination between host-range and plaque-type mutants of bacteriophage in single bacterial cells. *Genetics* **34:** 44–71.

Hershey A.D., Burgi E., and Ingraham L. 1963. Cohesion of DNA molecules isolated from phage lambda. *Proc. Natl. Acad. Sci.* **49:** 748–755.

Hershey A.D., Kamen M.D., Kennedy J.W., and Gest H. 1951. The mortality of bacteriophage containing assimilated radioactive phosphorus. *J. Gen. Physiol.* **34:** 305–319.

Visconti N. and Delbrück M. 1953. The mechanism of recombination in phage. *Genetics* **38:** 5–33.

SECTION III

Science

Hershey *the philosopher*

Individual members of the Phage Group had the chance to read various of Al's numerous unpublished essays. He was accustomed to disciplining his thoughts by enlisting his colleagues to critique his written summaries of them. The rest of the world got only a glimpse of Hershey the philosopher. Those glimpses, consisting of a Harvey Lecture, an essay in *Nature*, and his Nobel Lecture, are here reprinted in full.

Bacteriophage T2: Parasite or Organelle?*

A.D. HERSHEY

*Department of Genetics, Carnegie Institution of Washington,
Cold Spring Harbor, New York*

$\mathcal{B}$ACTERIOPHAGES HAVE ENTRANCED BIOLOGISTS since these viruses were discovered forty years ago. Here were dead things that grew. Speculation about them ranged over every conceivable alternative, and a few inconceivable ones. Lwoff[1] has summarized early and current thought about what used to be called, by the circumspect, the Twort-d'Herelle phenomenon.

Lwoff's review deals principally with a phase of viral infection during which phage and bacterium multiply in concert. Many indications suggest that in this phase the multiplying virus (prophage) forms an integral part of the hereditary equipment of the bacterium. If so, the prophage itself probably multiplies in the form of naked genetic material by whatever process is supposed to account for reproduction of other genetic materials. This inference comes exclusively from genetic experiments and, for the present at least, must remain somewhat formal.

I shall deal with a rather different example of viral growth, as revealed by experiments with bacteriophage T2. In the form that we know it, T2 always destroys the cells it infects. For this reason the experimental methods and results I shall refer to are quite unlike those yielding information about prophage. It is therefore remarkable, and reassuring, that the two lines of attack lead to rather similar conclusions about multiplying phage. The work with T2 promises to go a little further, pointing not only to a novel view of viral infection, but also toward a better understanding of the structure and function of hereditary material in general.

I shall begin my history by summarizing information available in 1950, by which time the nature of the problem was clear. In that year the basic facts about T2 added up to four items.

1. Resting phage particles are tadpole-shaped structures containing about 2×10^{-10} µg of deoxyribonucleic acid (DNA), a similar quantity of protein, and little else.[2] Their stability, as well as more direct tests, shows that they are metabolically inactive.

---

* Lecture delivered April 26, 1956.
Reprinted, with permission, from *The Harvey Lectures* (1957) **LI:** 229-239 (copyright Academic Press).

2. Growth of phage in the infected bacterium is linked to bacterial metabolism, including some rate-limiting processes (respiration, for example) already operating at maximal capacity at the time of infection.[3]

3. Viral inheritance can be referred to typical chromosome-like structures possessing conservative properties that are independent of genetic structure in the host cell.[4]

4. For a short time after infection, infected cells do not contain infective phage particles. The first infective particles to reappear are probably not the parents of particles formed later on.[5]

Doermann's discoveries (item 4) could be interpreted in only one way.[6-8] The infecting virus particle, on entering the cell, is converted into some noninfective, vegetative form. In this form it multiplies. The products of multiplication are then converted back into resting phage particles by a process that terminates growth.

The next major problem was very clear: What does multiplying phage look like? The information we possessed about it was paradoxical. On the one hand, multiplying phage manifested genetic properties directing attention to the virus as the unit of action; indeed, Luria[6,9] had just shown that phage multiplication produced genetically marked clones whenever mutations occurred, as if multiplication itself were a geometric process, in which each multiplying phage particle fathered others. On the other hand, metabolic studies[3] pointed to the infected bacterium as the unit of action; growth of phage and synthesis of phage nucleic acid occurred at linear rates determined, apparently, by the metabolism of the cell as a whole. Cohen[10] preferred to abandon the terminology used to describe growth (multiplication, self-reproduction), and speak of phage production by the cell.

Needless to say these were not very deep-seated contradictions: they could be reconciled by either of two equally satisfactory hypotheses concerning the structure of multiplying phage.[7]

First, by analogy with other parasites, one could suppose that the infecting virus particle undergoes minimal modifications (softening of surface structures, loss of ability to survive in the naked state, elaboration of enzymes) preparatory to growth in size and multiplication by fission. Loss of infectivity[5] and the main facts of genetic recombination[11] were thus accounted for, and linkage to bacterial metabolism could be explained by dependence on one or more specialized substrates normally manufactured by the cell. This analogy seemed perfectly satisfactory, as indeed it was, to the majority of virologists.

Second, abandoning analogy altogether, one could suppose that the infecting virus particle breaks down to yield a functionally unitary substance, vaguely described as a template, that would serve to reorganize the synthetic pathways of the cell along virus-specific channels.[6,7,10] Structurally, such a substance might be recognizable as a single chemical individual, and its function would be imaginable only in terms of a linear, or otherwise extended, molecule.[7] In a word, multiplying phage would possess exclusively the properties of hypothetical genetic materials in general, autocatalytic and heterocatalytic.[12]

I have recalled these ideas for two reasons. First, their mere formulation as potentially distinguishable hypotheses called for a prolonged and concerted effort, both in and out of the laboratory, on the part of all concerned. Second, it is doubtful whether the human energy for the next phase of research would have been forthcoming if the second hypothesis hadn't offered its irresistible attractions.[7]

Given these ideas, the initial strategy required to discriminate between them was fairly obvious. As Luria pointed out,[6] methods (and caution) were needed to distinguish between genetic and nongenetic materials. I asked[7]: Does or does not the replication of phage-specific materials occur inside a phage-specific membrane? Fortunately, a series of providential discoveries soon furnished the chemical tools essential to this program.

Anderson[13] found that phage particles could be inactivated by osmotic shock, leaving an intact but empty "ghost." Herriott[14] showed that the ghosts were protein structures that could be centrifuged down to leave a supernatant containing almost pure nucleic acid. Since the ghosts could adsorb to bacteria but were noninfective, Herriott correctly suspected that he had made a start toward separating phage particles into genetic and nongenetic parts.

A similar suspicion, laboriously arrived at following rather different leads,[15,16] together with Herriott's discoveries, literally forced Hershey and Chase[17] to perform the following experiment.

Phage particles were labeled in their proteins with radiosulfate. The labeled particles were allowed to attach to bacteria, and the suspension was spun vigorously in a Waring mixer. It was found that 80 per cent of the protein portion of the phage particles could be readily stripped from the cells by this means. A repetition of the experiment, starting with phage particles that had been labeled in their nucleic acid with radiophosphate, showed that only 20 per cent of the nucleic acid portion of the phage particles could be stripped from the cells. Evidently attachment of virus to bacterium was followed by a separation of viral protein and nucleic acid, most of the former remaining at the cell surface, most of the latter penetrating inside. Furthermore, stripped cells remained competent to yield phage, indicating that most of the parental viral protein was not directly concerned with intracellular growth of the virus.

These and other facts about the initial stages of infection can be summarized as follows. The virus particle contains a core of DNA enclosed in a coat of protein. It attaches to the cell wall by the tip of its tail.[18,19] The attachment is specific and passes through reversible to irreversible stages.[20] After attachment, nucleic acid is released from the virus particle and enters the cell. The protein coat remains at the cell surface, and does not undergo extensive chemical changes.[17] Some of these steps have been revealed microscopically.[21]

What do these facts say about multiplying phage? First, that its immediate precursor, at least, is not enclosed in a protein membrane. Second, its genetic specificity is determined, directly or indirectly, by all or any part of 2×10^{-10} µg of DNA of at least 90 per cent purity.[22] Third, it must acquire a protein membrane either before or after multiplication starts. These facts do not say that genetic specificity is determined by

DNA structure to the exclusion of other substances, nor how much DNA is intimately involved.[16,23] They only permit more or less justifiable assumptions about these questions.[23,24]

We can now ask our original question in more specific form: Is viral DNA synthesized inside a virus-specific protein membrane or not?

Attempts to answer this question by direct methods yielded the following information. If infected bacteria are broken open before their normal lysis time, the only characteristic materials one finds, besides phage particles, are structures recognizable as phage membranes and their component parts.[25] These are always empty of DNA, and, indeed, the characteristic viral DNA not yet incorporated into phage particles is found free.[26,27]

These results suggest that viral DNA and protein are formed separately and then put together.[25] Unfortunately, one cannot exclude the alternative possibility that multiplying virus is a complex particle so fragile that it always breaks apart when taken out of its intracellular environment.[25,28]

To avoid this difficulty, tracer methods for analyzing intracellular precursors in their natural state were worked out in my laboratory. These methods borrowed heavily from the earlier tracer experiments of Cohen, Putman, Kozloff, Graham, Maaløe, Watson, Stent, and their collaborators. Cohen had already shown that DNA started to form before phage particles in infected bacteria.[3] Levinthal and Fisher, Rountree, DeMars *et al.*, Lanni, and Maaløe and Symonds had shown the same for morphologically and serologically specific phage proteins.[25]

The description of viral growth that can be arrived at by kinetic tracer analysis alone is now rather complete.[23] The results summarized below refer specifically to cultures fed on glucose and ammonia, maintained at 36°C. Previously published results refer to various other conditions. These conditions affect numerical constants but not conclusions.

Within 10 minutes after infection, bacterial DNA begins to disappear from infected bacteria, and phage precursor DNA begins to form. The amount of precursor builds up to about 25 phage-particle equivalents per bacterium by the fifteenth minute, at which time the first phage particle is finished. After this time DNA synthesis continues at the constant rate of 2 to 3 phage equivalents per bacterium per minute, but does not accumulate as such because it is incorporated into phage particles at a similar rate. The DNA in the precursor pool is drawn on at random to make phage particles. For example, DNA of parental viral origin and DNA containing phosphorus recently assimilated from the culture medium compete on an equal basis for entry into phage particles.[26] Moreover, nearly all the intracellular DNA (excepting a small amount of bacterial DNA not yet destroyed) is contained in the viral precursor pool. This is so because radioactive phosphorus introduced into the pool is diluted by the entire amount of DNA before its entry into phage particles and eventually passes completely into particles, except for small losses easily accounted for by technical complications.

Under the same conditions, phage precursor protein accumulates to only about 13 phage equivalents per bacterium. All or most of this is already serologically specific at the time of synthesis, and it is incorporated into finished phage particles very soon after it is formed.[23]

These results showed that phage precursor protein and phage precursor DNA formed pools of unequal size and therefore could not share a single common pool. They suggested that synthesis of phage precursor DNA might precede synthesis of phage precursor protein, but were not very helpful otherwise. For example, one could still imagine that phage precursor DNA was formed inside particles lacking tails, or particles having protein membranes thinner than those of the finished particles. To test this possibility, it appeared promising to investigate the relation between nucleic acid and protein synthesis by the use of specific inhibitors.

Cohen and Fowler[29] had indicated that inhibition of protein synthesis by 5-methyltryptophan prevented synthesis of nucleic acid in infected bacteria. Burton[30] clarified this result in the following way. He found that blocking protein synthesis in bacteria at the time of infection prevented nucleic acid synthesis. Nucleic acid synthesis was not inhibited, however, if the block was imposed 5 or 10 minutes after infection. In the meantime chloramphenicol had been identified as a specific inhibitor of protein synthesis, and Burton's results were readily confirmed by Melechen and Tomizawa using this substance.[31]

In order to apply the methods of precursor analysis already outlined to cultures in which protein synthesis was inhibited, it was essential that all the infected bacteria remain capable of yielding phage when the inhibition was reversed. After some preliminary difficulties, Melechen and I found that chloramphenicol-inhibited bacteria behaved very well in this respect. The following schedule is typical of several experimental conditions that were employed successfully.[23]

Bacteria were infected with phage, and protein synthesis was reduced to about 5 per cent of the normal rate by adding chloramphenicol after 11 minutes. Under these conditions nucleic acid synthesis, stopped by the infection, is resumed about 10 minutes later. By the sixtieth minute 100 phage equivalents per bacterium of characteristic viral DNA have accumulated, none of it in phage particles nor in particles that sediment like phage particles after artificial lysis of the cells. A considerable amount of protein is also formed during the first 60 minutes, mostly before the addition of the antibiotic. This early protein was expected not to be phage precursor material.[32]

Actual precursor relationships are ascertained by removing chloramphenicol at the sixtieth minute. After this time DNA synthesis continues, protein synthesis resumes, and phage growth begins, all to continue at linear and almost normal rates for the next 2 or 3 hours. The appropriate isotope experiments showed that only 4 to 10 phage equivalents of the previously formed protein, but virtually all the previously formed DNA, are incorporated into phage particles.

These qualitative results leave several questions unanswered concerning the functional competence of phage precursor DNA synthesized in the presence of chloramphenicol. Our present experiments seem to answer one of these, as follows.

Consider the bacteria in which phage growth is just beginning after the removal of chloramphenicol. The cells contain 100 phage equivalents of labeled DNA made in the presence of chloramphenicol. At the time the first crop of phage can be isolated for specific radioactivity measurements, about 15 phage equivalents of additional unlabeled DNA have been formed. What do we expect to find in the phage particles? We can imagine four alternatives.

First, if the normal DNA synthesis accompanying formation of phage particles were occurring inside the coats of future virus particles, it would be nonradioactive, except perhaps for a little isotope made available by the breakdown of the "abnormal" labeled DNA in the cells. Second, if for some other reason the labeled DNA were incompletely functional, the maturing phage particles would have to have some method for discriminating against it, in which case also the virus particles first formed would be mostly nonradioactive. Third, maturing virus particles might convert some generalized form of DNA into its specific pattern as fast as they received it. This is virtually equivalent to asserting that precursor DNA has no specific function at all. It leads to the same prediction as the following. Fourth, if the labeled DNA made in the presence of chloramphenicol and the unlabeled DNA made after its removal were functionally equivalent, and shared a common precursor pool, the first sample of phage to be isolated should contain 100 parts radioactive to 15 parts nonradioactive DNA.

The prediction corresponding to the last alternative is fulfilled by experiment. I consider this result to prove that phage precursor DNA is not formed inside phage-precursor protein membranes. These must be put on afterward. It also tends to prove that DNA made in the presence of chloramphenicol, and therefore probably not containing protein as an integral part of its structure, is normally functional. To this second conclusion one must make certain reservations along the lines indicated by my third alternative, and further experiments are required to confirm it.

For the present, then, we can conclude only that the observed autonomy of synthesis of viral precursor DNA is quite unlike anything that might have been expected on the basis of analogy to typical intracellular parasites. On the contrary, we should recognize that viral infection is a phenomenon without familiar parallel, and expect from it in the future only what we find.

In a lecture of this kind one is permitted, perhaps expected, to conclude on a note of speculative inference. To this end I propose the following hypothesis. When T2 multiplies it does so in the form of naked genetic material composed of DNA. Other specific biosyntheses, though essential to the perpetuation of the virus, are superfluous to this basic process. They must, in fact, be secondary to it, though this thought is still unpleasantly vague.

A clear separation of the multiplicative phase from the particle-forming phase during growth of T2 is possible only with the aid of inhibitors. In other instances of viral infection this separation of phases may be the natural state of affairs, as when a bacterium is infected by a temperate bacteriophage and survives to start a lysogenic culture. Here, by analogy with T2, one may imagine that only the genetic material of the virus multiplies along with the bacterium; the complete virus re-forms only in occasional cells among the bacterial offspring,[1] and once formed it is completely useless to the population that gave it birth. Moreover, as long as one focuses attention on the bacterial component of the lysogenic culture, it is not possible to distinguish the genetic material of the bacterium from that of the virus; indeed, without additional biological reagents one cannot even recognize that the culture is lysogenic. To cite a dramatic example, the "mutation" from non-toxigenic to toxigenic diphtheria bacilli has been observed only after infection with a temperate bacteriophage. Apparently, the prophage forms part of the bacterial hereditary apparatus essential to toxin production.[33]

The two types of bacteriophage infection, interpreted in this way, have much in common. Neither gains anything from analogies to conventional examples of parasitism. Acquired toxigenesis in the diphtheria bacillus must be referred to a special kind of infective heredity: the bacteriophage is transformed into a functioning bacterial organelle, operationally part of a chromosome.[34] That it can alternatively regenerate bacteriophage does not alter this fact. So must the brusque, invariably lethal, intrusion of genetic material from T2 be thought of in genetic terms, in this instance a complete chemical and functional substitution.[16] Luria's dictum,[6] "parasitism at the genetic level," describes the unprecedented nature of both instances. Even this phrase, I suspect, contains nonessentials to be eliminated in the future.

I prefer to ascribe these and many other biological mysteries to universal properties of genetic material, among which opportunism and single-mindedness seem to be the only guiding principles. In different situations, genetic materials exhibit a few characteristic tendencies that we recognize but do not understand: to congregate, to substitute, to recombine, to change, all superimposed on the basic tendency to multiply conservatively. And, since each bit of genetic material derives from its own inscrutable history, different bits introduced into similar situations display the common tendencies in wholly unpredictable ways. In a word, genetic materials are quite as disorderly as the organisms that harbor them, refusing any classification beyond what suits the momentary whim of the taxonomist.

This limitation is absolute to the student of viruses, who cannot call on witnesses from antiquity. Viruses exist only today; what happened yesterday is already rumor. Fortunately, our ignorance of history is not a severe limitation to the goal of understanding properties-of viruses or of genetic materials in general.

Today, a bacterial virus is a parasitic microbe in ecology, a bacterial organelle during its existence as prophage, a marker on the bacterial chromosome in breeding experiments with lysogenic bacteria,[34] a subgamete for which names are lacking when it transmits lysogeny, a vector of unrelated bacterial organelles, including other prophages,[35] in transduction experiments,[36] and the inciter of an explosive disease of nucleic acid metabolism when it mimics T2. Bacteriophages are all these things, and probably more to be discovered. To ask which is the correct view is to ask what is the proper function of a window: to admit light, to let in air, to keep out wind, to exclude rain, to frame a pleasing landscape, or to pique the peeping Tom.

References

1. Lwoff, A. 1953. *Bacteriol. Revs.* **17,** 269.
2. Taylor, A. 1946. *J. Biol. Chem.* **165,** 271; Herriott, R., and Barlow, J. 1952. *J. Gen. Physiol.* **36,** 17.
3. Cohen, S. 1949. *Bacteriol. Revs.* **13,** 1.
4. Hershey, A. 1953. *Advances in Genet.* **5,** 89.
5. Doermann, A., and Dissosway, C. 1949. *Carnegie Inst. Wash. Year Book* **48,** 170.
6. Luria, S. 1950. *Science* **111,** 507.
7. Hershey, A. 1952. *Intern. Rev. Cytol.* **1,** 119.
8. Doermann, A. 1953. *Cold Spring Harbor Symposia Quant. Biol.* **18,** 3.
9. Luria, S. 1951. *Cold Spring Harbor Symposia Quant. Biol.* **16,** 463.
10. Cohen, S. 1947. *Cold Spring Harbor Symposia Quant. Biol.* **12,** 35.
11. Visconti, N., and Delbrück, M. 1953. *Genetics* **38,** 5.
12. Wright, S. 1941. *Physiol. Revs.* **21,** 487.
13. Anderson, T. 1953. *Ann. inst. Pasteur* **84,** 5.
14. Herriott, R. 1951. *J. Bacteriol.* **61,** 752.
15. Lesley, S., French, R., and Graham, A. 1950. *Arch. Biochem.* **28,** 149.
16. Hershey, A. 1953. *Cold Spring Harbor Symposia Quant. Biol.* **18,** 135.
17. Hershey, A., and Chase, M. 1952. *J. Gen. Physiol.* **36,** 39.
18. Weidel, W. 1953. *Cold Spring Harbor Symposia Quant. Biol.* **18,** 155.
19. Anderson, T. 1951. *Trans. N.Y. Acad. Sci.* **13,** 130.
20. Puck, T. 1953. *Cold Spring Harbor Symposia Quant. Biol.* **18,** 149.
21. Kellenberger, E., and Arber, W. 1955. *Z. Naturforsch.* **10b,** 698.
22. Hershey, A. 1955. *Virology* **1,** 108.
23. Hershey, A. 1956. *In* "Mutations," Brookhaven Symposia in Biology No. 8, page 6, Office of Technical Services, Dept. of Commerce, Washington, D.C.
24. Benzer, S. 1955. *Proc. Natl. Acad. Sci. U.S.* **41,** 344.
25. Levinthal, C., and Fisher, H. 1952. *Biochim. et Biophys. Acta* **9,** 419; Rountree, P. 1951. *Brit. J. Exptl. Pathol.* **32,** 341; DeMars, R., Luria, S., Fisher, H., and Levinthal, C. 1953. *Ann. inst. Pasteur* **84,** 113; Maaløe, O., and Symonds, N. 1953. *J. Bacteriol.* **65,** 177; Lanni, Y. 1954. *Ibid.* **67,** 640; DeMars, R. 1955. *Virology* **1,** 83.
26. Hershey, A. 1953. *J. Gen. Physiol.* **37,** 1.
27. Watanabe, I., Stent, G., and Schachman, H. 1954. *Biochim. et Biophys. Acta* **15,** 38.
28. Hershey, A. 1953. *Ann. inst. Pasteur* **84,** 99.
29. Cohen, S., and Fowler, C. 1947. *J. Exptl. Med.* **85,** 771.
30. Burton, K. 1955. *Biochem. J.* **61,** 473.
31. Melechen, N. 1955. *Genetics* **40,** 584; Tomizawa, J., and Sunakawa, S. 1956. *J. Gen. Physiol.* **39,** 553. Hahn, F., Wisseman, C., and Hopps, H. 1954. *J. Bacteriol.* **67,** 674.
32. Hershey, A., Garen, A., Fraser, D., and Hudis, J. 1954. *Carnegie Inst. Wash. Year Book* **53,** 210.
33. Freeman, V. 1951. *J. Bacteriol.* **61,** 675.
34. Lederberg, E., and Lederberg, J. 1953. *Genetics* **38,** 51; Wollman, E. 1953. *Ann. inst. Pasteur* **84,** 281.
35. Lennox, E. 1955. *Virology* **1,** 190; Jacob, F. 1955. *Ibid.* **1,** 207.
36. Zinder, N., and Lederberg, J. 1952. *J. Bacteriol.* **64,** 679.

Idiosyncrasies of DNA Structure

A.D. Hershey

*I*N 1958, WHAT HAD BEEN LEARNED about the genetic structure of phage particles presented a paradox. On the one hand, genetic crosses revealed only one linkage group.[1] On the other hand, physical evidence suggested that phage particles contained more than one DNA molecule and, probably, more than one species of DNA molecule. The paradox need not be dwelt on here, for it turned out that the physical evidence was mistaken: phage particles contain single DNA molecules that are species specific.

Seeking to resolve the paradox of 1958, my colleagues and I had to start at the beginning by learning how to extract, purify, and characterize DNA molecules. As it happened, our inexperience was not a severe handicap because the existing techniques were still primitive. They were primitive for good reason: until virus particles could be taken apart, nobody had ever seen a solution of uniform DNA molecules. Without knowing it, we were entering one of those happy periods during which each technical advance yields new information.

Joseph Mandell and I began by attempting to make chromatography of DNA work.[2] We succeeded, as had many chromatographers before us, more by art than by theory.[3] The first application of our method, by Elizabeth Burgi and me, yielded the following results.[4]

1. DNA extracted from phage T2 proved to be chromatographically homogeneous.
2. Subjected to a critical speed of stirring, the DNA went over by a single-step process to a second chromatographic species. The second species formed a single band that was not chromatographically homogeneous. We guessed that it consisted of half-length fragments produced by single breaks occurring preferentially near the centers of the original molecules.

The author is director of the Genetics Research Unit, Carnegie Institution of Washington, Cold Spring Harbor, New York 11724. This article is the lecture he delivered in Stockholm, Sweden, 12 December 1969, when he received a Nobel Prize in Physiology or Medicine, which he shared with Dr. Salvador Luria and Dr. Max Delbrück. It is published here with the permission of the Nobel Foundation and will also be included in the complete volume of *Les Prix Nobel en 1969* as well as in the series Nobel Lectures (in English) published by the Elsevier Publishing Company, Amsterdam and New York. Dr. Luria's lecture appeared in the 5 June issue of *Science,* page 1166, and that of Dr. Delbrück in the 12 June issue, page 1312.

3. A single chromatographic fraction of the half-length fragments, when subjected to a higher critical speed of stirring, went over in a single step to a third chromatographic species that we called quarter-length fragments.

4. Unfractionated half-length fragments subjected to stirring could be altered in a gradual but not a stepwise manner, presumably because the fragments of various lengths broke at various characteristic speeds of stirring.

5. These results showed that chromatographic behavior and fragility under shear depended on molecular length, and that our starting material was uniform with respect to length by both criteria.

Burgi and I verified the above results by sedimentation analysis and took pains to isolate precise molecular halves and quarters.[5]

At this point we believed we had characteristic DNA molecules in our hands but lacked any method of weighing or measuring them. Fortunately, Irwin Rubenstein and C. A. Thomas, Jr. had a method of measurement but were experiencing difficulties in preparing materials. We joined forces to measure by radiographic methods the phosphorus content of T2 DNA molecules and their halves and quarters.[6] We found a molecular weight of 130 million for the intact DNA. Moreover, since the DNA molecule and the phage particle contained equal amounts of phosphorus, there could be only one molecule per particle. Evidently T2 possessed a unimolecular chromosome.

With the materials derived from T2 available as standards, Burgi and I worked out conditions under which sedimentation rates in sucrose could be used as measures of molecular weight.[7] We found the useful relation

$$D_2/D_1 = (M_2/M_1)^{0.35} \qquad (1)$$

in which D means distance sedimented, M means molecular weight, and the subscripts refer to two DNA species. The relation serves to measure the molecular weight of an unknown DNA from that of a known DNA when the two are sedimented in mixture. By this method the DNA of phage λ, for instance, shows a molecular weight of 31 million.

Of course, Eq. 1 holds only for typical bihelical DNA molecules. As a check for equivalent structures, we measured fragility under hydrodynamic shear, which also depends on molecular weight and molecular structure.

While the rudiments of T2 DNA structure were being worked out as described above, genetic analysis of the chromosome was generating its own paradox. By this time T4 had largely replaced T2 for experimental purposes, but the two phages are so closely related that information gained from either one usually applies to both.

The paradox appeared in the work of Doermann and Boehner[9] who found, in effect, that T4 heterozygotes could replicate without segregating and were somehow polarized in structure. These properties were incompatible with the heteroduplex model for heterozygotes and eventually led Streisinger and his colleagues to postulate two radical features of T4 DNA structure: circular permutation and terminal repetition.[10] I shall come back to these features presently, and note here only that both have been confirmed by physical analysis.[11] I turn now to a rather different DNA, that of phage λ.

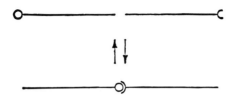

Fig. 1. Reversible joining of molecular halves of λ DNA. The upper part of the figure shows a schematic version of a single molecule cut in two. The lower part of the figure shows the halves rejoined through their terminal cohesive sites.

From the first, our preparations of λ DNA proved refractory in that they refused to pass through our chromatographic column and failed to yield reasonable boundaries in the ultracentrifuge. Only after we achieved sedimentation patterns of high resolution in sucrose did our results begin to make sense.[12] Then we could see in suitable preparations four distinct components. Three of these sedimented at the proper rates according to Eq. 1 for linear structures with length ratios 1: 2: 3. We provisionally called them monomers, dimers, and trimers. The fourth component sedimented faster than the monomer but slower than the dimer. We called it a closed or folded monomer.

Further analysis was possible when we found that heating to about 75°C (insufficient to cause denaturation) converted everything into linear monomers. Then we could show that heating dilute solutions at 55°C converted monomers entirely into the closed form. Alternatively, heating concentrated solutions at 55°C yielded dimers, trimers, and larger aggregates. Moreover, the fragility of the various structures under shear decreased in the proper order: trimers, dimers, closed monomers, then linear monomers. Finally, closure and aggregation were clearly competitive processes, suggesting that each molecule possessed two cohesive sites that were responsible for both processes.

We tested our model by examining molecular halves, which by hypothesis should carry one cohesive site each, and should be able to join only in pairs (Fig. 1). This proved to be correct: by thermal treatment we could reversibly convert molecular halves into structures sedimenting at the rate of unbroken linear molecules. Furthermore, the rejoined halves exhibited a buoyant density appropriate to paired right and left molecular halves, not to paired right halves or paired left halves.[13] Therefore the cohesive sites were complementary in structure, not just sticky spots. Our results were also consistent with terminally situated cohesive sites, though clear proof of this came from the electron microscopists. The structures corresponding to closed and open forms of λ DNA are diagrammed in Fig. 2.

Phages T4 and λ do not exhaust the modalities of phage DNA structure[14] but they cover much of the ground. To bring together what they teach us I show in Fig. 3 idealized DNA molecules of three types.

Structure I represents the Watson-Crick double helix in its simplest form. Actually, structure I is not known to exist, perhaps because replication of the molecular ends would be mechanically precarious.

Structure II represents the linear form of λ DNA, as we have seen. Note that on paper it derives from structure I without gain or loss of nucleotides. In nature, struc-

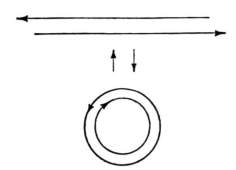

Fig. 2. Interconversion of open and closed forms of λ DNA. The bihelical DNA molecule is indicated by parallel lines of opposite polarity.

ture II cannot derive from I because it contains the sequence *za* not present in I. In structure II, *A* and *a* may be called joining sequences, and they need only be long enough to permit specific base pairing. In λ, the joining sequences each contain about 20 nucleotides.[15] Presumably phage λ circumvents the difficulty of replicating ends by abolishing them (Fig. 2).

Structure III represents T4 DNA. It derives from structure I, on paper, by cyclic permutation followed by the addition of repeats at one end. T4 DNA molecules are said to be circularly permuted, meaning that the 200,000 possible cyclic permutations occur with equal frequency. The terminal repetitions in T4 DNA are relatively long: about 1 percent of the molecular length containing about two genes. Thus T4 DNA solves the problem of replication of ends by making them dispensable. The same feature protects gene function during permutation, because the severed genes in a particular DNA molecule are always present in a second intact copy.

Structure III has three genetic consequences. First, molecules arising by recombination are heterozygous for markers situated in the terminal repetitions.[16] Second, such heterozygotes are likely to appear one-ended and can replicate before segregation.[9] Third, the circular permutation gives rise to a circular genetic map.[10]

T4 cannot generate its permuted DNA molecules by cutting specific internucleotide bonds, and seemingly must somehow cut them to size. In principle, it could do this by measuring either total length or the lengths of the terminal repetitions. Many years ago, George Streisinger and I looked for a shortening of T4 DNA molecules as a result of genetic deletions. We didn't find any shortening. The reason is that T4 cuts its chromosome by measuring overall size. Thus a reduction of the genomic length just increases the lengths of the terminal repetitions.[17]

Structure II as seen in λ DNA also has genetic consequences. Here the chromosome ends are clearly generated by cuts at specific internucleotide bonds, with the result that deletions necessarily shorten the DNA molecule.[18] The cohesive sites in λ DNA, though they do not give rise to a circular genetic map, do show up as a joint in the center of the prophage, thus giving to λ a genetic map with two cyclic permutations. In fact, Campbell[19] had foreseen the need for potential circularity of the λ chromosome at a time when the DNA could only be described as goo.

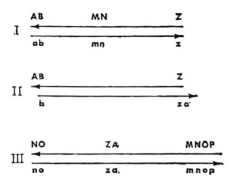

Fig. 3. Three bihelical DNA molecules represented as information diagrams. Each capital letter signifies a nucleotide sequence of arbitrary length; each small letter the corresponding complementary sequence.

Summary

The study of two phage species led to three generalizations probably valid for all viruses.

1. Virus particles contain single molecules of nucleic acid.
2. The molecules are species specific and, with interesting exceptions, are identical in virus particles of a single species.
3. Different viral species contain nucleic acids that differ not only in length and nucleotide sequence but in many unexpected ways as well. I have described only two examples: λ DNA, characterized by terminal joining sequences, and T4 DNA, exhibiting circular permutation and terminal repetition of nucleotide sequences.

Epilogue

In the foregoing account I have deliberately pursued a single line of thought, neglecting both parallel developments in other laboratories and work performed in my own laboratory in which I didn't directly participate. A few of these omissions I feel obliged to repair.

Several people studied breakage of DNA by shear before we did. Davison[20] first noticed the extreme fragility of very long DNA molecules, and he and Levinthal[21] pursued the theory of breakage. However, we first noticed stepwise breakage at critical rates of shear. That observation was needed both to substantiate theory and to complete our evidence for molecular homogeneity.

Davison *et al.*[22] also showed that particles of phage T2 contain single DNA molecules, though they didn't attempt direct measurements of molecular weight.

Physical studies of DNA had of course been under way for some years before analysis of virus particles began. For instance, Doty, McGill, and Rice[23] had observed a relation equivalent to our Eq. 1, containing the exponent 0.37. Their data covered a range of molecular weights below seven million. Larger molecules were not known at the time of their work and could not have been studied by the existing methods anyway.

Our work on sedimentation of DNA in sucrose would have been considerably eased if we had known of the earlier work on sedimentation of enzymes by Martin and Ames.[24]

The first example of a circular DNA,[25] as well as the first evidence for one DNA molecule per phage particle,[26] came from Sinsheimer's work with phage ϕX174. Its DNA comes in single strands that weigh only 1.7 million daltons, which is small enough to permit light-scattering measurements.

Elizabeth Burgi[18] demonstrated reductions of molecular weight of DNA in deletion mutants of phage λ already shown by G. Kellenberger *et al.*[27] to contain reduced amounts of DNA per phage particle.

References and Notes

1. A. D. Kaiser, *Virology* **1**, 424 (1955); G. Streisinger and V. Bruce, *Genetics* **45**, 1289 (1960).
2. J. D. Mandell and A. D. Hershey, *Anal. Biochem.* **1**, 66 (1960).
3. I recall a discouraging conversation with Paul Doty, in which he pointed out and I agreed that large DNA molecules could hardly he expected to reach equilibrium with the bed material during passage through a column. But at least Mandell and I were dealing with ionic interactions. A couple of years later Doty and Marmur and their colleagues were to face a real theoretical crisis: having discovered renaturation of DNA, how explain the remarkable speed of the process?
4. A. D. Hershey and E. Burgi, *J. Mol. Biol.* **2**, 143 (1960).
5. E. Burgi and A. D. Hershey, *ibid.* **3**, 458 (1961).
6. I. Rubenstein, C. A. Thomas, Jr., A. D. Hershey, *Proc. Nat. Acad. Sci. U.S.* **47**, 1113 (1961).
7. E. Burgi and A. D. Hershey, *Biophys. J.* **3**, 309 (1963).
8. A. D. Hershey, E. Burgi, L. Ingraham, *ibid.* **2**, 424 (1962).
9. A. H. Doermann and L. Boehner, *Virology* **21**, 551 (1963).
10. G. Streisinger, R. S. Edgar, G. Harrar Denhardt, *Proc. Nat. Acad. Sci. U.S.* **51**, 775 (1964).
11. C. A. Thomas, Jr., and I. Rubenstein, *Biophys. J.* **4**, 94 (1964); L. A. MacHattie, D. A. Ritchie, C. A. Thomas, Jr., C. C. Richardson, *J. Mol. Biol.* **23**, 355 (1967).
12. A. D. Hershey, E. Burgi, L. Ingraham, *Proc. Nat. Acad. Sci. U.S.* **49**, 748 (1963).
13. A. D. Hershey and E. Burgi, *ibid.* **53**, 325 (1965).
14. C. A. Thomas, Jr., and L. A. MacHattie, *Ann. Rev. Biochem.* **36**, 485 (1967).
15. R. Wu and A. D. Kaiser, *J. Mol. Biol.* **35**, 523 (1968).
16. J. Séchaud, G. Streisinger, J. Emrich, J. Newton, H. Lanford, H. Reinhold, M. Morgan Stahl, *Proc. Nat. Acad. Sci. U.S.* **54**, 1333 (1965).
17. G. Streisinger, J. Emrich, M. Morgan Stahl, *ibid.* **57**, 292 (1967).
18. E. Burgi, *ibid.* **49**, 151 (1963).
19. A. Campbell, *Advan. Genet.* **11**, 101 (1962).
20. P. F. Davison, *Proc. Nat. Acad. Sci. U.S.* **45**, 1560 (1959).
21. C. Levinthal and P. F. Davison, *J. Mol. Biol.* **3**, 674 (1961).
22. P. F. Davison, D. Freifelder, R. Hede, C. Levinthal, *Proc. Nat. Acad. Sci. U.S.* **47**, 1123 (1961).

23. P. Doty, B. McGill, S. Rice, *ibid.* **44,** 432 (1958).
24. R. G. Martin and B. N. Ames, *J. Biol. Chem.* **236,** 1372 (1961).
25. W. Fiers and R. L. Sinsheimer, *J. Mol. Biol.* **5,** 408 (1962).
26. R. L. Sinsheimer, *ibid.* **1,** 43 (1959).
27. C. Kellenberger, M. L. Zichichi, J. Weigle, *ibid.* **3,** 399 (1961).

Genes and Hereditary Characteristics

A. D. HERSHEY

Genetics Research Unit, Carnegie Institution, Cold Spring Harbor, New York

*I*NFLUENTIAL IDEAS ARE ALWAYS SIMPLE. Since natural phenomena need not be simple, we master them, if at all, by formulating simple ideas and exploring their limitations. The notion that genes determine the characteristics of biological individuals and species is exceptional among simple ideas: its limitations have consistently diminished with the passage of time. In this article I shall consider both the simplicity and the limitations of ideas about genetic determination.

The early students of heredity were forced to distinguish between the genetic constitution of an animal or plant (its genotype) and the expression of its genes in visible characters (its phenotype).[1] The distinction is particularly clear in an individual that received dissimilar genes from its two parents: the dual genotype clearly gives rise to a single phenotype. But even an individual with a single set of genes has a phenotype that depends on its stage of development, environmental influences and accidental factors.

The notion that genotype controls phenotype is pure tautology in the typical breeding experiment in which genotype means differences between genotypes, and phenotype means differences between phenotypes. Mendel brought the tautology to light by showing that inheritance depends on unitary factors. Morgan and his students traced these factors to hypothetical genes at known locations in visible chromosomes and created the science of formal genetics.[2] The generality of Mendel's discovery can be summarized by saying that we know of no biological characters, including developmental patterns, that are immune to gene mutations. The importance of his discovery lay in showing that inheritance could be analysed.

Of necessity, formal genetics largely ignored questions about gene function. "As to the nature of 'genes' it is as yet of no value to propose any hypothesis; but that the notion 'gene' covers a reality is evident from Mendelism." This statement, quoted from Johannsen[1], sufficed pretty well for three decades. E. B. Wilson in his famous book expressed a more reckless opinion.[3] "No one ... now conceives the chromosomes as exclusive agents of heredity. They may be considered as modifiers in a reaction system. ... Chromosomes are to be regarded as *differential factors of heredity*, rather than central governing elements. ..." The relation between genotype and phenotype for many years seemed infinitely complicated, as indeed it was in the prechemical era of genetics.

This appraisal of ideas about genetic determination is based on Dr. Hershey's report in the sixty-eighth *Year Book* of the Carnegie Institution of Washington, issued last February.

Help from Neurospora

In 1940 Beadle and Tatum turned new light on Mendel's unit factors when they described their first experiments with the bread mould *Neurospora*. This organism was unique at the time in permitting both genetic and nutritional experiments. With it Srb and Horowitz,[4] for example, could show by mutational analysis that each biochemical step in the synthesis of the amino-acid arginine is the province of a single gene functioning in the production of a single enzyme. Results of this sort, epitomized in the phrase "one gene-one enzyme," obviously conflicted with what everybody knew: that the relation between genotype and phenotype was infinitely complicated. Chiefly because of this conflict, and partly too because the meaning of the phrase was not precisely defined, the new hypothesis met with strenuous opposition.

Actually, the hypothesis consisted of two parts: only one gene functions specifically in the synthesis of a single enzyme; and one gene functions specifically in the synthesis of only one enzyme (if any). Both parts were necessarily a little vague because the significance of the word "specifically" was not clarified for another decade or more. (We now say that one gene determines the amino-acid sequence of one enzyme.)

I do not suppose it is possible to assign a date to the eventual acceptance of the one gene-one enzyme hypothesis, but I remember clearly the Cold Spring Harbor Symposium of 1951 at which the controversy reached its climax. At that meeting Horowitz and Leupold[5] presented their paper entitled "Some Recent Studies bearing on the One Gene-One Enzyme Hypothesis," which would be better known today if it had been called "Confirmation of the One Gene-One Enzyme Hypothesis by the Use of Conditional Lethal Mutations." In the stormy discussion that followed I was unable to grasp the issues, but it was clear to me that nearly everyone was quarrelling with Beadle and Tatum, who were absent, and ignoring the results just presented.

Horowitz and Leupold offered rather subtle arguments in support of the hypothesis that only one gene functions "in a direct manner" in the synthesis of a single enzyme. They also showed that there are relatively few genes concerned with functions common to synthesis of proteins in general, and proposed that the immediate precursors of proteins are single amino-acids or their derivatives, not polypeptides. Their paper is historic both for cogency of argument and because it describes the first systematic use of temperature-sensitive mutants.

Chemical Genetics

In the context of the one gene-one enzyme hypothesis, subsequent developments were chiefly the discovery of two smaller classes of genes: regulator genes,[6] the products of which in the well known examples are proteins interacting directly with DNA to interfere with the expression of other genes[7,8]; and genes in which the structures of ribosomal and transfer RNAs are encoded.[9] At present there are no clear indications that more classes remain to be discovered, although, of course, not all proteins determined by genes are properly called enzymes.

Genetics was a perfected theoretical discipline some time before chemical genetics was invented, and no real conflict between them was possible. (Read, for example, Wright's review,[10] then Beadle's.[11]) The initial aim of chemical genetics called for elucidation of the functional basis of the determination of phenotype by genotype. This aim has been achieved in large measure, largely through elucidation of the structure of DNA[12]. The chief element of surprise, I think, was the simplicity of the denouement, anticipated in part by Beadle and Tatum.

If the overall plan is simple, it ought to be possible to put it into a few words, which I attempt as follows. First, the genotype resides in DNA—more importantly, in the linear sequence of the four nucleotides in single DNA strands. Second, nucleotide sequences in single DNA strands are transcribable into complementary sequences according to simple one-to-one rules: the four bases form only two interstrand pairs, guanine-cytosine and adenine-thymine. This code or its equivalent is used for DNA replication, gene transcription, and synthesis of ribosomal and transfer RNAs. It also regulates the structure of typical double stranded DNA molecules.

Third, sequences in one of the two complementary strands, transcribed into messenger RNA, are translatable into amino-acid sequences in proteins according to a second code unrelated to the first. This is a non-overlapping triplet code (three bases per amino-acid) usually called the genetic code.

Fourth, the phenotype, to the extent that it is understood at all, depends on the structures of species-specific proteins, structures that depend in turn on linear sequences of amino-acids. This conclusion is a necessary inference from what is known of enzyme control of metabolism, on the one hand, and the nature of the genetic code, on the other. That amino-acid sequences determine protein structure might have been guessed years ago from the fact that thermal inactivation of enzymes is often reversible. It is now being proved by artificial synthesis of enzymes. Moreover, gross biological structure can be directly determined by subunit structure, as seen in the reconstitution of virus particles from their parts.[13,14] These successes demonstrate that three dimensional structures can be represented in one dimensional messages, and thus verify one of the most important implications of the Watson-Crick model of DNA—an implication that familiarity with human speech had scarcely prepared us for.

My summary is necessarily abstract. Perhaps I should mention one concrete fact. Rüst and Sinsheimer[15] have shown that either complementary strand of DNA from the phage φX can infect bacterial cells to give rise to identical viral progeny. Therefore the two strands contain the same genetic information: they encode the same genotype. If these two strands could be examined in detail, they would be seen to differ from each other in a systematic way: each adenine residue in one is matched by a thymine residue at the corresponding position in the other, and similarly for the pair guanine and cytosine. This is certain, even though nucleotide sequences cannot be read directly. How we know is fairly simple, too, but the evidence cannot be put into a few words.

What Critics Say

So far I have made what I consider to be factual statements. Perhaps the best way to assess my judgment in this matter is to look at the criticisms raised by people who see things differently. Carl Lindegren, a perceptive man with a brave disregard for the rules of debate, once pointed out that the city of Chicago existed for some time before it became dependent on electricity.[16] He inferred that if DNA happened to be a late-comer on the biological scene, contemporary research would be hard put to discover that fact. Lindegren's analogy is not very apt, for towns do not acquire public utilities by inheritance. Nevertheless, it is fair to ask why biologists should believe in something not visible in the historical record. The answer, to which molecular genetics has contributed something, can be given as follows.

If we assume kinship of living things, we at once imagine an evolutionary family tree stemming from an aboriginal branch point that represents an event of singular importance. That event, the invention of hereditary differences between lines destined to persist, I take to be conceptually equivalent to the origin of life. If we find DNA in both aboriginal branches, we conclude either that DNA function antedates the origin of life or that DNA was independently created two or more times. In fact, biochemists find a common genetic code exploited in all members of a reasonable sample of biological species. To most people, the hypothesis of unique origin provides the only economical explanation. (My weak statement reflects a certain lack of rigour in the argument. Strictly speaking, every homological explanation has an analogical equivalent.)

The same issues are raised in extreme form when experimental evolutionists suggest that life is being continually recreated. There are really two issues. On the one hand, the unity of biology speaks for unique origin. On the other hand, the complexity of the simplest forms of life suggests an inordinately low frequency of spontaneous generation: only once on Earth as far as we know and perhaps not at all in an unexplored universe. I am aware that these arguments are logically ambiguous. I am also aware that they were advanced before the phrase "molecular genetics" was coined. But I wish to make two less obvious points. First, in spite of the simplicity of the overall plan of inheritance, recent advances in molecular genetics serve only to augment, not diminish, our appreciation of biological complexity. For example, Peter Lengyel recently added up 130 known macromolecular components necessary just for the synthesis of protein. Second, without evidence for the unity of biochemistry, now clearly evident in the universality of the genetic code, the doctrine of the unity of biology would be insecure indeed, particularly with respect to the simpler forms of life. In short, life exists and we infer that it had a beginning. To find anything credible in either the fact or the inference is to miss the point.

Barry Commoner has devoted more thought to the search for weaknesses in molecular biology than anyone else (except, of course, molecular biologists). Recently[17] he cited the discovery by Speyer that mutations affecting Kornberg's DNA polymerase (an enzyme concerned with DNA synthesis) can influence the rate of further mutation. This fact shows, according to Commoner, that heritable characteristics are determined in part by nucleotide sequences in DNA, in part by the enzymes that reproduce those sequences. He has also remarked somewhere, presumably to defend his inference against obvious forms of attack, that it has not been shown that information theory is applicable to biological problems. Both these statements strike me as irresponsible: they are not untrue, just obtuse and misleading. Because Commoner is an eloquent writer,[18] and has used such phrases as "theoretical crisis," "illusory successes of molecular biology," and "collision course" to manufacture conflict between "two kinds of biology," it seems worthwhile to try to straighten out the technical basis of his argument. Fleischman[19] has made one attempt. My own is a little simpler.

Long before 1953 it was evident that general mutation rates are themselves determined by genes. Geneticists described the situation as follows. The characteristics of organisms are determined by genes received from their parents and transmitted to their progeny, genes that must, therefore, be duplicated in each cell generation. But genes are not perfectly stable, or are not duplicated with perfect fidelity, whence arise the mutations that serve evolutionary purposes. Thus it is necessary to distinguish between a mechanism of duplication on which inheritance depends, and occasional mistakes made by that mechanism, few of which can be put to use. The frequency of error depends on all sorts of things—external radiation, exposure to various chemicals, temperature, ionic milieu, as well as genetic constitution itself—including, we now know, the gene-determined structure of at least one enzyme. Thus information theory, insofar as it is needed in this connexion, is just the common sense of geneticists, who saw fit to distinguish between speech and noise. I am forced to conclude that Commoner is quarrelling with the principles of genetics, and that he has not said anything about molecular biology.

Reply to the Critics

To separate my counter-criticism from the arbitrary meanings of words, I repeat it in a form that avoids loaded phrases. I have noted that agencies external to the cell, such as sources of radiation, affect mutation rates. According to Commoner's reasoning we ought, on that account, to abandon the notion that living cells can reproduce themselves.

There are, I think, more interesting limitations to current biological principles than anything pointed out by the carpers. The discovery by early geneticists that a unique set of genes determines the characteristics of the individual at once raised the question whether or not biological traits are determined exclusively by those genes. This question has never been answered and is difficult to phrase intelligibly. The demonstration by molecular biologists that a linear genetic message can be translated into three dimensional structure, as in the multiplication of virus particles, showed that in principle the known mechanisms of inheritance could be the only mechanisms. (The example of the viruses is important because there one can observe the regeneration of quite different viral species in the same cellular milieu, depending only on the kind of DNA molecule introduced at the start.) The inference that all three dimensional structure is encoded in nucleotide sequences does not necessarily follow, however. I shall call that inference the unwritten dogma, for it must be shared at least by those biologists who consider molecular biology all but finished. I turn now to some problems of cellular organization that seem to lie beyond the reach of current molecular principles. Here, if anywhere, one might expect to enter the province of biological theory.

The cell theory, dating from 1839 or earlier, engenders lengthy discussions in textbooks, discussions that are interesting but do not, I think, succeed very well in stating a theory. The central notions are: that living things come in cellular form, that cells arise only from pre-existing cells, and that all cells are homologous. Without the last proviso, the cell theory is only a historical remnant of controversy about facts. The last proviso itself comes from Darwin, not cell theorists. Discussion of the theory commonly ends with the statement that the cell is "the structural unit of life," or even "life's minimum unit."[20] Such statements serve chiefly as a reminder that scientific theories cannot be profitably discussed outside their proper experimental context.

According to Tartar,[21] the pertinent experiments are nearly as venerable as the cell theory. For example, Gruber (1885), pursuing earlier work, showed that a single cell of the protozoan genus *Stentor* could be cut into three parts, from each of which a complete animal would regenerate. In subsequent experiments with other ciliates, pieces as small as 1/80 of the cell volume were found capable of regeneration. Such experiments show that the cell is a homoeostatic unit of life, not a minimum unit, and raise the question, what is the minimum unit? By suppressing this question, the cell theory in effect keeps living things out of the laboratory.

More useful theories divide the cell into conceptually distinct parts in the attempt to identify partial functions. Here the successes, notably enzymology and genetics, should not be allowed to overshadow the failures, such as analysis of cellular development, which also has a long history. Tartar ascribes to Prowazek (1904, 1913) what I take to be the traditional view: "Guidance of the elaboration of formed parts in the cytoplasm is to be sought in neither a nucleus of unprescribed location nor a flowing endoplasm but in the most solid portion of the cell, namely, the ectoplasm."[21] This simplifying hypothesis divides the cell into two parts: on the one side, genes and gene products; on the other side, a cell cortex or skeleton that seems to manifest supramolecular properties. Strictly speaking, the hypothesis says nothing about the origin of the guidance system, which could in principle be recreated every time an egg is fertilized, and whenever a protozoon emerges from its cyst. But some sort of primary guiding principle is usually assumed. For example, Luria[22] remarked, in effect, that just as biological evolution is separate from cultural evolution, so is DNA evolution separate from the evolution of cellular organization. This I suspect is as close as it is possible to come to a contemporary statement of the cell theory. Like all good theories, it admits alternatives: for example, the unwritten dogma, according to which biological evolution is solely the evolution of nucleotide sequences. These alternatives were presented in essentially the same terms by Sonneborn.[23]

The hypothesis of bipartite inheritance arose naturally, I think, from the fact that cellular differentiation can be reversible or not to any degree. The reversible changes seemed to imply extragenic determination and the irreversible changes suggested extragenic inheritance, that is, differentiations dating from the origins of the species. The force of this argument is now weakened by the evidence that developmental changes sometimes can be explained in terms of selective control of gene action.[24] But not all persistent changes in cell structure and behaviour can be so explained.[24,25] Before turning to the experiments, I call attention to the purely logical difficulty of imagining an experimental result requiring the hypothesis of bipartite cellular inheritance.

Acquired Traits Retained

There can be no question about evidence that certain acquired traits can persist in a cell lineage. This is not new but improved evidence, which should not be allowed to confuse us. To say that some acquired traits are inherited may be literal truth, but it sacrifices the purity of language that comes to us from Mendel: all traits are acquired, no trait is inherited, not even DNA structure is inherited, only hereditary principles are inherited, for example, nucleotide sequences. All this was explained by Johannsen,[1] questioned by Wilson,[3] and justified when it was found how genes really work. Only then could one invoke information theory to distinguish between what is inherited and what is merely indispensable. The possibility that cells utilize extragenic sources of information should, of course, be kept open, but it must be admitted that no one has found two biological species with identical genomes, nor has anybody proposed a plausible model for extragenic inheritance. (The notion of the primer,[26] that is, a structure that is both substrate and product of enzyme action, is promising but rudimentary.)

If cells draw on an extragenic source of information, a second abstraction must be invoked, another vital principle superimposed on the genotype. A likely candidate already exists in what is usually called cell polarity, which tradition places in a rigid ectoplasm for good reason—it is a spatial principle, and as such requires mystical language. Seemingly independent of the visible structures that respond to it, polarity pervades the cell much as a magnetic field pervades space without help from the iron filings that bring it to light. Biological fields are species specific, as seen in the various patterns and symmetries of growing things.[27] The existence of polar fields is best seen in experiments with *Stentor*, whose ectoplasm contains figurative iron filings in the form of pigmented granules. The following description is condensed from Tartar's book.[21]

Stentor coeruleus is an aquatic one-celled animal bearing feeding organs at its head end and a hold-fast at its tail end. The entire surface is marked by longitudinal stripes of two kinds: clear stripes carrying cilia, alternating with granular stripes without cilia. During the life of the animal, its cortical stripes grow both in width and in number.

The splitting of old (wide) stripes into young (narrow) ones occurs on the ventral surface and proceeds asymmetrically, producing a circumferential (left-right) gradient of stripe widths meeting as a visible boundary on the median ventral surface.

Before each cell division, and during regeneration of decapitated animals, new mouth parts start to form in the region of stripe multiplication on the ventral surface. (The parts afterwards migrate to their proper position.) These observations define an oral primordium site lying near the junction between wide and narrow stripes. However, this site does not contain an indispensable structure, because the dorsal half of a longitudinally bisected animal regenerates a new primordium that appears in the newly created junction between wide and narrow stripes. Apparently the oral primordium develops when required at the poles of a left-right gradient, and near the equator of a longitudinal gradient.

Likewise the hold-fast comes not from a specific hereditary primordium but from something generated at the posterior pole of a longitudinal gradient. Thus if the hold-fast is tucked forward by surgical means, a second one develops at the newly created posterior pole. Animals with two mouths and one tail, or with two tails and one mouth, once created by surgical interference, reproduce as biotypes with varying degrees of stability.

A piece of cortex grafted in reverse orientation into an animal may rotate to restore normal polarities, may degenerate and disappear, or may develop its own mosaic stripe pattern. Similarly, an animal bisected transversely, and reconstructed with the head portion rotated 180° with respect to the tail, may regain its normal stripe pattern either by rotatory slippage of the two halves with respect to each other or by replacement of stripes in one of the halves through outgrowth from the other.

These experiments, like comparable experiments with *Paramecium*,[24] show that polarity resides in all parts of the cell cortex. Perhaps, as many people think, polarity represents something that was invented only once and has evolved since on its own.

The well known experiments of Hämmerling on *Acetabularia*[27] set limits to the hypothesis of bipartite inheritance. In this unicellular alga the single nucleus lies at the base of an erect stalk, well separated from the growing tip. At the tip of the stalk a cap eventually forms as part of an elaborate reproductive process. The structure of the cap is species specific, and interspecies grafts show that cap structure is determined by materials originating in the nucleus or rhizoids and migrating up the stalk, to the exclusion of outgrowths from the tip. Cortical inheritance, if it exists in *Acetabularia*, is apparently not species specific. Perhaps, contrary to what many people think, polarity resides exclusively in the gene-determined structures of polar molecules, and only genes evolve. If so, the only cell theory worthy of the name is wrong.

In *Stentor* and certain other cells, all reasonably large pieces of the cell cortex are equipotent with respect to regeneration of cortical patterns—patterns that are, moreover, subject to metastable variations. In bacteriophages, supramolecular patterns do not persist as such but recur, apparently residing exclusively in the gene-determined structures of individual molecules. *Acetabularia* presents a tantalizing example somewhere in between. Taken together, the facts encourage us to see in cortical polarity a historical invention that ought to be analysable in terms of structure and process. They do not encourage us to think that the task of molecular biology is finished, even at the cellular level. In Tartar's words,[21] "Our greatest lack and most fruitful opportunity in biology lies in conceiving and testing the nature and capabilities of persistent supramolecular patterns."

References

1. Johannsen, W., *Amer. Naturalist,* **45,** 129 (1911).
2. Sturtevant, A. H., *History of Genetics* (Harper and Row, New York, 1965).
3. Wilson, E. B., *The Cell in Development and Heredity,* third ed., 916 (Macmillan Company, New York, 1925).
4. Srb, A. M., and Horowitz, N. H., *J. Biol. Chem.,* **154,** 129 (1944).
5. Horowitz, N. H., and Leupold, U., *Cold Spring Harbor Symp. Quant. Biol.,* **16,** 65 (1951).
6. Jacob, F., and Monod, J., *CR Acad. Sci.,* **249,** 1282 (1959).
7. Gilbert, W., and Müller-Hill, B., *Proc. US Nat. Acad. Sci.,* **56,** 1891 (1966).
8. Ptashne, M., *Proc. US Nat. Acad. Sci.,* **57,** 306 (1967).
9. Yanofaky, S. A., and Spiegelman, S., *Proc. US Nat. Acad. Sci.,* **48,** 1069 (1962).
10. Wright, S., *Physiol. Rev.,* **21,** 487 (1941).
11. Beadle, G. W., *Chem. Rev.,* **37,** 15 (1945).
12. Watson, J. D., and Crick, F. H. C., *Nature,* **171,** 737 (1953).
13. Fraenkel-Conrat, H., and Williams, R. C., *Proc. US Nat. Acad. Sci.,* **41,** 69 (1955).
14. Edgar, R. S., and Wood, W. B., *Proc. US Nat. Acad. Sci.,* **55,** 498 (1966).
15. Rüst, P., and Sinsheimer, R. L., *J. Mol. Biol.,* **23,** 545 (1967).
16. Lindegren, C. C., *Nature,* **176,** 1244 (1955).
17. Commoner, B., *Nature,* **220,** 334 (1968).
18. Commoner, B., *Science and Survival* (Viking Press, New York, 1967).
19. Fleischman, P., *Nature,* **225,** 30 (1970).
20. Simpson, G. G., and Beck, W. S., *Life* (Harcourt, Brace, and World, New York, 1969).
21. Tartar, V., *The Biology of Stentor,* 376, 378 (Pergamon Press, Oxford, 1961).
22. Luria, S. E., *Macromolecular Metabolism,* Suppl., *J. Gen. Physiol.,* **49,** 330 (1966).
23. Sonneborn, T. M., in *The Nature of Biological Diversity* (McGraw-Hill, New York, 1963).
24. Sonneborn, T. M., *Proc. US Nat. Acad. Sci.,* **51,** 915 (1964).
25. Beisson, J., and Sonneborn, T. M., *Proc. US Nat. Acad. Sci.,* **53,** 275 (1965).
26. Robbins, P. W., Wright, A., and Dankert, M., *Macromolecular Metabolism,* Suppl., *J. Gen. Physiol.,* **49,** 331 (1966).
27. Ebert, J. D., *Interacting Systems in Development* (Holt, Rinehart and Winston, New York, 1965).

DNA-mediated transformation of capsular antigens in Pneumococcus provided strong evidence that DNA was the hereditary material. However, the interpretation of that evidence was in doubt for a variety of reasons, one of which was that the determination of capsular antigens might not be typical. The "Waring Blendor" paper by Al and his assistant, Martha Chase, provided evidence that most, maybe all, of the hereditary properties of phage T2 were DNA-determined. In his essay in this book, Rollin Hotchkiss discusses the dynamics between the two, complementary lines of inquiry.

[Reprinted from THE JOURNAL OF GENERAL PHYSIOLOGY, September 20, 1952;
Vol. 36, No. 1, pp. 39–56]
Printed in U.S.A.

INDEPENDENT FUNCTIONS OF VIRAL PROTEIN AND NUCLEIC ACID IN GROWTH OF BACTERIOPHAGE*

BY A. D. HERSHEY AND MARTHA CHASE

(*From the Department of Genetics, Carnegie Institution of Washington, Cold Spring Harbor, Long Island*)

(Received for publication, April 9, 1952)

The work of Doermann (1948), Doermann and Dissosway (1949), and Anderson and Doermann (1952) has shown that bacteriophages T2, T3, and T4 multiply in the bacterial cell in a non-infective form. The same is true of the phage carried by certain lysogenic bacteria (Lwoff and Gutmann, 1950). Little else is known about the vegetative phase of these viruses. The experiments reported in this paper show that one of the first steps in the growth of T2 is the release from its protein coat of the nucleic acid of the virus particle, after which the bulk of the sulfur-containing protein has no further function.

Materials and Methods.—Phage T2 means in this paper the variety called T2H (Hershey, 1946); T2*h* means one of the host range mutants of T2; UV-phage means phage irradiated with ultraviolet light from a germicidal lamp (General Electric Co.) to a fractional survival of 10^{-5}.

Sensitive bacteria means a strain (H) of *Escherichia coli* sensitive to T2 and its *h* mutant; resistant bacteria B/2 means a strain resistant to T2 but sensitive to its *h* mutant; resistant bacteria B/2*h* means a strain resistant to both. These bacteria do not adsorb the phages to which they are resistant.

"Salt-poor" broth contains per liter 10 gm. bacto-peptone, 1 gm. glucose, and 1 gm. NaCl. "Broth" contains, in addition, 3 gm. bacto-beef extract and 4 gm. NaCl.

Glycerol-lactate medium contains per liter 70 mM sodium lactate, 4 gm. glycerol, 5 gm. NaCl, 2 gm. KCl, 1 gm. NH$_4$Cl, 1 mM MgCl$_2$, 0.1 mM CaCl$_2$, 0.01 gm. gelatin, 10 mg. P (as orthophosphate), and 10 mg. S (as MgSO$_4$), at pH 7.0.

Adsorption medium contains per liter 4 gm. NaCl, 5 gm. K$_2$SO$_4$, 1.5 gm. KH$_2$PO$_4$, 3.0 gm. Na$_2$HPO$_4$, 1 mM MgSO$_4$, 0.1 mM CaCl$_2$, and 0.01 gm. gelatin, at pH 7.0.

Veronal buffer contains per liter 1 gm. sodium diethylbarbiturate, 3 mM MgSO$_4$, and 1 gm. gelatin, at pH 8.0.

The HCN referred to in this paper consists of molar sodium cyanide solution neutralized when needed with phosphoric acid.

---

* This investigation was supported in part by a research grant from the National Microbiological Institute of the National Institutes of Health, Public Health Service. Radioactive isotopes were supplied by the Oak Ridge National Laboratory on allocation from the Isotopes Division, United States Atomic Energy Commission.

Adsorption of isotope to bacteria was usually measured by mixing the sample in adsorption medium with bacteria from 18 hour broth cultures previously heated to 70°C. for 10 minutes and washed with adsorption medium. The mixtures were warmed for 5 minutes at 37°C., diluted with water, and centrifuged. Assays were made of both sediment and supernatant fractions.

Precipitation of isotope with antiserum was measured by mixing the sample in 0.5 per cent saline with about 10^{11} per ml. of non-radioactive phage and slightly more than the least quantity of antiphage serum (final dilution 1:160) that would cause visible precipitation. The mixture was centrifuged after 2 hours at 37°C.

Tests with DNase (desoxyribonuclease) were performed by warming samples diluted in veronal buffer for 15 minutes at 37°C. with 0.1 mg. per ml. of crystalline enzyme (Worthington Biochemical Laboratory).

Acid-soluble isotope was measured after the chilled sample had been precipitated with 5 per cent trichloroacetic acid in the presence of 1 mg./ml. of serum albumin, and centrifuged.

In all fractionations involving centrifugation, the sediments were not washed, and contained about 5 per cent of the supernatant. Both fractions were assayed.

Radioactivity was measured by means of an end-window Geiger counter, using dried samples sufficiently small to avoid losses by self-absorption. For absolute measurements, reference solutions of P^{32} obtained from the National Bureau of Standards, as well as a permanent simulated standard, were used. For absolute measurements of S^{35} we relied on the assays (±20 per cent) furnished by the supplier of the isotope (Oak Ridge National Laboratory).

Glycerol-lactate medium was chosen to permit growth of bacteria without undesirable pH changes at low concentrations of phosphorus and sulfur, and proved useful also for certain experiments described in this paper. 18-hour cultures of sensitive bacteria grown in this medium contain about 2×10^9 cells per ml., which grow exponentially without lag or change in light-scattering per cell when subcultured in the same medium from either large or small seedings. The generation time is 1.5 hours at 37°C. The cells are smaller than those grown in broth. T2 shows a latent period of 22 to 25 minutes in this medium. The phage yield obtained by lysis with cyanide and UV-phage (described in context) is one per bacterium at 15 minutes and 16 per bacterium at 25 minutes. The final burst size in diluted cultures is 30 to 40 per bacterium, reached at 50 minutes. At 2×10^8 cells per ml., the culture lyses slowly, and yields 140 phage per bacterium. The growth of both bacteria and phage in this medium is as reproducible as that in broth.

For the preparation of radioactive phage, P^{32} of specific activity 0.5 mc./mg. or S^{35} of specific activity 8.0 mc./mg. was incorporated into glycerol-lactate medium, in which bacteria were allowed to grow at least 4 hours before seeding with phage. After infection with phage, the culture was aerated overnight, and the radioactive phage was isolated by three cycles of alternate slow (2000 G) and fast (12,000 G) centrifugation in adsorption medium. The suspensions were stored at a concentration not exceeding 4 μc./ml.

Preparations of this kind contain 1.0 to 3.0×10^{-12} μg. S and 2.5 to 3.5×10^{-11} μg. P per viable phage particle. Occasional preparations containing excessive amounts of sulfur can be improved by absorption with heat-killed bacteria that do not adsorb

A. D. HERSHEY AND MARTHA CHASE   41

the phage. The radiochemical purity of the preparations is somewhat uncertain, owing to the possible presence of inactive phage particles and empty phage membranes. The presence in our preparations of sulfur (about 20 per cent) that is precipitated by antiphage serum (Table I) and either adsorbed by bacteria resistant to phage, or not adsorbed by bacteria sensitive to phage (Table VII), indicates contamination by membrane material. Contaminants of bacterial origin are probably negligible for present purposes as indicated by the data given in Table I. For proof that our principal findings reflect genuine properties of viable phage particles, we rely on some experiments with inactivated phage cited at the conclusion of this paper.

The Chemical Morphology of Resting Phage Particles.—Anderson (1949) found that bacteriophage T2 could be inactivated by suspending the particles in high concentrations of sodium chloride, and rapidly diluting the suspension with water. The inactivated phage was visible in electron micrographs as tadpole-shaped "ghosts." Since no inactivation occurred if the dilution was slow

TABLE I

Composition of Ghosts and Solution of Plasmolyzed Phage

Per cent of isotope|	Whole phage labeled with		Plasmolyzed phage labeled with	
	P^{32}	S^{35}	P^{32}	S^{35}
Acid-soluble.............................	—	—	1	—
Acid-soluble after treatment with DNase.......	1	1	80	1
Adsorbed to sensitive bacteria................	85	90	2	90
Precipitated by antiphage..................	90	99	5	97

he attributed the inactivation to osmotic shock, and inferred that the particles possessed an osmotic membrane. Herriott (1951) found that osmotic shock released into solution the DNA (desoxypentose nucleic acid) of the phage particle, and that the ghosts could adsorb to bacteria and lyse them. He pointed out that this was a beginning toward the identification of viral functions with viral substances.

We have plasmolyzed isotopically labeled T2 by suspending the phage (10^{11} per ml.) in 3 M sodium chloride for 5 minutes at room temperature, and rapidly pouring into the suspension 40 volumes of distilled water. The plasmolyzed phage, containing not more than 2 per cent survivors, was then analyzed for phosphorus and sulfur in the several ways shown in Table I. The results confirm and extend previous findings as follows:—

1. Plasmolysis separates phage T2 into ghosts containing nearly all the sulfur and a solution containing nearly all the DNA of the intact particles.

2. The ghosts contain the principal antigens of the phage particle detectable by our antiserum. The DNA is released as the free acid, or possibly linked to sulfur-free, apparently non-antigenic substances.

42   VIRAL PROTEIN AND NUCLEIC ACID IN BACTERIOPHAGE GROWTH

3. The ghosts are specifically adsorbed to phage-susceptible bacteria; the DNA is not.

4. The ghosts represent protein coats that surround the DNA of the intact particles, react with antiserum, protect the DNA from DNase (desoxyribonuclease), and carry the organ of attachment to bacteria.

5. The effects noted are due to osmotic shock, because phage suspended in salt and diluted slowly is not inactivated, and its DNA is not exposed to DNase.

TABLE II

Sensitization of Phage DNA to DNase by Adsorption to Bacteria

Phage adsorbed to		Phage labeled with	Non-sedimentable isotope, *per cent*	
			After DNase	No DNase
Live bacteria.............................		S^{35}	2	1
" "		P^{32}	8	7
Bacteria heated before infection...............		S^{35}	15	11
" " " "		P^{32}	76	13
Bacteria heated after infection................		S^{35}	12	14
" " " "		P^{32}	66	23
Heated unadsorbed phage: acid-soluble P^{32}	70°.......	P^{32}	5	
	80°.......	P^{32}	13	
	90°.......	P^{32}	81	
	100°.......	P^{32}	88	

Phage adsorbed to bacteria for 5 minutes at 37°C. in adsorption medium, followed by washing.

Bacteria heated for 10 minutes at 80°C. in adsorption medium (before infection) or in veronal buffer (after infection).

Unadsorbed phage heated in veronal buffer, treated with DNase, and precipitated with trichloroacetic acid.

All samples fractionated by centrifuging 10 minutes at 1300 G.

Sensitization of Phage DNA to DNase by Adsorption to Bacteria.—The structure of the resting phage particle described above suggests at once the possibility that multiplication of virus is preceded by the alteration or removal of the protective coats of the particles. This change might be expected to show itself as a sensitization of the phage DNA to DNase. The experiments described in Table II show that this happens. The results may be summarized as follows:—

1. Phage DNA becomes largely sensitive to DNase after adsorption to heat-killed bacteria.

2. The same is true of the DNA of phage adsorbed to live bacteria, and then

heated to 80°C. for 10 minutes, at which temperature unadsorbed phage is not sensitized to DNase.

3. The DNA of phage adsorbed to unheated bacteria is resistant to DNase, presumably because it is protected by cell structures impervious to the enzyme.

Graham and collaborators (personal communication) were the first to discover the sensitization of phage DNA to DNase by adsorption to heat-killed bacteria.

The DNA in infected cells is also made accessible to DNase by alternate freezing and thawing (followed by formaldehyde fixation to inactivate cellular enzymes), and to some extent by formaldehyde fixation alone, as illustrated by the following experiment.

Bacteria were grown in broth to 5×10^7 cells per ml., centrifuged, resuspended in adsorption medium, and infected with about two P^{32}-labeled phage per bacterium. After 5 minutes for adsorption, the suspension was diluted with water containing per liter 1.0 mM $MgSO_4$, 0.1 mM $CaCl_2$, and 10 mg. gelatin, and recentrifuged. The cells were resuspended in the fluid last mentioned at a concentration of 5×10^8 per ml. This suspension was frozen at $-15°C$. and thawed with a minimum of warming, three times in succession. Immediately after the third thawing, the cells were fixed by the addition of 0.5 per cent (v/v) of formalin (35 per cent HCHO). After 30 minutes at room temperature, the suspension was dialyzed free from formaldehyde and centrifuged at 2200 G for 15 minutes. Samples of P^{32}-labeled phage, frozen-thawed, fixed, and dialyzed, and of infected cells fixed only and dialyzed, were carried along as controls.

The analysis of these materials, given in Table III, shows that the effect of freezing and thawing is to make the intracellular DNA labile to DNase, without, however, causing much of it to leach out of the cells. Freezing and thawing and formaldehyde fixation have a negligible effect on unadsorbed phage, and formaldehyde fixation alone has only a mild effect on infected cells.

Both sensitization of the intracellular P^{32} to DNase, and its failure to leach out of the cells, are constant features of experiments of this type, independently of visible lysis. In the experiment just described, the frozen suspension cleared during the period of dialysis. Phase-contrast microscopy showed that the cells consisted largely of empty membranes, many apparently broken. In another experiment, samples of infected bacteria from a culture in salt-poor broth were repeatedly frozen and thawed at various times during the latent period of phage growth, fixed with formaldehyde, and then washed in the centrifuge. Clearing and microscopic lysis occurred only in suspensions frozen during the second half of the latent period, and occurred during the first or second thawing. In this case the lysed cells consisted wholly of intact cell membranes, appearing empty except for a few small, rather characteristic refractile bodies apparently attached to the cell walls. The behavior of intracellular P^{32} toward DNase, in either the lysed or unlysed cells, was not significantly different from

44   VIRAL PROTEIN AND NUCLEIC ACID IN BACTERIOPHAGE GROWTH

that shown in Table III, and the content of P^{32} was only slightly less after lysis. The phage liberated during freezing and thawing was also titrated in this experiment. The lysis occurred without appreciable liberation of phage in suspensions frozen up to and including the 16th minute, and the 20 minute sample yielded only five per bacterium. Another sample of the culture formalinized at 30 minutes, and centrifuged without freezing, contained 66 per cent of the P^{32} in non-sedimentable form. The yield of extracellular phage at 30 minutes was 108 per bacterium, and the sedimented material consisted largely of formless debris but contained also many apparently intact cell membranes.

TABLE III

Sensitization of Intracellular Phage to DNase by Freezing, Thawing, and Fixation with Formaldehyde

	Unadsorbed phage frozen, thawed, fixed	Infected cells frozen, thawed, fixed	Infected cells fixed only
Low speed sediment fraction			
Total P^{32}	—	71	86
Acid-soluble	—	0	0.5
Acid-soluble after DNase	—	59	28
Low speed supernatant fraction			
Total P^{32}	—	29	14
Acid-soluble	1	0.8	0.4
Acid-soluble after DNase	11	21	5.5

The figures express per cent of total P^{32} in the original phage, or its adsorbed fraction.

We draw the following conclusions from the experiments in which cells infected with P^{32}-labeled phage are subjected to freezing and thawing.

1. Phage DNA becomes sensitive to DNAse after adsorption to bacteria in buffer under conditions in which no known growth process occurs (Benzer, 1952; Dulbecco, 1952).

2. The cell membrane can be made permeable to DNase under conditions that do not permit the escape of either the intracellular P^{32} or the bulk of the cell contents.

3. Even if the cells lyse as a result of freezing and thawing, permitting escape of other cell constituents, most of the P^{32} derived from phage remains inside the cell membranes, as do the mature phage progeny.

4. The intracellular P^{32} derived from phage is largely freed during spontaneous lysis accompanied by phage liberation.

We interpret these facts to mean that intracellular DNA derived from phage is not merely DNA in solution, but is part of an organized structure at all times during the latent period.

Liberation of DNA from Phage Particles by Adsorption to Bacterial Fragments.—The sensitization of phage DNA to specific depolymerase by adsorption to bacteria might mean that adsorption is followed by the ejection of the phage DNA from its protective coat. The following experiment shows that this is in fact what happens when phage attaches to fragmented bacterial cells.

TABLE IV

Release of DNA from Phage Adsorbed to Bacterial Debris

	Phage labeled with	
	S^{35}	P^{32}
Sediment fraction		
Surviving phage..................................	16	22
Total isotope....................................	87	55
Acid-soluble isotope..............................	0	2
Acid-soluble after DNase..........................	2	29
Supernatant fraction		
Surviving phage..................................	5	5
Total isotope....................................	13	45
Acid-soluble isotope..............................	0.8	0.5
Acid-soluble after DNase..........................	0.8	39

S^{35}- and P^{32}-labeled T2 were mixed with identical samples of bacterial debris in adsorption medium and warmed for 30 minutes at 37°C. The mixtures were then centrifuged for 15 minutes at 2200 G, and the sediment and supernatant fractions were analyzed separately. The results are expressed as per cent of input phage or isotope.

Bacterial debris was prepared by infecting cells in adsorption medium with four particles of T2 per bacterium, and transferring the cells to salt-poor broth at 37°C. The culture was aerated for 60 minutes, M/50 HCN was added, and incubation continued for 30 minutes longer. At this time the yield of extracellular phage was 400 particles per bacterium, which remained unadsorbed because of the low concentration of electrolytes. The debris from the lysed cells was washed by centrifugation at 1700 G, and resuspended in adsorption medium at a concentration equivalent to 3×10^9 lysed cells per ml. It consisted largely of collapsed and fragmented cell membranes. The adsorption of radioactive phage to this material is described in Table IV. The following facts should be noted.

46   VIRAL PROTEIN AND NUCLEIC ACID IN BACTERIOPHAGE GROWTH

1. The unadsorbed fraction contained only 5 per cent of the original phage particles in infective form, and only 13 per cent of the total sulfur. (Much of this sulfur must be the material that is not adsorbable to whole bacteria.)

2. About 80 per cent of the phage was inactivated. Most of the sulfur of this phage, as well as most of the surviving phage, was found in the sediment fraction.

3. The supernatant fraction contained 40 per cent of the total phage DNA (in a form labile to DNase) in addition to the DNA of the unadsorbed surviving phage. The labile DNA amounted to about half of the DNA of the inactivated phage particles, whose sulfur sedimented with the bacterial debris.

4. Most of the sedimentable DNA could be accounted for either as surviving phage, or as DNA labile to DNase, the latter amounting to about half the DNA of the inactivated particles.

Experiments of this kind are unsatisfactory in one respect: one cannot tell whether the liberated DNA represents all the DNA of some of the inactivated particles, or only part of it.

Similar results were obtained when bacteria (strain B) were lysed by large amounts of UV-killed phage T2 or T4 and then tested with P^{32}-labeled T2 and T4. The chief point of interest in this experiment is that bacterial debris saturated with UV-killed T2 adsorbs T4 better than T2, and debris saturated with T4 adsorbs T2 better than T4. As in the preceding experiment, some of the adsorbed phage was not inactivated and some of the DNA of the inactivated phage was not released from the debris.

These experiments show that some of the cell receptors for T2 are different from some of the cell receptors for T4, and that phage attaching to these specific receptors is inactivated by the same mechanism as phage attaching to unselected receptors. This mechanism is evidently an active one, and not merely the blocking of sites of attachment to bacteria.

Removal of Phage Coats from Infected Bacteria.—Anderson (1951) has obtained electron micrographs indicating that phage T2 attaches to bacteria by its tail. If this precarious attachment is preserved during the progress of the infection, and if the conclusions reached above are correct, it ought to be a simple matter to break the empty phage membranes off the infected bacteria, leaving the phage DNA inside the cells.

The following experiments show that this is readily accomplished by strong shearing forces applied to suspensions of infected cells, and further that infected cells from which 80 per cent of the sulfur of the parent virus has been removed remain capable of yielding phage progeny.

Broth-grown bacteria were infected with S^{35}- or P^{32}-labeled phage in adsorption medium, the unadsorbed material was removed by centrifugation, and the cells were resuspended in water containing per liter 1 mM $MgSO_4$, 0.1 mM $CaCl_2$, and 0.1 gm. gelatin. This suspension was spun in a Waring

A. D. HERSHEY AND MARTHA CHASE   47

blendor (semimicro size) at 10,000 R.P.M. The suspension was cooled briefly in ice water at the end of each 60 second running period. Samples were removed at intervals, titrated (through antiphage serum) to measure the number of bacteria capable of yielding phage, and centrifuged to measure the proportion of isotope released from the cells.

The results of one experiment with each isotope are shown in Fig. 1. The data for S^{35} and survival of infected bacteria come from the same experiment, in which the ratio of added phage to bacteria was 0.28, and the concentrations

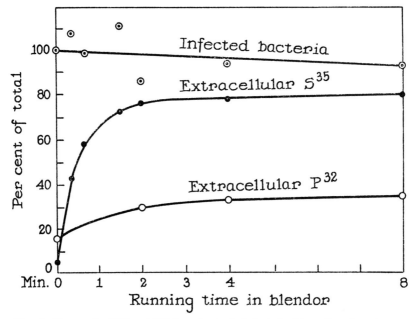

FIG. 1. Removal of S^{35} and P^{32} from bacteria infected with radioactive phage, and survival of the infected bacteria, during agitation in a Waring blendor.

of bacteria were 2.5×10^8 per ml. infected, and 9.7×10^8 per ml. total, by direct titration. The experiment with P^{32}-labeled phage was very similar. In connection with these results, it should be recalled that Anderson (1949) found that adsorption of phage to bacteria could be prevented by rapid stirring of the suspension.

At higher ratios of infection, considerable amounts of phage sulfur elute from the cells spontaneously under the conditions of these experiments, though the elution of P^{32} and the survival of infected cells are not affected by multiplicity of infection (Table V). This shows that there is a cooperative action among phage particles in producing alterations of the bacterial membrane which weaken the attachment of the phage. The cellular changes detected in this way may be related to those responsible for the release of bacterial components from infected bacteria (Prater, 1951; Price, 1952).

A variant of the preceding experiments was designed to test bacteria at a later stage in the growth of phage. For this purpose infected cells were aerated in broth for 5 or 15 minutes, fixed by the addition of 0.5 per cent (*v/v*) commercial formalin, centrifuged, resuspended in 0.1 per cent formalin in water, and subsequently handled as described above. The results were very similar to those already presented, except that the release of P^{32} from the cells was slightly less, and titrations of infected cells could not be made.

The S^{35}-labeled material detached from infected cells in the manner described possesses the following properties. It is sedimented at 12,000 G, though less completely than intact phage particles. It is completely precipitated by

TABLE V

Effect of Multiplicity of Infection on Elution of Phage Membranes from Infected Bacteria

Running time in blendor	Multiplicity of infection	P^{32}-labeled phage		S^{35}-labeled phage	
		Isotope eluted	Infected bacteria surviving	Isotope eluted	Infected bacteria surviving
min.		*per cent*	*per cent*	*per cent*	*per cent*
0	0.6	10	120	16	101
2.5	0.6	21	82	81	78
0	6.0	13	89	46	90
2.5	6.0	24	86	82	85

The infected bacteria were suspended at 10^9 cells per ml. in water containing per liter 1 mM $MgSO_4$, 0.1 mM $CaCl_2$, and 0.1 gm. gelatin. Samples were withdrawn for assay of extracellular isotope and infected bacteria before and after agitating the suspension. In either case the cells spent about 15 minutes at room temperature in the eluting fluid.

antiphage serum in the presence of whole phage carrier. 40 to 50 per cent of it readsorbs to sensitive bacteria, almost independently of bacterial concentration between 2×10^8 and 10^9 cells per ml., in 5 minutes at 37°C. The adsorption is not very specific: 10 to 25 per cent adsorbs to phage-resistant bacteria under the same conditions. The adsorption requires salt, and for this reason the efficient removal of S^{35} from infected bacteria can be accomplished only in a fluid poor in electrolytes.

The results of these experiments may be summarized as follows:—

1. 75 to 80 per cent of the phage sulfur can be stripped from infected cells by violent agitation of the suspension. At high multiplicity of infection, nearly 50 per cent elutes spontaneously. The properties of the S^{35}-labeled material show that it consists of more or less intact phage membranes, most of which have lost the ability to attach specifically to bacteria.

2. The release of sulfur is accompanied by the release of only 21 to 35 per

cent of the phage phosphorus, half of which is given up without any mechanical agitation.

3. The treatment does not cause any appreciable inactivation of intracellular phage.

4. These facts show that the bulk of the phage sulfur remains at the cell surface during infection, and takes no part in the multiplication of intracellular phage. The bulk of the phage DNA, on the other hand, enters the cell soon after adsorption of phage to bacteria.

Transfer of Sulfur and Phosphorus from Parental Phage to Progeny.—We have concluded above that the bulk of the sulfur-containing protein of the resting phage particle takes no part in the multiplication of phage, and in fact does not enter the cell. It follows that little or no sulfur should be transferred from parental phage to progeny. The experiments described below show that this expectation is correct, and that the maximal transfer is of the order 1 per cent

Bacteria were grown in glycerol-lactate medium overnight and subcultured in the same medium for 2 hours at 37°C. with aeration, the size of seeding being adjusted nephelometrically to yield 2×10^8 cells per ml. in the subculture. These bacteria were sedimented, resuspended in adsorption medium at a concentration of 10^9 cells per ml., and infected with S^{35}-labeled phage T2. After 5 minutes at 37°C., the suspension was diluted with 2 volumes of water and resedimented to remove unadsorbed phage (5 to 10 per cent by titer) and S^{35} (about 15 per cent). The cells were next suspended in glycerol-lactate medium at a concentration of 2×10^8 per ml. and aerated at 37°C. Growth of phage was terminated at the desired time by adding in rapid succession 0.02 mM HCN and 2×10^{11} UV-killed phage per ml. of culture. The cyanide stops the maturation of intracellular phage (Doermann, 1948), and the UV-killed phage minimizes losses of phage progeny by adsorption to bacterial debris, and promotes the lysis of bacteria (Maaløe and Watson, 1951). As mentioned in another connection, and also noted in these experiments, the lysing phage must be closely related to the phage undergoing multiplication (*e.g.*, T2H, its *h* mutant, or T2L, but not T4 or T6, in this instance) in order to prevent inactivation of progeny by adsorption to bacterial debris.

To obtain what we shall call the maximal yield of phage, the lysing phage was added 25 minutes after placing the infected cells in the culture medium, and the cyanide was added at the end of the 2nd hour. Under these conditions, lysis of infected cells occurs rather slowly.

Aeration was interrupted when the cyanide was added, and the cultures were left overnight at 37°C. The lysates were then fractionated by centrifugation into an initial low speed sediment (2500 *G* for 20 minutes), a high speed supernatant (12,000 *G* for 30 minutes), a second low speed sediment obtained by recentrifuging in adsorption medium the resuspended high speed sediment, and the clarified high speed sediment.

50   VIRAL PROTEIN AND NUCLEIC ACID IN BACTERIOPHAGE GROWTH

The distribution of S^{35} and phage among fractions obtained from three cultures of this kind is shown in Table VI. The results are typical (except for the excessively good recoveries of phage and S^{35}) of lysates in broth as well as lysates in glycerol-lactate medium.

The striking result of this experiment is that the distribution of S^{35} among the fractions is the same for early lysates that do not contain phage progeny, and later ones that do. This suggests that little or no S^{35} is contained in the mature phage progeny. Further fractionation by adsorption to bacteria confirms this suggestion.

Adsorption mixtures prepared for this purpose contained about 5×10^9 heat-killed bacteria (70°C. for 10 minutes) from 18 hour broth cultures, and

TABLE VI

Per Cent Distributions of Phage and S^{35} among Centrifugally Separated Fractions of Lysates after Infection with S^{35}-Labeled T2

Fraction	Lysis at $t = 0$ S^{35}	Lysis at $t = 10$ S^{35}	Maximal yield	
			S^{35}	Phage
1st low speed sediment......................	79	81	82	19
2nd " " "	2.4	2.1	2.8	14
High speed "	8.6	6.9	7.1	61
" " supernatant.....................	10	10	7.5	7.0
Recovery.................................	100	100	96	100

Infection with S^{35}-labeled T2, 0.8 particles per bacterium. Lysing phage UV-killed *h* mutant of T2. Phage yields per infected bacterium: <0.1 after lysis at $t = 0$; 0.12 at $t = 10$; maximal yield 29. Recovery of S^{35} means per cent of adsorbed input recovered in the four fractions; recovery of phage means per cent of total phage yield (by plaque count before fractionation) recovered by titration of fractions.

about 10^{11} phage (UV-killed lysing phage plus test phage), per ml. of adsorption medium. After warming to 37°C. for 5 minutes, the mixtures were diluted with 2 volumes of water, and centrifuged. Assays were made from supernatants and from unwashed resuspended sediments.

The results of tests of adsorption of S^{35} and phage to bacteria (H) adsorbing both T2 progeny and *h*-mutant lysing phage, to bacteria (B/2) adsorbing lysing phage only, and to bacteria (B/2*h*) adsorbing neither, are shown in Table VII, together with parallel tests of authentic S^{35}-labeled phage.

The adsorption tests show that the S^{35} present in the seed phage is adsorbed with the specificity of the phage, but that S^{35} present in lysates of bacteria infected with this phage shows a more complicated behavior. It is strongly adsorbed to bacteria adsorbing both progeny and lysing phage. It is weakly adsorbed to bacteria adsorbing neither. It is moderately well adsorbed to bacteria adsorbing lysing phage but not phage progeny. The latter test shows that the S^{35} is not contained in the phage progeny, and explains the fact that the S^{35} in early lysates not containing progeny behaves in the same way.

The specificity of the adsorption of S^{35}-labeled material contaminating the phage progeny is evidently due to the lysing phage, which is also adsorbed much more strongly to strain H than to B/2, as shown both by the visible reduction in Tyndall scattering (due to the lysing phage) in the supernatants of the test mixtures, and by independent measurements. This conclusion is further confirmed by the following facts.

TABLE VII

Adsorption Tests with Uniformly S^{35}-Labeled Phage and with Products of Their Growth in Non-Radioactive Medium

Adsorbing bacteria	Per cent adsorbed				
	Uniformly labeled S^{35} phage		Products of lysis at $t = 10$	Phage progeny (Maximal yield)	
	+ UV-*h*	No UV-*h*			
	S^{35}	S^{35}	S^{35}	S^{35}	Phage
Sensitive (H).....................	84	86	79	78	96
Resistant (B/2)...................	15	11	46	49	10
Resistant (B/2*h*)................	13	12	29	28	8

The uniformly labeled phage and the products of their growth are respectively the seed phage and the high speed sediment fractions from the experiment shown in Table VI.

The uniformly labeled phage is tested at a low ratio of phage to bacteria: +UV-*h* means with added UV-killed *h* mutant in equal concentration to that present in the other test materials.

The adsorption of phage is measured by plaque counts of supernatants, and also sediments in the case of the resistant bacteria, in the usual way.

1. If bacteria are infected with S^{35} phage, and then lysed near the midpoint of the latent period with cyanide alone (in salt-poor broth, to prevent readsorption of S^{35} to bacterial debris), the high speed sediment fraction contains S^{35} that is adsorbed weakly and non-specifically to bacteria.

2. If the lysing phage and the S^{35}-labeled infecting phage are the same (T2), or if the culture in salt-poor broth is allowed to lyse spontaneously (so that the yield of progeny is large), the S^{35} in the high speed sediment fraction is adsorbed with the specificity of the phage progeny (except for a weak non-specific adsorption). This is illustrated in Table VII by the adsorption to H and B/2*h*.

It should be noted that a phage progeny grown from S^{35}-labeled phage and containing a larger or smaller amount of contaminating radioactivity could not be distinguished by any known method from authentic S^{35}-labeled phage, except that a small amount of the contaminant could be removed by adsorption to bacteria resistant to the phage. In addition to the properties already mentioned, the contaminating S^{35} is completely precipitated with the phage by antiserum, and cannot be appreciably separated from the phage by further fractional sedimentation, at either high or low concentrations of electrolyte. On the other hand, the chemical contamination from this source would be very small in favorable circumstances, because the progeny of a single phage particle are numerous and the contaminant is evidently derived from the parents.

The properties of the S^{35}-labeled contaminant show that it consists of the remains of the coats of the parental phage particles, presumably identical with the material that can be removed from unlysed cells in the Waring blendor. The fact that it undergoes little chemical change is not surprising since it probably never enters the infected cell.

The properties described explain a mistaken preliminary report (Hershey *et al.*, 1951) of the transfer of S^{35} from parental to progeny phage.

It should be added that experiments identical to those shown in Tables VI and VII, but starting from phage labeled with P^{32}, show that phosphorus is transferred from parental to progeny phage to the extent of 30 per cent at yields of about 30 phage per infected bacterium, and that the P^{32} in prematurely lysed cultures is almost entirely non-sedimentable, becoming, in fact, acid-soluble on aging.

Similar measures of the transfer of P^{32} have been published by Putnam and Kozloff (1950) and others. Watson and Maaløe (1952) summarize this work, and report equal transfer (nearly 50 per cent) of phosphorus and adenine.

A Progeny of S^{35}-Labeled Phage Nearly Free from the Parental Label.—The following experiment shows clearly that the obligatory transfer of parental sulfur to offspring phage is less than 1 per cent, and probably considerably less. In this experiment, the phage yield from infected bacteria from which the S^{35}-labeled phage coats had been stripped in the Waring blendor was assayed directly for S^{35}.

Sensitive bacteria grown in broth were infected with five particles of S^{35}-labeled phage per bacterium, the high ratio of infection being necessary for purposes of assay. The infected bacteria were freed from unadsorbed phage and suspended in water containing per liter 1 mM $MgSO_4$, 0.1 mM $CaCl_2$, and 0.1 gm. gelatin. A sample of this suspension was agitated for 2.5 minutes in the Waring blendor, and centrifuged to remove the extracellular S^{35}. A second sample not run in the blendor was centrifuged at the same time. The cells from both samples were resuspended in warm salt-poor broth at a concentration of 10^8 bacteria per ml., and aerated for 80 minutes. The cultures were then lysed by the addition of 0.02 mM HCN, 2×10^{11} UV-killed T2, and 6 mg. NaCl per ml. of culture. The addition of salt at this point causes S^{35} that would otherwise be eluted (Hershey *et al.*, 1951) to remain attached to the bacterial debris. The lysates were fractionated and assayed as described previously, with the results shown in Table VIII.

The data show that stripping reduces more or less proportionately the S^{35}-content of all fractions. In particular, the S^{35}-content of the fraction containing most of the phage progeny is reduced from nearly 10 per cent to less than 1 per cent of the initially adsorbed isotope. This experiment shows that the bulk of the S^{35} appearing in all lysate fractions is derived from the remains of the coats of the parental phage particles.

Properties of Phage Inactivated by Formaldehyde.—Phage T2 warmed for 1 hour at 37°C. in adsorption medium containing 0.1 per cent (v/v) commercial formalin (35 per cent HCHO), and then dialyzed free from formalde-

TABLE VIII

Lysates of Bacteria Infected with S^{35}-Labeled T2 and Stripped in the Waring Blendor

Per cent of adsorbed S^{35} or of phage yield:	Cells stripped		Cells not stripped	
	S^{35}	Phage	S^{35}	Phage
Eluted in blendor fluid...............	86	—	39	—
1st low-speed sediment...............	3.8	9.3	31	13
2nd " " "	(0.2)	11	2.7	11
High-speed "	(0.7)	58	9.4	89
" " supernatant...............	(2.0)	1.1	(1.7)	1.6
Recovery...............	93	79	84	115

All the input bacteria were recovered in assays of infected cells made during the latent period of both cultures. The phage yields were 270 (stripped cells) and 200 per bacterium, assayed before fractionation. Figures in parentheses were obtained from counting rates close to background.

hyde, shows a reduction in plaque titer by a factor 1000 or more. Inactivated phage of this kind possesses the following properties.

1. It is adsorbed to sensitive bacteria (as measured by either S^{35} or P^{32} labels), to the extent of about 70 per cent.

2. The adsorbed phage kills bacteria with an efficiency of about 35 per cent compared with the original phage stock.

3. The DNA of the inactive particles is resistant to DNase, but is made sensitive by osmotic shock.

4. The DNA of the inactive particles is not sensitized to DNase by adsorption to heat-killed bacteria, nor is it released into solution by adsorption to bacterial debris.

5. 70 per cent of the adsorbed phage DNA can be detached from infected cells spun in the Waring blendor. The detached DNA is almost entirely resistant to DNase.

These properties show that T2 inactivated by formaldehyde is largely incapable of injecting its DNA into the cells to which it attaches. Its behavior in the experiments outlined gives strong support to our interpretation of the corresponding experiments with active phage.

DISCUSSION

We have shown that when a particle of bacteriophage T2 attaches to a bacterial cell, most of the phage DNA enters the cell, and a residue containing at least 80 per cent of the sulfur-containing protein of the phage remains at the cell surface. This residue consists of the material forming the protective membrane of the resting phage particle, and it plays no further role in infection after the attachment of phage to bacterium.

These facts leave in question the possible function of the 20 per cent of sulfur-containing protein that may or may not enter the cell. We find that little or none of it is incorporated into the progeny of the infecting particle, and that at least part of it consists of additional material resembling the residue that can be shown to remain extracellular. Phosphorus and adenine (Watson and Maaløe, 1952) derived from the DNA of the infecting particle, on the other hand, are transferred to the phage progeny to a considerable and equal extent. We infer that sulfur-containing protein has no function in phage multiplication, and that DNA has some function.

It must be recalled that the following questions remain unanswered. (1) Does any sulfur-free phage material other than DNA enter the cell? (2) If so, is it transferred to the phage progeny? (3) Is the transfer of phosphorus (or hypothetical other substance) to progeny direct—that is, does it remain at all times in a form specifically identifiable as phage substance—or indirect?

Our experiments show clearly that a physical separation of the phage T2 into genetic and non-genetic parts is possible. A corresponding functional separation is seen in the partial independence of phenotype and genotype in the same phage (Novick and Szilard, 1951; Hershey *et al.*, 1951). The chemical identification of the genetic part must wait, however, until some of the questions asked above have been answered.

Two facts of significance for the immunologic method of attack on problems of viral growth should be emphasized here. First, the principal antigen of the infecting particles of phage T2 persists unchanged in infected cells. Second, it remains attached to the bacterial debris resulting from lysis of the cells. These possibilities seem to have been overlooked in a study by Rountree (1951) of viral antigens during the growth of phage T5.

SUMMARY

1. Osmotic shock disrupts particles of phage T2 into material containing nearly all the phage sulfur in a form precipitable by antiphage serum, and capable of specific adsorption to bacteria. It releases into solution nearly all

A. D. HERSHEY AND MARTHA CHASE   55

the phage DNA in a form not precipitable by antiserum and not adsorbable to bacteria. The sulfur-containing protein of the phage particle evidently makes up a membrane that protects the phage DNA from DNase, comprises the sole or principal antigenic material, and is responsible for attachment of the virus to bacteria.

2. Adsorption of T2 to heat-killed bacteria, and heating or alternate freezing and thawing of infected cells, sensitize the DNA of the adsorbed phage to DNase. These treatments have little or no sensitizing effect on unadsorbed phage. Neither heating nor freezing and thawing releases the phage DNA from infected cells, although other cell constituents can be extracted by these methods. These facts suggest that the phage DNA forms part of an organized intracellular structure throughout the period of phage growth.

3. Adsorption of phage T2 to bacterial debris causes part of the phage DNA to appear in solution, leaving the phage sulfur attached to the debris. Another part of the phage DNA, corresponding roughly to the remaining half of the DNA of the inactivated phage, remains attached to the debris but can be separated from it by DNase. Phage T4 behaves similarly, although the two phages can be shown to attach to different combining sites. The inactivation of phage by bacterial debris is evidently accompanied by the rupture of the viral membrane.

4. Suspensions of infected cells agitated in a Waring blendor release 75 per cent of the phage sulfur and only 15 per cent of the phage phosphorus to the solution as a result of the applied shearing force. The cells remain capable of yielding phage progeny.

5. The facts stated show that most of the phage sulfur remains at the cell surface and most of the phage DNA enters the cell on infection. Whether sulfur-free material other than DNA enters the cell has not been determined. The properties of the sulfur-containing residue identify it as essentially unchanged membranes of the phage particles. All types of evidence show that the passage of phage DNA into the cell occurs in non-nutrient medium under conditions in which other known steps in viral growth do not occur.

6. The phage progeny yielded by bacteria infected with phage labeled with radioactive sulfur contain less than 1 per cent of the parental radioactivity. The progeny of phage particles labeled with radioactive phosphorus contain 30 per cent or more of the parental phosphorus.

7. Phage inactivated by dilute formaldehyde is capable of adsorbing to bacteria, but does not release its DNA to the cell. This shows that the interaction between phage and bacterium resulting in release of the phage DNA from its protective membrane depends on labile components of the phage particle. By contrast, the components of the bacterium essential to this interaction are remarkably stable. The nature of the interaction is otherwise unknown.

8. The sulfur-containing protein of resting phage particles is confined to a

56   VIRAL PROTEIN AND NUCLEIC ACID IN BACTERIOPHAGE GROWTH

protective coat that is responsible for the adsorption to bacteria, and functions as an instrument for the injection of the phage DNA into the cell. This protein probably has no function in the growth of intracellular phage. The DNA has some function. Further chemical inferences should not be drawn from the experiments presented.

REFERENCES

Anderson, T. F., 1949, The reactions of bacterial viruses with their host cells, *Bot. Rev.*, **15**, 464.

Anderson, T. F., 1951, *Tr. New York Acad. Sc.*, **13**, 130.

Anderson, T. F., and Doermann, A. H., 1952, *J. Gen. Physiol.*, **35**, 657.

Benzer, S., 1952, *J. Bact.*, **63**, 59.

Doermann, A. H., 1948, *Carnegie Institution of Washington Yearbook, No. 47*, 176.

Doermann, A. H., and Dissosway, C., 1949, *Carnegie Institution of Washington Yearbook, No. 48*, 170.

Dulbecco, R., 1952, *J. Bact.*, **63**, 209.

Herriott, R. M., 1951, *J. Bact.*, **61**, 752.

Hershey, A. D., 1946, *Genetics*, **31**, 620.

Hershey, A. D., Roesel, C., Chase, M., and Forman, S., 1951, *Carnegie Institution of Washington Yearbook, No. 50*, 195.

Lwoff, A., and Gutmann, A., 1950, *Ann. Inst. Pasteur*, **78**, 711.

Maaløe, O., and Watson, J. D., 1951, *Proc. Nat. Acad. Sc.*, **37**, 507.

Novick, A., and Szilard, L., 1951, *Science*, **113**, 34.

Prater, C. D., 1951, Thesis, University of Pennsylvania.

Price, W. H., 1952, *J. Gen. Physiol.*, **35**, 409.

Putnam, F. W., and Kozloff, L., 1950, *J. Biol. Chem.*, **182**, 243.

Rountree, P. M., 1951, *Brit. J. Exp. Path.*, **32**, 341.

Watson, J. D., and Maaløe, O., 1952, *Acta path. et microbiol. scand.*, in press.

When the Cold Spring Harbor laboratories were reorganized in the early 1960s, Hershey accepted the Directorship of the newly created Genetics Research Unit, made up of Barbara McClintock and himself and their associates. As Director, Al contributed Introductions to the annual reports of the Unit. These reports, with some editorial omissions, are reprinted below. The Introduction for the year 1968-69 was reprinted with minor changes by *Nature*. Since the changes presumably reflect Al's more considered statement, the Introduction for that year is here omitted, and the reprint is to be found with Al's published essays.

Reprinted, with permission, from the Carnegie Institution of Washington, Washington, D.C.

In spite of numerous distractions connected with the reorganization of our own group and the formation of the new Laboratory of Quantitative Biology, we in the Genetics Research Unit have enjoyed a lively year of research. Individual reports will be found in the following pages.

Barbara McClintock, serving as a Trustee of the Laboratory of Quantitative Biology, played an important role in its organization. Through that organization the Genetics Research Unit has gained tangible assets, notably a welcome new colleague in John Cairns, Director of the new laboratory.

The Unit has also suffered a loss. Margaret McDonald, a member of the former Department of Genetics since 1943, is leaving to continue her research at the Waldemar Medical Research Foundation. We shall miss her and wish her well.

As a sequel to her past work in South America, sponsored by the Rockefeller Foundation, Dr. McClintock and Dr. William L. Brown are supervising the research of four Latin American Fellows now studying at North Carolina State College. The general purpose is to encourage research in maize genetics and cytology, research that is equally important in the Americas for botanical, historical, and economic reasons.

The Drosophila Educational Project, under Jennie S. Buchanan, Curator, continues to thrive. During the current year about 3500 requests for cultures, and 950 for copies of the *Drosophila Guide*, have been filled. The cultures go mainly to college and high school teachers, who use them in classroom work.

The Library continued under the capable management of Mrs. G. C. Smith, who makes it possible for our small group to enjoy the near-luxury of a reasonably complete reference source in biology and biochemistry.

All members of the staff benefited through visits to and visitors from other laboratories within and without the Institution. Notable quantitatively was the annual Phage Meeting, which brought to Cold Spring Harbor some dozens of scientists interested in a field of research still neither too wide for intimacy nor too narrow for mutual stimulus among its practitioners. The meeting was organized by Elizabeth Burgi, with advice from George Streisinger and Frank Stahl of the University of Oregon.

Our justification for existence as a Unit, however, resides in the value of our research. We like to think that much of that value is as unstatable and as durable as other human produce that cannot be sold. Some can be put on paper, however. That we offer with the usual human mixture of pride and diffidence.

Current activities at the Genetics Research Unit are related to two areas of research that are being explored intensively in a number of laboratories at the present time. Our situation in this respect is somewhat novel, partly because the laboratory at Cold Spring Harbor has traditionally fostered unpopular research, partly because of the increased number of laboratories now devoted to biological research of all kinds. The change must be counted a welcome one, none the less because it calls for some soul-searching on the part of all concerned.

McClintock's present interests, in particular, were generated almost exclusively by her own work. In 1945 she began studying a number of unstable genes in maize. By 1955 she had discovered and analyzed several control systems responsible for the behavior of such genes. Each system consists of two chromosomal elements: one that determines by propinquity which gene is to be affected and the nature of the effect, and a second that transmits from a distance the controlling signals. The particular systems amenable to study in maize play no obvious role in the economy of the plant. McClintock interpreted them as models applicable to situations where control at the genetic level is essential, as in tissue differentiation.

In microbes, certain metabolic control systems are accessible to analysis owing to the rather direct relation between the nutrients supplied in the culture medium and the metabolic tasks the culture has to perform. Metabolic control in microorganisms has been studied for many years, and particularly since it was found that not only the activity of enzymes but also their biosynthesis is subject to environmental influences. Several microbial systems in which synthesis of enzymes can be switched on or off according to the prevailing metabolic demand proved to resemble the maize systems described by McClintock. They also consist of a chromosomal element (operator or receptor) that lies in or adjacent to the gene or genes subject to control, and a second element (regulator or emittor gene) that causes the production of a substance (repressor), probably a cytoplasmic protein, which seems to react directly with the operator to determine the functional status of the operator-linked genes. Chemical signals from the environment intervene in this system by interacting specifically with the repressor, which exhibits therefore two separate but interdependent specificities: affinity for a given metabolite, and affinity for a given operator, the second affinity depending on whether or not the first is satisfied. The duality of the system takes advantage of the fact that only proteins have the required specificity, variety, and flexibility of structure (as seen in the enzymes) to function as repressors.

One of the most unexpected features of the system described is that it operates apparently at the genetic level to control superficial metabolic reactions: for example, hydrolysis of lactose to monosaccharides. This encourages the belief, as McClintock pointed out long ago, that similar systems direct fundamental biological processes in which control at the gene level seems necessary.

The general scheme discussed above, first recognized by Jacob and Monod, represents one outcome of many years of intensive study of inducible and repressible enzymes, on the one hand, and of microbial genetics on the other, both in their laboratories and elsewhere. The scheme is far from complete and probably represents but one of many types of control systems yet to be discovered, but it does provide a powerful tool for further exploration. The exploration is, in fact, under way in numerous laboratories throughout the world, including those of Umbarger and Margolin at Cold Spring Harbor. McClintock herself is pursuing one of the important unanswered questions, namely, how metabolic timing is achieved, as she relates in the factual part of this report.

The second area under investigation in the Genetics Research Unit, by the writer and his colleagues, might be described as the relation between DNA structure and chromosomal function in the phages. That direction of inquiry, too, was in part an outgrowth of earlier work at Cold Spring Harbor and was evidently a logical choice, since the same thought occurred to others. Our own progress since the inception of the program in 1958 is summarized in this and preceding Year Books. Here I shall describe briefly some important contributions from other laboratories.

One of the most remarkable physical clues to the structure and function of DNA, evolving from some years of study of thermal denaturation in Doty's laboratory at Harvard, came from the demonstration by Marmur and Doty and their colleagues that denaturation at sufficiently high temperatures is accompanied by the literal separation of the complementary strands of the helix and yet is largely reversible. Re-formation of helical structures from the separated strands is observed during a second period of heating at a suitably lowered temperature. It depends strongly on the concentration of complementary strands and the nature of the solvent. What happens under optimum conditions is an exchange of partners according to the base-pairing rules of Watson and Crick, which means that the rejoining is specific for long base sequences. Thus renaturation is nearly as complicated mechanically as DNA replication itself, and the work of Marmur and Doty did much to make that process credible. The same work also formed the basis for several tests of genetic homology developed since, by Marmur, Spiegelman, Bautz, Hall, Bolton, and their colleagues.

A second remarkable physical technique, that of equilibrium density-gradient centrifugation in cesium chloride and other salt solutions, was invented by Meselson, Stahl, and Vinograd at the California Institute of Technology and used by them to confirm the principle of DNA replication proposed by Watson and Crick. Another application of the method, introduced by Weigle and others, is described in connection with Mosig's work in this report. It served, in the experiments of Meselson and Weigle, and of Kellenberger, Zichichi, and Weigle, to demonstrate for the first time that genetic recombination can be brought about by the literal fracture and rejoining of DNA molecules.

The most definitive answer to questions about chromosome structure in phage comes from the work of Kaiser and Hogness at Stanford University. They began, like some other people, by trying to develop a biological assay for phage DNA. Unlike many other people, including us until recently, they succeeded. In their system, employing lambda phage, it was clear that all the markers of the phage chromosome are carried by a single piece of native DNA, a conclusion nicely complementing ours that a phage particle contains only one piece of DNA. Moreover, when the DNA solution was subjected to a critical rate of shear, which according to our results breaks DNA molecules into halves, the genetic markers separated into two groups, corresponding, within the limitations of their analysis, to markers situated in the right and left halves of the genetic map, respectively. In further work along the same lines, Hogness and Simmons succeeded in separating the two molecular halves with which the proper biological activities were associated. Their result, though partly anticipated by some experiments of Jacob and Wollman on bacterial mating and by much earlier correlations of genetic and cytological analyses of chromosomes in *Drosophila* by Sturtevant and Bridges, is the first clear demonstration that genetic crosses do in fact yield information about molecular structure. This principle Niels Bohr chose to question a few years ago on philosophical grounds. It is still not established with great precision.

Important gains have been made by comparative study of nucleic acids from different viral species. The phage SP8 was found, in Marmur's laboratory at Brandeis University, to contain DNA with a sufficient bias in composition of the two strands of the helix to permit their separation by chromatographic methods. With these materials, and making use of the homology tests of Spiegelman and of Bolton and McCarthy, Marmur and his colleagues showed that in the infected cell only the denser (pyrimidine-rich) strand of the DNA is transcribed into the complementary RNA that presumably directs synthesis of phage proteins. The possibility of isolating the two individual strands of SP8 and a few other selected phage DNA's may be compared with the opportunity afforded by lambda DNA to isolate several of the molecular segments generated by transverse cuts, as described in the experimental part of this report.

Benefits to be gained from a happy choice of experimental material are also illustrated by the study in Sinsheimer's laboratory at the California Institute of Technology of a phage, called ø · X · 174, that was chosen because of its small size. This phage proved to contain a unique DNA consisting of a single-stranded ring that represents, moreover, only one of the strands of the more usual base-paired structure. The double-stranded equivalent, also ring shaped, was isolated from the infected bacterium, where it appeared to be the immediate precursor and product of replication. Sinsheimer's results, together with Spiegelman's work with the same phage at the University of Illinois, suggest further that the eventual regeneration of single-stranded rings required to make the offspring phage particles takes place in a manner analogous to the transcription from double-stranded DNA to single-stranded RNA, a process by which all DNA's presumably function. Work with ø · X · 174 provided the first definite indication that DNA molecules could be circular. It demonstrated, too, the potential biological competence of one strand of a base-paired helix, an important feature of the Watson-Crick structural model for DNA.

Another small-particle phage, discovered and studied chiefly in Zinder's laboratory at the Rockefeller Institute, proved to contain ribonucleic acid, like several plant and animal viruses. This phage apparently dispenses altogether with DNA throughout its life cycle.

We may speculate about the significance of some of the specializations seen among the phages. DNA has obvious advantages as a genetic material in its remarkable stability, maintained by bonds that are nevertheless easily opened. The duplicated information carried in its two strands, now a well authenticated biological fact, protects further against accidental damage and, as suggested by recent work of Setlow and Carrier and of Boyce and Howard-Flanders, perhaps serves also in a cross-checking and error-correcting mechanism that operates during replication. Packing a relatively stiff double-stranded DNA molecule into a phage particle, however, presents a formidable mechanical problem. That problem is sometimes solved by the introduction of periodic cuts in one of the two strands of the molecule, according to work by Davison and Freifelder.

In choosing to gain other advantages conferred by small size, both the DNA- and the RNA-containing phages could dispense with the double structure entirely, probably because, as is frequently seen in the laboratory, the hazards to which nucleic acid molecules are exposed must be reckoned per unit length, so that a short molecule is intrinsically more stable than a long one.

If RNA can also assume a double-stranded structure, as apparently happens during replication of RNA viruses, why is DNA needed at all? Perhaps because the bonding between RNA strands, for chemical reasons not understood, is too strong, tending to diminish the specificity of complementary pairing on the one hand, and to interfere with the separation of complementary strands during replication on the other. The second difficulty could be surmounted, as Sinsheimer's work shows, by a replication scheme (double-stranded template to single-stranded product) not calling for frank separation of complementary strands at all. This scheme cannot apply, though, to some viruses recently discovered to contain double-stranded RNA. As usual in such discussions, it must be concluded that different organisms exploit different ways of doing things for a reason best described as historical accident.

How pertinent are these remarks to the report of our small research unit? For one thing, we can point out that, if important scientific gains are often made in laboratories other than our own, we have nevertheless often contributed to them in small ways. More important, it is in the light of such accomplishments that we can ourselves best understand, and best explain to others, the guiding spirit of our work, which is scarcely evident in a day-to-day record of it. More important still, the account of any successful phase of research in biology always contains a lesson about the communal nature of scientific effort, which has its moral and esthetic qualities, to be sure, but is first of all a practical necessity arising out of the constitution of living things.

For biological laws (unlike physical ones, perhaps) do not emerge from the elucidation of single mechanisms, however intellectually satisfying that may be. The single mechanism, and especially that part of it decipherable in a given species, is usually a truncated, even in a sense a pathological, variation on a theme. As such it is at best a clue to that theme, at worst a senseless diversion. To discover the theme many clues, including the false, must be run to ground. Just so anatomy became a meaningful science in terms of comparative anatomy, and biochemistry in terms of comparative biochemistry. Indeed, if the position of the pessimists who used to deny the possibility of mechanistic explanations of important biological phenomena is no longer tenable, they can fall back to a second stand: that it is not yet quite clear that the common denominators of ascertainable mechanisms will ever be found.

Hence the dicta that seem to emerge from these paragraphs: that duplication of effort in biological research is a virtually nonexistent bugbear; that recent progress in certain areas of biology, while sometimes happily jading the senses, by no means diminishes the number of valid questions; and that laboratories, as diverse in size and character as possible, should be encouraged to multiply as fast as the multiplication of dedicated researchers permits.

Work in the laboratory at Cold Spring Harbor can be described at various levels. First, it demonstrates a philosophy: not, to be sure, a moral principle consciously chosen, but a pattern that an observer might discern in our activities. We have the freedom and the obligation to look at natural things. We look at them with vision that is both sharpened and restricted by the tools at our disposal and the direction our curiosity has taken. What we first see is an element of novelty, without which nature and art cannot charm. Perceiving that element, we have already in a measure succeeded. But we should like to prolong our surprise, to savor it fully, to communicate it. So we formulate a question, and find ourselves trying to solve a problem. It is always the same sort of question: Why is this object constructed as it is? And always the same kind of problem: How can our question be framed and answered in conveyable terms? In its particular example, the problem may be solved or not, with varying degrees of success or failure. In any event, excepting the worst failures, we have prolonged and illuminated our surprise and can hope to share our experience with others.

Research, including ours, exemplifies other philosophies as well. For instance, it can start with a problem that appeals to us mainly because we have an idea how to solve it, or mainly because a solution is urgently needed. Our research, if different from anybody's, differs in exhibiting as a major ingredient the philosophy first stated.

To place the element of novelty at the center of endeavor may strike some readers as frivolous. Though at a tangent to the purpose of these paragraphs, I would go further. I would argue that the sophistication and costliness of contemporary research are necessary because the threshold of human surprise is rising. We don't exhaust our natural surroundings; we exhaust our ability to respond to them. This may be the secret and real price of the scientific vitality of our time. Except for the element of novelty, ritual could take the place of research.

At a second level of description, work in our laboratory can be identified by subject matter. The writer and his colleagues are studying the structure and function of the nucleic acids of phage particles. Barbara McClintock is exploring mechanisms of control of gene action in maize. These two areas of research have some aims in common, notably to understand cooperation among biological elements, but they differ greatly in materials and methods.

At a third level of description, we present in the following pages edited versions of nine laboratory notebooks.

1965

The enduring goal of scientific endeavor, as of all human enterprise, I imagine, is to achieve an intelligible view of the universe. One of the great discoveries of modern science is that its goal cannot be achieved piecemeal, certainly not by the accumulation of facts. To understand a phenomenon is to understand a category of phenomena or it is nothing. Understanding is reached through creative acts.

The universe presents an infinite number of phenomena. The faith of the scientist, if he has faith, is that these can be reduced to a finite number of categories. Even so he tends to consider the path toward his goal as endless. Not too discontentedly, either, because human history is replete with glorious paths, not goals achieved.

To speak of goals at all is to speak in unscientific terms. One cannot measure progress toward the goal of understanding. Various peoples at various times have thought they had an intelligible view of the universe and, so thinking, had in fact. Most of us today, in spite of much talk about contemporary spiritual malaise, are complacent enough intellectually. If understanding is reached through creative acts, they are partly acts of faith.

These are large questions. They are pondered by professional thinkers, who evidently believe in the power of abstract thought. If that power is efficacious, it behooves the scientist to exercise it now and then when his experiments flag. Otherwise he risks failing a personal goal: to see his work in selfless perspective.

Because nearly all living things exhibit sexuality, one is led to believe that sex plays nearly indispensable roles in biological economy. Among diploid organisms, whose physiology depends in part on the continuous interaction of two unlike genomes, the roles of sexual processes are multiple and diverse. Among organisms that reproduce in the haploid phase, the single intent of sex may be to ensure opportunities for genetic recombination. Having reached that conclusion, a perceptive biologist might have anticipated what has been found, namely, that among haploid organisms, opportunities for genetic recombination are often provided by asexual means. The diversity of means itself proves to be a remarkable discovery, which I shall recount here in an appreciative, as opposed to authoritative, manner.

Until 1946, bacteria were defined as asexual creatures. In that year Lederberg and Tatum described for *Escherichia coli* what is still one of the few known examples of sexual recombination in bacteria. Not the least remarkable feature of their experiment was that it turned up, against considerable odds, just what they were looking for, in violation of the rule of serendipity that prevailed for some years afterwards in microbial genetics. Even so, more than a decade of intensive work was necessary before the general features of sexual recombination in *E. coli* could be appreciated.

Sexual differentiation in *E. coli* depends on the status of the cells with respect to a fertility factor called F, first found in lines of *E. coli* called F⁺ and absent in other lines called F⁻. The crosses of Lederberg and Tatum happened to be F⁺ x F⁻ and yielded few recombinants because, as it turned out, F⁺ cultures owe their fertility to a third category of cell, called Hfr, present in small numbers. The name Hfr, which stands for high-frequency recombination, is of historical significance only. Hfr cells are males (more strictly, donors), and F⁻ cells are females (recipients), in a process of genetic transfer.

F⁺ cultures may be called vector lines in that they carry F as an extrachromosomal factor, demonstrable in extracts as a small piece of DNA, and transfer it with high frequency at cell contact to F⁻ cells, converting them to F⁺. The F factor can be lost, whereupon F⁺ cells revert to F⁻.

F⁺ cells also give rise to Hfr (male) lines with low frequency. Male lines no longer carry F in infectious form, the F factor having been inserted into the bacterial chromosome. Male cells can mate with female (F⁻) cells by forming cell pairs connected by a cytoplasmic bridge through which DNA passes from male to female, not, apparently, by passive transfer but by a process associated with DNA replication in the male. The female cell thus becomes a zygote, and shortly reverts to the haploid condition, preserving the female genotype except as modified by genetic recombination that may have occurred in the zygote.

Clues to male function came rapidly with detailed genetic study of Hfr x F⁻ crosses. The transfer of genes proved to be a linear process characterized by a temporal order that could be determined experimentally by interrupting mating at various times. Several methods were used for this, the most effective being to separate conjugal pairs by stirring the cell suspension. Distance between genes came to be measured as difference between times required for transfer of those genes. Complete chromosome transfer proved to take about 100 minutes after formation of conjugal pairs.

The next step called for isolation and comparison of several male lines. Then it was found that different males transfer chromosomal markers in different orders that are cyclic permutations of one another and, moreover, can be either forward (ABC) or reverse (CBA) in direction. The male determinant itself is exceptional, and is always the last to be transferred. Therefore males differ with respect to the point of insertion of F into the chromosome, the genetic map is circular, and the chromosome appears to be a ring opened by F.

It follows that F functions in two ways: first, to modify the cell exterior so as to facilitate conjugation; second, to direct the transfer of chromosomal material. These two functions are clearly distinct, since either F^+ or Hfr cells can conjugate with F^-, but only Hfr cells transfer chromosomal genes during conjugation.

Why such a complicated process? Given the decision of *E. coli* to mate by DNA transfer rather than by cell fusion, one answer is clear. The mating pair is a precarious structure, which seldom persists until chromosomal transfer is complete. Thus, if all genes are to be transferred with equal frequency, a necessary condition if purposes of genetic recombination are to be achieved, a regiment of specialized males is needed.

The process described is sexual in two senses: it involves two mating types and, potentially, it allows interaction between two intact chromosomes. In fact, though, the transfer of genetic material is usually fragmentary and not very different in effect from other processes of genetic transfer that are clearly asexual.

Asexual recombination was named "transduction" by Zinder and Lederberg, who discovered one of the well-known examples and quite properly wished to emphasize the differences between sexual and asexual processes. Somewhat ironically, it turns out that the fertility factor itself engages in transduction in *E. coli*. Hfr lines can revert to F^+, and the revertants usually carry wild-type F factor. Occasionally, however, an F factor recovered from an Hfr chromosome proves to contain one or another small chromosomal fragment of bacterial origin. Elements of this sort, called F′ factors, function like F in most respects but in addition carry their extra load of genetic determinants with them, determinants that accordingly behave as contagious elements or as authentic chromosomal duplications, depending on their status, extrachromosomal or chromosomal, in the carrier culture.

The conjugation system in *E. coli* is not strictly species specific. F′ factors can be transferred from *E. coli* to *Proteus* and *Serratia* species, for instance. In the foreign species they are carried as extrachromosomal elements whose persistence is recognizable by the functioning of the particular genes they contain. Turning this process around, one can imagine that certain "transfer factors" found in *E. coli* are F′ factors that originated in another species. At any rate, some colicine determinants and drug-resistance factors behave in *E. coli* as do F′ factors from *E. coli* in *Proteus*. Their relation to F is indicated by their capacity for transfer by cell contact, and by their maintenance as extrachromosomal elements in carrier cell lines. Their failure to interact directly with the *E. coli* chromosome suggests a foreign origin. Whatever their origin, they function in *E. coli* rather like viruses, carrying genetic determinants not recognizable as chromosomal genes, and pursuing evolutionary ends of their own. They can be called neither sex factors nor transducing elements.

Transduction, as a plausible natural means of genetic exchange among bacteria, was first discovered in *Salmonella*, where it is brought about by bacteriophage P22. Only a few transducing phages have been studied, but those few clearly belong to two quite dissimilar classes.

One class of transducing phage, exemplified by the coliphage P1, works by picking up more or less at random a phage-sized piece of bacterial DNA and incorporating it, in place of a P1 chromosome, into what is otherwise a phage particle. As a result, that particle can transduce its content of bacterial genes by virtue of phage-specific mechanisms of attachment to bacteria. This process, called generalized transduction by phage, mimics sex at the population level, since all bacterial genes are transduced with similar frequencies. Here no mechanism is provided for maintenance of the transduced genes in the recipient cell. That requires genetic exchange between the transduced fragment and the bacterial chromosome. Phage P22 also is a generalized transducing phage, but hasn't been studied as thoroughly as P1.

A very different kind of transduction is carried out by phage λ, as described in detail in later pages of this report. Phage λ specifically transduces genes responsible for galactose and biotin metabolism in *E. coli,* and provides both a mechanism for cell-to-cell transfer and a mechanism for insertion of the transduced genes into the chromosome of the recipient cell. The related phage ø80 transduces in a similar way genes concerned with tryptophan synthesis. By extension, this highly specialized transduction mechanism may also serve all the purposes by which sex can be imagined to benefit *E. coli.* Note that the specialized transducing phages combine features of F′ and P1 transduction.

The classical example of pure transducing agent is of course DNA itself, though DNA transduction doesn't seem to work for *E. coli* and, to be sure, hardly seems necessary to that species. In the pneumococcus and related bacteria, according to work of Hotchkiss and others, DNA transduction plays an authentic role in nature and promises to display its own adaptive specialties.

In short, at least seven mechanisms provide opportunities for exchange of hereditary determinants in bacteria. None is typically sexual, though one closely resembles sexual processes and several involve mating between differentiated cell types. Genes passed from cell to cell by the various mechanisms may differ with respect to origin, manner of transfer, and mode of replication in the carrier-cell line. In the face of such modalities, the meaning of many words useful in other contexts dissolves—sex, fertility, conjugation, transduction, meromixis, episome, virus—a forceful reminder that living things embrace opportunities without much regard for ideological categories.

Bacteria and phages, more literally but not more truly than other organisms, are not seen through the eye of the beholder, to which only the raw materials of creation are directly visible. *Escherichia coli,* for instance, beset by the dual transience of mortality and shifting opportunities, exists as we know it against overwhelming odds. Its story is therefore trivial, unless it invites reflection. What we see on reflection is an intimation of history, another witness interrogated as to what living things are about. What we see is also a human construction, given us by the dozen or so people who looked with more than their eyes.

Elucidation of the structure of DNA has had consequences of two sorts. As all interested people know, it set off a numerous and continuing train of successes in the analysis of the living machine. It also led to conclusions of more general significance: notably, that natural phenomena do not present immovable obstacles to human comprehension.

One of our less machinelike attributes, as humans, is our tendency to ask whether problems are soluble in principle. The answer has never been clear. In fact, human history can be epitomized in terms of modes of seeking the answer. The scientific mode, impressive as its record is in many ways, engenders its own grounds for reasoned doubt.

J. D. Watson, in his Nobel Lecture, stated part of the difficulty very clearly. "In pessimistic moods," he wrote, "we often worried that the correct structure [of DNA] might be dull-that is, that it would suggest absolutely nothing. . . ." Indeed, both the history of speculation about DNA structure, and the record of chemical pharmacodynamics as a whole, encouraged such pessimism. As it turned out, the correct structure suggested a great deal, and it is this result, not the structure itself, that illuminates.

I once publicly expressed a similar pessimism in connection with my own work on genetic recombination. Phage research, I complained, is beset with the following difficulty. One notices a phenomenon that looks interesting. After suitable thought and labor, one performs an experiment that ought to be instructive. What it actually does is to turn up a second phenomenon not related in a simple way to the first. This sequence of events makes for lively research. It does not provide explanations.

As practicing scientists we don't have to be much concerned about the search for explanations, because the discovery and description of natural phenomena are ends in themselves. As philosophers, though, we have to consider the possibility that the sequence I have described may prove endless, giving an indefinitely complicated description of nature without making us, in human terms, appreciably wiser. To an outsider, this seems to be the state of affairs in research on the structure of matter at both ends of the scale of magnitude. For somewhat similar reasons, what goes on in atomic nuclei, outer space, and living cells might prove equally refractory to explanation. This must be roughly what Watson and Crick dreaded in their pessimistic moments.

What do we seek in a scientific explanation? Our universe is necessarily described in terms of conceptual fictions: mass, field, electromagnetic wave, energy quantum, genome, or, for that matter, gross national product. These we combine into relations that have predictive value. At this stage we have a successful theory but hardly an explanation, for Apollo's chariot serves as well as Newtonian principles in appropriate areas of prediction. According to P. W. Bridgman, we arrive at explanations by establishing, among items of experience, connections that are independent of the operations by which our fictions are defined. Apollo is manifest only in the sun, which places him in the unscientific category of hypotheses beyond dispute. By contrast, Bouguer found gravity in a Peruvian mountain, and rotational inertia must have been noticed by the inventor of the wheel.

There are not many explanations of the sort exemplified by classical mechanics, certainly not many in biology. Most biological explanations necessarily borrow their elements from mechanics and chemistry. The exceptions, such as the metabolic role of enzymes and the evolutionary origin of species, are interesting in the present connection and share common features. Darwin's hypothesis is perfectly satisfactory as an explanation, since its elements, hereditary variation and selection, are demonstrable independently of the process they explain. It fails as a theory because, like all historical schemes, it deals with unique events, says nothing about direction, and lacks predictive value. Similar qualifications apply to metabolic principles.

In short, the DNA revolution is one of a small class of successful scientific revolutions, and should renew significantly our confidence that the structure of the universe is eventually decipherable. As Albert Einstein wrote in an earlier context, "... [scientific] endeavors are based on the belief that existence should have a completely harmonious structure. Today we have less ground than ever before for allowing ourselves to be forced away from this wonderful belief."

Note that a scientific explanation has a negative aspect: it must not introduce any new conceptual fictions. (If it does, the explanation rests in abeyance until the new fictions can be independently authorized. Hence Bouguer's search for Newton's action at a distance.) Gunther Stent, in a charming and luminous essay recently published in *Science*, has stressed this negative aspect of the DNA revolution, which uncovered no paradox unique to living things.* According to the point of view I have presented, his essay might be described as a lament for success. Of course, one may argue that discovery of primary concepts is more important than explanation, or even that explanations stale the freshness of the human outlook. But these are untestable hypotheses.

Barbara McClintock, now Distinguished Service Member of our group, continues her work on controlling mechanisms in maize. Her new status carries the privilege of aperiodic publication. Readers of this report will miss her written contribution, but may know that her influence is there just the same.

Rudolf Werner's work on replication of T4 DNA, referred to in previous *Year Books*, has been published in the *Journal of Molecular Biology*. In March he moved next door, to the Laboratory of Quantitative Biology. He no longer reports in these pages, but he too remains one of us.

Phyllis Bear left our group in December 1967 to join the faculty of the University of Wyoming.

The work described in the following pages was partly supported by a research grant, GM15876, from the National Institute of General Medical Sciences, U.S. Public Health Service.

Hershey *the scientist*

Al communicated with members of the Phage Group by sending them reprints of his annual research reports. These reports were eagerly read and discussed. Their meaning was debated everywhere, but, although they were not peer-reviewed, their reliability was never doubted. In these reports, Al was free to omit details that might interrupt the flow of the narrative, revealing Hershey the Scientist more clearly than do many of his published writings.

This, Al's first research report from his new position at Cold Spring Harbor, announced the existence of individual T2 phage particles that carry two different alleles. That was surprising for a creature thought to be haploid, i.e., to have but one set of chromosomes. No explanation was possible until the structure of phage chromosomes was worked out. In 1953, the duplex nature of DNA was proposed by Watson and Crick, providing an obvious basis for heterozygosity. Later, another structural basis for heterozygosity proved to be the unimaginable terminal redundancy of a circularly permuted chromosome.

The insight into the nature of viral growth provided by the observation of phenotypic mixing was historic.

The section on transfer of phosphorus and sulfur isotopes from parent to progeny phage preceded, but did not presage, the analyses that follow in the next two reports.

The section on bispecific antibodies probably summarized Al's last work in immunology.

GROWTH AND INHERITANCE IN BACTERIOPHAGE

A. D. Hershey, Catherine Roesel, Martha Chase, and Stanley Forman

During the past year we have studied, in the coliphage T2H, the transfer of phosphorus and sulfur from viral parent to progeny, and have begun work on two new types of intracellular interaction between particles. These experiments have called for both isotopic and genetic labeling of bacteriophage particles. P32 and S35 have proved to be suitable radioactive tracers for our purposes. The genetic markers we have used are previously studied mutant genes, a few of which are described below.

Dr. Roesel has made use of the phages T2H and T5, which may be thought of as individually labeled antigenic particles, to investigate the structure of antibody molecules.

Among the genetically marked stocks of T2H, the r, or rapidly lysing, mutants are particularly useful. They form large, clear plaques on agar plates seeded with sensitive bacteria, as contrasted with the smaller, fuzzy plaques of the wild-type. A number of mapped r genes are available. Phage stocks lacking an r gene are called r^+, irrespective of other details of genetic constitution.

A particular mutation in another gene confers the ability to lyse certain bacteria that are resistant to the wild-type phage. Lines carrying the mutant gene are called h, meaning altered host-specificity. Lines carrying the normal allele of this gene are called h^+.

The h and r genes recombine, in genetic crosses, to yield four kinds of phage: wild-type (meaning here h^+r^+), h (meaning here hr^+), r (meaning here h^+r), and hr. This same pattern of recombination is observed between h and any of the r genes. We speak of closely linked, distantly linked, or unlinked pairs, depending on the frequency of recombination in crosses involving h and the several r genes.

Heterozygous Phage Particles

A full report of the work summarized in the following paragraphs will appear in a paper by Hershey and Chase in the *Cold Spring Harbor Symposia on Quantitative Biology,* volume 16 (1951).

When bacteria are infected with both *r* and *r*[+] phage, and the progeny liberated at lysis are plated out on sensitive bacteria, the majority of the plaques are pure clones of one parental type or the other. About 2 per cent of the plaques unexpectedly contain mixtures of the two types, which can be recognized because of the visible mottling of the plaques. These mottled plaques might originate (1) from clumps containing one or more particles of each kind, (2) from rapidly mutating variants, or (3) from rapidly segregating heterozygous particles, that is, particles containing two alleles of the same gene that separate from each other when the particles multiply.

The explanation (1) in terms of clumping was excluded by inactivation tests with antiserum, heat, ultraviolet light, and beta rays. Mottling clumps should be inactivated more rapidly than single particles because the inactivation of either one of the two components of the clumps would destroy the mottling property. This expectation was verified with artificial clumps prepared by agglutinating phage mixtures with antibody. The mottling phage particles, on the contrary, were inactivated like single particles by the agents mentioned.

Explanation (2) could also be excluded, chiefly for the following reason. Phage stocks containing a factor for instability of the *r* character are known. They give rise to mixtures of stable *r* and unstable *r*[+] progeny, never to stable *r*[+]. The mottling particles, on the other hand, yield approximately equal numbers of stable *r* and stable *r*[+] phage, with only traces of mottling phage. The latter can best be explained as newly formed mottling particles, rather than direct descendants of the original particles.

For these reasons we conclude that the mottling particles are heterozygotes.

The frequency of heterozygosis was found to be 2 per cent for five different *r* genes tested. By a special method the frequency of heterozygosis with respect to the *h* gene could be roughly measured. It proved likewise to be about 2 per cent. This shows that the doubling of individual genes does not depend on structural peculiarities resulting from the mutations.

Further experiments were designed to answer the following question: If bacteria are infected with pairs of phage particles carrying both *h* and *r* markers, will the heterozygotes with respect to *h* and the heterozygotes with respect to *r* be the same particles, or different ones? The facts bearing on this and related questions can be summarized as follows.

(1) Crosses involving both *h* and *r* markers yield some particles heterozygous for *h* only, some heterozygous for *r* only, and some heterozygous for both markers. We call the first two classes single heterozygotes, and the third class double heterozygotes.

(2) The single heterozygotes form four classes of approximately equal size, segregating into the pairs *h, hr; h,* wild; *r, hr;* and *r,* wild. Thus single heterozygotes yield one parental type and one recombinant type with respect to the original cross.

(3) When the two markers are linked, the double heterozygotes segregate to yield the two parental types of the original cross, never the two recombinants, and never three kinds of phage. When the two markers are unlinked, these alternatives cannot be distinguished.

DEPARTMENT OF GENETICS

(4) When the two markers are unlinked or distant, heterozygosis for one marker is almost independent of heterozygosis for the second, and the doubly heterozygous class amounts to only 3 per cent of the total number of heterozygotes. Thus in the crosses $hr7 \times$ wild-type and $hr1 \times$ wild-type, which yield respectively 20 and 40 per cent of recombinants, the pooled heterozygotes segregate to yield about 48 per cent of recombinants.

(5) In the cross $hr13 \times$ wild-type (closely linked markers) the doubly heterozygous class makes up about 59 per cent of the total number of heterozygotes. This cross yields about 2 per cent recombinants, but the pooled heterozygotes segregate to yield about 20 per cent recombinants.

Our results indicate that if bacteria could be infected with two phage particles carrying many markers, every progeny particle would be heterozygous for one or more of the markers, but in general particles heterozygous for one marker would not be heterozygous for the others. This situation can be pictured in either of two ways. Phage particles may regularly carry small extra segments of genetic material, which are substituted for the homologous segments in some of the very early progeny of the particles. Or phage particles may be diploid, in which case some special mechanism is needed to explain the fact that a particle tends to receive most of its gene complement from one parent, and very little of it from the other. No experimental choice between these alternatives can be made at present.

PHENOTYPIC MIXING

Hershey and Chase have made a preliminary study of what appears to be a physiological, as opposed to genetic, interaction between phage particles. When a bacterium is infected with both h and h^+ phage particles, the progeny particles exhibit characteristics different from those of either parent. We speak of these abnormal progeny as confused particles. Genetically, the confused particles are normal; that is, they multiply in pure clones to form h or h^+ particles like one of the parents. Phenotypically, they are intermediate between the two parents with respect to heat sensitivity, and with respect to adsorption to the selective host for the h mutant.

This situation suggests that the characteristics of a given phage particle may be determined not by the genes of that particle, but by the joint action of all the viral genes present in the cell in which the particle is formed.

TRANSFER OF PHOSPHORUS AND SULFUR FROM PARENT TO PROGENY PHAGE

Putnam and Kozloff (1950) showed that when bacteria growing in nonradioactive medium are infected with phage labeled with radioactive phosphorus, only a fraction of the labeled atoms appear in the progeny particles. This and other findings (Hershey and others, 1951) suggested that the transfer might consist of special parts of the phage particle that are conserved during growth. S. S. Cohen proposed a two-cycle growth experiment as a test of this hypothesis. If the labeled atoms were transferred in the form of special hereditary material, the progeny of a first cycle of growth from radioactive seed would contain radioactive atoms principally in this special material. During a second cycle of growth, therefore, radioactivity should be more efficiently conserved.

Hershey and Forman, as well as A. H. Doermann, J. D. Watson, and O. Maaløe, have made this test. The results from different laboratories, obtained under dissimilar conditions of experiment, are in essential agreement. Our experiments are summarized in the following paragraphs. The experiments with S35 are the first attempt to label the protein fraction of the phage selectively.

Radioactive phage T2H was cultivated in a glucose-ammonia medium containing about 20 microcuries of radioactive isotope per milliliter, in the form either of P32-phosphate of specific activity 1.0, or of S35-sulfate of specific activity 10 μc./μg. The phage particles were washed by centrifugation and suspended in dilute buffer. The preparations contained not more than 3×10^{-11} μg of phosphorus per phage particle, or showed an extinction coefficient (uncorrected for scattering) of less than 10^{-11} cm.^{2} per phage particle at wave length 2600 A.

First-cycle transfer experiments were carried out according to the following scheme. Broth-grown bacteria (10^9 per ml.) were infected in buffer, 97 per cent adsorption of phage being obtained in 5 minutes. The mixture was diluted with water and centrifuged, and the supernatant fluid was decanted (fraction I). The sedimented bacteria were transferred to salt-free broth at 37° C. and aerated 60 to 80 minutes, lysis being complete in about 50 minutes. The lysate was centrifuged at low speed to obtain a bacterial sediment (II), and at high speed to throw down the phage. The latter was fractionated into a low-speed sediment (IIIA), and a high-speed sediment (IIIB). The pooled high-speed supernatants and washings are called fraction IV.

Second-cycle experiments were done in the same way, starting with phage progeny from a first-cycle experiment. The unusual feature of the procedure described is the use of salt-free broth, which prevents losses of phage progeny by readsorption to unlysed bacteria and debris.

Fraction I contains some radioactive material that is not unadsorbed phage. Fraction II contains unlysed bacteria and debris. Fractions IIIA and IIIB contain roughly equal amounts of phage, comprising 90 per cent of the yield. The B portion contains slightly less isotope per phage than the A portion, and is used for measuring the extinction coefficient, and as starting material in second-cycle experiments. Fraction IV contains about 10 per cent of the phage, plus nonsedimentable material. There is no significant loss of phage titer during fractionation.

The results of several experiments are given in table 3. They show that the transfer of either isotope is about 35 per cent in both the first and the second cycle of growth. This means that neither phosphorus nor sulfur is transferred from parent to progeny in the form of special hereditary parts of the phage particles. It does not mean necessarily that the transfer takes an indirect metabolic route. Additional experiments that may clarify this point are under way.

Demonstration of Bispecific Antibody Molecules

Classic experiments have established that animals injected with two or more unrelated antigens produce a corresponding number of antibodies, each reacting with one of the antigens. More recently, the question has been raised whether any of the molecules can react with more than one antigen. This question is related to the general problem of structure and origin of antibody molecules.

Roesel has explored the possibility that some bispecific antibody may be formed in response to injections of a mixture of two bacteriophages. For this purpose she worked out the following method:

Specific precipitates of two kinds were prepared: (1) Serum from rabbits injected

DEPARTMENT OF GENETICS

with mixtures of T2H and T5 was precipitated with T5. (2) Sera from pairs of rabbits injected respectively with T2H and T5 were mixed, and the mixtures were precipitated with T5. The adsorption of

be in addition some antibody molecules specific for both antigens. Any systematic difference in the adsorption of T2H to the two kinds of precipitate would have to be due to molecules of antibody capable

TABLE 3

DISTRIBUTION OF RADIOACTIVITY AMONG PRODUCTS OF LYSIS AFTER ONE OR TWO CYCLES OF GROWTH FROM RADIOACTIVE SEED

EXPERIMENT NO.	MULTI-PLICITY OF INFECTION	BURST SIZE	PERCENTAGE OF INPUT RADIOACTIVITY				$E260^* \times 10^{-12}$ cm.^{2}
			Unadsorbed (I)	Bacterial sediment (II)	Phage (IIIA+ IIIB)	Nonsedimentable (IV)	
FIRST CYCLE; P32-LABELED							
3A............	4.5	200	10	?	40	32	11
3B............	10.0	280	12	11	37	40	10
3D............	10.0	350	9	9	32	37	7
3F............	1.3	244	10	15	30	34	12
3H............	0.5	270	9	21	27	25	?
SECOND CYCLE; P32-LABELED							
3C............	5.3	206	23	8	32	37	11
3E............	6.5	250	24	7	35	41	7
FIRST CYCLE; S35-LABELED							
4A............	5.9	276	7	17	38	15	7
4C............	7.1	310	13	18	34	15	8
SECOND CYCLE; S35-LABELED							
4B............	8.1	406	†	†	43	†	7
4D............	8.3	300	†	†	39	†	6

* Extinction coefficient of phage yield per phage particle at wave length 2600 A, not corrected for scattering.
† Fraction too dilute for assay.

T2H to washed precipitates of the two kinds was then measured under carefully controlled conditions.

Both kinds of precipitate would be expected to contain T5, anti-T5, and a small amount of nonspecifically occluded anti-T2. In precipitates of the first kind, there might

of combining successively with the two antigens.

The method proved to be adequately sensitive. A large number of sera had to be studied because (1) wide variations are observed in the properties of sera from different animals; (2) the adsorption of

T2H to precipitates of the second kind is strongly dependent on the amounts of the two antisera used to form the precipitates; (3) heat-labile constituents present in both normal and immune sera strongly influence the result.

A consistent difference in adsorption of T2H to the two kinds of precipitate was found. This means that a small amount of bispecific antibody is produced by animals simultaneously injected with the two antigens used.

The work on bacteriophage was supported in part by a grant from the National Institutes of Health, U. S. Public Health Service.

The bipartite nature of the mature phage particle emerged, and the blendor experiment, demonstrating the genetic role of DNA, was briefly reported.

The co-discovery with other groups that T2 DNA contains a nitrogenous base other than cytosine was exciting and offered the possibility of monitoring phage DNA replication unobscured by a background of host cell DNA.

INITIAL STEPS IN THE REPRODUCTION OF BACTERIOPHAGE

A. D. HERSHEY AND MARTHA CHASE

The idea has developed during the past few years that intracellular viruses, bacterial and animal, differ in fundamental ways from the extracellular particles that have heretofore been subject to biochemical and biological study. This was most clearly shown by Doermann (Year Book No. 47, 1947–1948), who found that no infective virus particles could be recovered by bursting open bacteria during the first 10 minutes after infection with phage. Inasmuch as virus is undoubtedly multiplying in the cells at this time, the existence of a masked, vegetative structure is clearly implied. Information about this structure has come chiefly from genetic and biochemical study; and at present we are concentrating on the latter. Our results may be summarized as follows.

It is generally agreed that extracellular particles of phage T2 consist chiefly of protein and desoxypentose nucleic acid. We have used S_{35} and P_{32} as convenient labels for these two components. When radioactive phage is suspended in a three-molar solution of sodium chloride, and the suspension is diluted very rapidly with water, the phage particles are disrupted, as can be shown in several ways. First, as T. F. Anderson found, the infectivity of the particles is lost, and electron micrographs reveal empty tadpole-shaped ghosts. Secondly, as Herriott showed, the ghosts can be freed of phosphorus by treatment with the enzyme desoxyribonuclease and centrifugal washing. The phosphorus-free particles are capable of attaching to phage-susceptible bacteria and lysing them. Finally, radiochemical analysis of osmotically shocked suspensions shows that nearly all the sulfur-containing protein of the phage can be adsorbed to bacteria, or precipitated with antiphage serum, leaving nearly all the phosphorus in solution. Analysis of these materials shows that phage particles consist of a protective membrane, chiefly protein, which is responsible for the specific attachment of virus to bacterium, and a nonantigenic core consisting chiefly or entirely of nucleic acid and a small amount of a free basic amino acid. The amino acid has not been identified, but is probably identical with the "fast arginine" found by Dent in extracts of various tissues.

The interaction of virus particles with phage-sensitive bacteria can also be studied by radiochemical and enzymatic methods. After phage attaches to heat-killed bacteria, the viral phosphorus can be digested out of the cells with desoxyribonuclease, leaving the viral sulfur behind. (Experiments of this type were first performed by A. F. Graham and collaborators.) Since whole phage is not affected by the enzyme, this result shows that the viral membrane is altered in some way by the attachment of phage to bacterial cells. A similar change can be demonstrated after the attachment of phage to live bacteria, but only if the infected cells are damaged in some way to permit the enzyme to get inside. This shows that the viral nucleic acid, at least, enters the cell on infection, and suggests the possibility that it is ejected from the viral membrane into the cell more or less as it is discharged into solution during osmotic shock.

The discharge of nucleic acid from the viral membrane following attachment to bacteria was demonstrated in two ways. One consisted in allowing isotopically labeled phage to attach to broken fragments of bacterial cells, and then centrifuging the mixture to separate the cellular debris from soluble materials. It was found that nearly all the phage membranes (sulfur) sedimented with the debris, whereas about half the phage nucleic acid (phosphorus) remained in solution.

The second method of demonstrating the discharge of viral nucleic acid was suggested by the recent electron micrographs of T. F. Anderson, which show that the primary attachment of virus to bacterium takes place by means of the exceedingly slender tail of the virus particle. We found, using isotopically labeled phage, that this attachment could be broken by spinning a suspension of infected cells in a Waring Blendor. This treatment strips the viral membranes from the infected cells, and leaves the viral nucleic acid inside. What is most remarkable, the ability of the infected cells subsequently to yield viral progeny is not affected by stripping. The viral nucleic acid thus becomes the center of interest in the investigation of intracellular phage, though further work is required to give this statement chemical precision.

Another indication of the importance of nucleic acid in viral growth is the following. If bacteria are infected with isotopically labeled phage, and the viral progeny are subsequently isolated and analyzed, it is found that nearly half the nucleic acid phosphorus of the parental virus has been transferred to the progeny (Putnam and Kozloff). The sulfur-containing amino acids of the parental viral membranes, on the other hand, remain metabolically inert.

The initial steps in viral growth—adsorption of virus to bacteria and injection of viral nucleic acid into the cells—occur

DEPARTMENT OF GENETICS

independently of bacterial metabolism. The multiplication of virus, however, is tied to cell metabolism. As a first approximation, we may think of the initial phases of this multiplication in terms of the synthesis of the specific viral nucleic acid. Corollary ideas are that the viral nucleic acid functions as the sole agent of genetic continuity in the virus, and that the synthesis of viral protein (likewise a specific substance foreign to the uninfected cell) occurs *de novo* as a late step in viral growth. These ideas may prove correct, incorrect, or untestable, but they are likely to orient research on viral growth for some time to come.

A fortunate circumstance will make it possible to study the economy of nucleic acid synthesis in infected cells. We find that bacterial nucleic acid contains one characteristic pyrimidine (cytosine) and viral nucleic acid another (unidentified) in addition to common constituents. The absence of cytosine from the nucleic acid of T2 had previously been reported by Marshak; and Cohen and Wyatt, independently of us, have also found the unidentified substance. We have developed a quantitative chromatographic method for analyzing mixtures of the two kinds of nucleic acid. Preliminary experiments show that the method can be applied to infected cells. It is to be hoped that this development may lead to a method for measuring the size of intracellular viral populations, and thus bring closer together genetical and biochemical questions about growth.

With the blendor result securely in hand, Al undertook separate analyses of the intracellular syntheses of DNA and of protein. He was not the first to make such measurements, but he was the first to address them to issues of heredity. The studies on DNA were testing the hypothesis that "naked" DNA corresponds to the genetically inferred "vegetative pool."

Measurements of the molecular weight of DNA isolated from phage particles indicated ten molecules per particle. This result was later shown to be an artifact introduced by breakage of the single DNA molecule during extraction and handling.

ROLE OF DESOXYRIBOSE NUCLEIC ACID IN BACTERIOPHAGE INFECTION

A. D. Hershey, June Dixon Hudis, and Martha Chase

Our report last year summarized evidence that infection of *Escherichia coli* by phage T2 starts off with the injection of the viral nucleic acid into the cell, leaving most of the viral protein on the cell surface in metabolically inert form. This situation seemed to offer a unique opportunity to learn something about the function of nucleic acid in viral infection, and, more specifically, to inquire into the chemical basis of viral inheritance. At the outset it was clear that this project would call for a number of different lines of investigation into the chemistry of viral growth. During the current year we have made an appreciable start along two of these lines.

The first question we have considered is whether any protein is injected into the cell along with the viral desoxyribose nucleic acid (DNA). This question calls for the use of isotopically labeled phage. We have prepared phage uniformly labeled by growth in a medium containing C_{14}-glucose of high specific activity. Our experiments with this material have been exploratory, and only provisional conclusions can

be drawn from them. They were of three types:

1. C_{14}-labeled phage was allowed to attach to sensitive bacteria, and the suspension was spun in a Waring Blendor for 2.5 minutes, and then separated into cellular and extracellular fractions by centrifugation. The infected cells remained competent to yield phage. Both fractions were then analyzed for labeled amino acids, purines, and pyrimidines by paper chromatography and radio assay. It was found that the cellular fraction contained 80 per cent of each of the purines and pyrimidines contained in the whole phage, but only 20 per cent of each of several amino acids, including lysine and arginine.

The proper interpretation of this result, which is in agreement with the earlier work with phage labeled with P_{32} and S_{35}, is as follows. The phage particles attach to the bacteria by the ends of their tails. All or most of the nucleic acid of most of the particles then passes into the bacterial cells, presumably through the tails of the particles. After this happens it is possible to strip the empty viral membranes from the cells, by spinning in the Waring Blendor, without affecting the subsequent course of the infection. This treatment apparently breaks the tails of the particles, leaving stubs attached to the cells, as judged by the properties and microscopic appearance (Levinthal) of the material stripped off. The stubs of the tails presumably account for the 20 per cent of protein, resembling the whole viral membranes in amino acid composition, left on the cells after stripping. It is evident, however, that this type of experiment cannot detect any protein differing in composition from the membranes and comprising appreciably less than 20 per cent of the total viral protein. Nor can it reveal anything about the function of the protein not removed by stripping.

2. In cells, and particularly in chromosomes of cells, desoxypentose nucleic acid is supposed to be associated with basic proteins rich in arginine or lysine. If this is true of the nucleic acid of T2, it should be possible to detect nonmembrane protein after disrupting the phage particles by osmotic shock.

We subjected C14-labeled T2 to osmotic shock, and digested the nucleic acid with the enzyme desoxyribonuclease. Most of the membrane protein was then removed, in one experiment by centrifugation at 30,000 g, in another experiment by precipitation with antiphage. In both experiments the supernatant fluids contained about 5 per cent of the total viral protein, and the unprecipitated fraction did not differ in amino acid composition from the remainder. The supernatant fluids also contained nearly all the viral nucleic acid (in degraded form), and one or both should have contained any protein originally associated with the nucleic acid. We conclude that the viral nucleic acid is not associated with a characteristic protein, or that, if it

is, the protein content of the complex is appreciably less than 5 per cent by weight.

Additional experiments of the same type with S35-labeled phage have shown that the membrane protein content of the supernatant fluid can be reduced to about 0.5 per cent by repeated centrifugation at 30,000 g. This suggests that the 5 per cent of protein analyzed above consisted of fragments of membranes, and shows that more sensitive tests for basic protein will be feasible if this source of contamination is reduced.

The apparently very low basic protein content of resting particles of T2 might be interpreted as an extreme example of the rule deduced by Mirsky and his collaborators from analyses of chromosomes of different types of cells, namely, that the content of basic protein is correlated with different levels of metabolic activity of the tissue of origin.

3. If all or part of the 20 per cent of viral protein that remains attached to infected cells after stripping in the Waring Blendor has any metabolic function in viral growth, it might be expected that some of the parental amino acids would be incorporated into the viral progeny, as is true of the constituents of the parental nucleic acid. Preliminary experiments by us, by Kozloff, and by French have failed to demonstrate transfer of amino acids, but the best use has not yet been made of the available techniques.

The analyses described above confirm, on the whole, the conjecture that viral nucleic acid rules viral reproduction, and hence viral inheritance. At the same time it is clear that static analysis will not lead to any really decisive proof of this idea. As a second approach, we have planned a comparative study of the times and rates of synthesis of viral protein and nucleic acid during viral growth in infected cells. The analysis of protein and nucleic acid synthesis represents two major projects,

DEPARTMENT OF GENETICS

which need to be co-ordinated not only with each other but also with related work already under way in other laboratories, particularly the immunological studies of viral growth being carried on by Luria and his collaborators. At the present time, only the work on nucleic acid synthesis has progressed very far. Accordingly, we shall summarize the results of this work without saying much about its bearing on the larger questions.

The questions we have asked about nucleic acid synthesis are very simple. We know from Doermann's work that infective phage particles begin to form in the infected cell about 10 minutes after infection, and that some masked form of the virus has already multiplied to a considerable extent before this time. We asked how much nucleic acid is formed during this period of blind multiplication, and whether such nucleic acid is in fact the nucleic acid of future virus particles.

One method of attack depended on the fortunate circumstance that the nucleic acid of T2 contains 5-hydroxymethyl cytosine, whereas the desoxyribonucleic acid of the bacterial host contains only the usual purines and pyrimidines. Making use of these facts, we were able to work out a quantitative chromatographic method of analysis for the two kinds of nucleic acid in infected bacteria. By this method we found that the bacterial nucleic acid disappears during the first 25 minutes of viral growth, and that viral nucleic acid begins to accumulate promptly after infection. At 10 minutes, when infective virus is just beginning to reappear, the characteristic viral nucleic acid already measures about 50 units (phage-particle equivalents) per bacterium. After this time, the synthesis of nucleic acid keeps pace with the formation of infective particles in such a way that the cell always

contains 50 to 100 units of surplus nucleic acid.

To determine whether this surplus nucleic acid is a precursor of the infective particles, as it should be if the hypothetical vegetative form of the virus contains nucleic acid, we made a rather thorough study of the kinetics of DNA synthesis and phage maturation, using P32 as a nucleic acid label.

The experimental principles are outlined in figure 5, which divides the source of

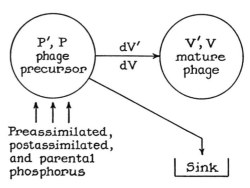

Fig. 5. Experimental arrangements for the analysis of precursor phosphorus. P', phage-precursor P32; P, phage-precursor phosphorus (P31 + P32); V', mature-phage P32; V, mature-phage phosphorus. (Reprinted from *Journal of General Physiology*, vol. 37, 1953, p. 2.)

phosphorus for viral growth into three parts: phosphorus assimilated by the bacteria before infection, which is first incorporated into bacterial DNA (Kozloff); phosphorus contained in the parental phage; and phosphorus assimilated by the bacteria after infection. Any one of these three parts can be separately labeled with P32, as was shown previously by Cohen, and by Putnam and Kozloff.

We first labeled phosphorus assimilated after infection, and particularly phosphorus assimilated during the first 4 minutes after infection. Atoms of this phosphorus were found to be incorporated into nucleic acid after spending an average of 8 or 9 minutes in the cell. The radioactive phosphorus in the newly synthesized nucleic acid was later incorporated almost completely into infective phage particles. The second step in the transport required an additional 7 or 8 minutes. These times are in good agreement with the results of Stent and Maaløe, who found that the transport from medium to phage required about 14 minutes.

These results showed that one of the precursors of the virus is nucleic acid synthesized after infection, presumably the surplus cytosineless nucleic acid previously detected by chromatographic analysis.

We next inquired whether this surplus nucleic acid is the sole or principal immediate precursor of the nucleic acid of mature virus. This is equivalent to saying that the precursor pool indicated in figure 5 is a unitary pool, and contains only cytosineless nucleic acid. Moreover, if it contains all the cytosineless nucleic acid, as indicated by the completeness of the transfer of early-assimilated phosphorus, the size of the pool is known from measurements already described: it contains 50 to 100 phage-particle equivalents of precursor nucleic acid.

This prediction could be tested independently of the assumption that the precursor is nucleic acid. We knew that the precursor is converted into phage by an irreversible process, because practically all the early-assimilated P32 eventually ends up in phage. Accordingly, we could define the specific radioactivity of phosphorus in the precursor pool (see fig. 5) by the relation $dV'/dV = P'/P$. According to this relation, the specific activity of the precursor pool (P'/P) can be measured by measuring the specific activity (dV'/dV) of a sample that has just been converted into mature phage.

Suppose, now, that a small amount of P32 is admitted into the precursor pool at the very start, by infecting the cells with labeled phage. There is, to be sure, a complication arising from the fact that not all the parental phosphorus enters the pool; about 60 per cent of it goes into a "sink" of nonprecursor material. Taking this into account, we shall expect that when the first phage particles are formed they will receive nucleic acid drawn from a pool containing about 40 per cent of the parental radioactive nucleic acid, diluted by about 50 units per bacterium of newly synthesized, nonradioactive nucleic acid. Thus the first virus particle formed in a bacterium infected with P32-labeled phage should contain about 0.8 per cent as much P32 as was present in the parental particles. A slight extension of this reasoning permits a very precise measure of the size of the precursor pool by analysis of the kinetics of transfer of parental phosphorus to progeny phage.

This analysis shows that the precursor pool contains phosphorus equivalent to 54 to 94 phage particles per bacterium, in agreement with direct measures of the amount of nucleic acid synthesized after infection. We conclude, therefore, that the precursor pool contains all its phosphorus in the form of nucleic acid, and contains all the nucleic acid synthesized after infection and not yet incorporated into infective particles.

A similar analysis of the preassimilated bacterial phosphorus shows that it is not present in the pool of immediate viral-precursor nucleic acid, but enters this pool gradually during viral growth. This result agrees with the fact, since established by Kozloff, that the precursor material is bacterial nucleic acid, and has to be converted into cytosineless nucleic acid before it is directly available for viral growth. We find that this conversion is completed in about 25 minutes.

As a whole, the tracer experiments show that viral nucleic acid is synthesized at the

DEPARTMENT OF GENETICS   227

rate to be expected if the precursor nucleic acid is contained in the vegetative phage particles identified by genetic methods. It remains to be seen whether the synthesis of viral-precursor protein obeys the same or different rules.

MOLECULAR CHARACTERISTICS OF THE DNA COMPONENT OF PHAGE T2

In view of the primary role played by the viral desoxyribose nucleic acid (DNA) in the multiplication of phage T2, the molecular characteristics of this material have been investigated by Garen in collaboration with Dr. E. Reichmann, of the Chemistry Department of Harvard University. The DNA of T2 was separated from its protein membrane by several shakings of an aqueous suspension of the phage in the presence of chloroform. This procedure removes most of the sulfur-containing protein (as indicated by the residual radioactivity when S35-labeled phage is used as a tracer) and recovers more than 50 per cent of the DNA. Preliminary measurements of the isolated DNA by light-scattering techniques indicate a molecular weight of ten million. From this value we can estimate that a mature particle of phage T2 contains around 10 molecules of DNA of this molecular weight. Further experiments are now in progress, using viscosity and rotary-diffusion techniques in addition to light scattering, to gain additional information about the size and shape of the molecules.

The principal focus of the report was on correlations between the behavior of DNA as monitored by various kinds of labels and the behavior of genetic information. The aim was to establish in a quantitative way the now accepted relationship between DNA and genes. For instance, what fraction of the DNA is dedicated to the encoding of genetic information?

The suggestion of recombination between the chromosome of phage T3 and that of its host foreshadowed reports to come that the *E. coli* chromosome contains segments of DNA that are homologous with segments of the chromosome of phage λ. We now call such segments of *E. coli* DNA "cryptic prophages."

The sources of viral components following infection by phage T2 were reported. T2, like the other T-even phages, not only stops syntheses of macromolecular host components, but also uses breakdown products of host nucleic acids to synthesize its own DNA.

The report concludes that the rate of protein synthesis following phage infection is the same as that before infection and proposes, incorrectly, that the infected cell continues to make most of the proteins it was making before. Al then muses that this is paradoxical in view of the fact that "typical bacterial enzymes" are not made in infected cells. This paradox was later resolved by the demonstration that T-even phage reprogram the bacterial ribosomes by blocking synthesis of bacterial mRNA while producing their own.

GROWTH AND INHERITANCE IN BACTERIOPHAGE

A. D. HERSHEY, ALAN GAREN, DOROTHY K. FRASER, AND JUNE DIXON HUDIS

Putnam and Kozloff first showed that isotopically labeled phage particles contribute some but not all of their labeled atoms to their progeny. Their work and that of their successors has given rise to the following well defined questions about the transferred atoms: (1) Origin—in random or special parts of the parental particles (Putnam and Kozloff)? (2) Location —in random or special parts of the offspring particles (Maaløe and Watson)? (3) Route—by way of large specific pieces or small nonspecific ones (Kozloff)? (4) Distribution—in one, several, or all the offspring particles (Hershey, Kamen, Kennedy, and Gest)? (5) Function—transferred independently, or not, of the transfer of genes (Kozloff)?

Questions (1) and (2) were partly answered by Hershey and Chase, and by French, who found that only the viral nucleic acid (DNA) contributes and receives conserved atoms. Question (1) remains unanswered with respect to parts of the DNA, however, because only about 40 per cent of its label is transferred. Question (2) was further answered by Maaløe and Watson, who showed that the DNA of the offspring of labeled particles is itself randomly labeled with respect to transfer during a second cycle of growth. Question (4) was partly answered by Watson and Maaløe, and by French, Graham, Lesley, and van Rooyen, who found that

only particles formed early receive parental atoms. As is described in this report, Garen has now shown that the transferred atoms originating from a single parental particle enter into several offspring particles. Most of the work to be summarized here, however, deals with attempts to answer questions (3) and (5).

Some of our experiments relate also to a question rather new to both genetics and virology. This involves the relation between virus and bacterial nucleus. Rather diverse lines of genetic evidence (Lederberg and Lederberg, Zinder, Bertani, Lwoff and his collaborators) show that some close relation exists in lysogenic bacteria. Dorothy Fraser has carried the idea of nuclear localization to its logical conclusion by looking for evidence of genetic recombination between phage and bacterium. In addition, we are exploring possible ways of bringing chemical methods to bear on the question, and some of our work with T2 has this objective.

As in the past, the work with bacteriophage was supported in part by a grant (C-2158) from the National Cancer Institute of the National Institutes of Health, U. S. Public Health Service. Isotopes were supplied by the Oak Ridge National Laboratory on allocation from the Atomic Energy Commission. Garen's participation in the work was made possible by his

210

fellowship from The National Foundation for Infantile Paralysis.

Biochemical Route from Parent to Offspring Phage

The chemistry of the transfer of atoms from parental to offspring phage interests us for two reasons. First, the atoms are those of DNA, primarily or exclusively, which suggests that DNA has a genetic function in T2. Second, it seems likely that most of the viral DNA is genetically potent, because the amount in a single particle of T2 (2×10^{-13} mg) is small as compared with the amount per nucleus in most cells. In fact, current radiobiological experiments (Stent, Doermann) seem to show that more than 20 per cent of it is genetic material, and that the ratio of genetic to nongenetic DNA remains constant as one goes to still smaller phages. For these reasons we feel justified in hoping that the conservation of materials during viral growth is a genetic problem, as well as a biochemical one.

The isotope-competition method offers one way of investigating the biochemical route from parental to progeny DNA. This approach is inseparable from the problem of mechanisms of synthesis of DNA, about which equally little is known. In this work we have benefited greatly from the generous advice of R. B. Roberts and E. T. Bolton of the Institution's Department of Terrestrial Magnetism, and from the active collaboration of A. F. Graham, L. Siminovitch, and S. M. Lesley at the University of Toronto, who first undertook experiments along the lines indicated here, and kindly permitted us to join them.

The experimental principle is as follows. Uniformly C14-labeled T2 is prepared by propagation in bacterial cultures containing radioactive glucose. The labeled virus is then allowed to multiply in cultures in which the parental viral material is the only source of C14. In these cultures the source of food for the synthesis of new viral material is glucose, with or without added precursors of DNA such as guanine. Supplements of this kind are called competitors because they can be expected to compete specifically with intracellular sources of radioactive guanine and related substances. That they do so in fact is readily checked by placing them in competition with radioactive glucose.

Whether the source of radioactivity is the parental viral material or the glucose in the culture medium, the analytical method is the same. One isolates the viral DNA, hydrolyzes it to its constituent bases, separates these on paper chromatograms, and assays the radioactivity in the spots located by inspection in ultraviolet light. This method has the advantage that bases are recovered quantitatively, and the disadvantage that not all the radioactivity in the spots is contained in the bases. It is being used in conjunction with more precise methods worked out by the group in Toronto.

Table 1 summarizes the results of some isotope-competition experiments in which the parental phage was the source of radioactivity appearing in the offspring. In these experiments, 0.5 mg per ml of competitor was added at the time of infection. Phage yields were isolated after artificial lysis 1 hour later. They contained 30 to 50 per cent of the parental DNA C14. The results show that parental guanine, adenine, hydroxymethyl cytosine (hmc), and thymine are used equally to make the DNA of the offspring phage. Furthermore, this equal transfer cannot be upset by supplementing the culture medium with the competitors indicated. This must mean that the parental DNA is not broken down to the level of these intermediates en route to progeny DNA. Within the rather severe limitations of the method, the findings suggest that DNA is transferred in large pieces, and therefore conceivably in the form of genetically functional pieces.

TABLE 1

Transfer of C14 from parental to progeny T2

MATERIAL ANALYZED	PER CENT DISTRIBUTION OF C14			
	Guanine	Hmc	Adenine	Thymine
Parental phage	18	18	31	34
Yield, no competitor	18	17	30	35
Yield, thymidine competing	18	17	32	34
Yield, guanine competing	18	18	29	34
Yield, adenine competing	18	19	31	33
Yield, thymine competing	18	18	30	34

Correlation between Genetic and Material Transfer of Viral Deoxyribonucleic Acid

A much more general clue to the meaning of the transfer of DNA from parents to offspring can be sought in terms of the correlation, or lack of it, between genetic and biochemical results. To this end we have studied the properties of phage inactivated by beta rays and ultraviolet light before infection, or self-inactivated after infection by attachment to previously infected bacteria. The principle of all the experiments is the same. One infects bacteria with two genetically and isotopically marked phages under conditions such that only one of them contributes genetic markers to the viral progeny. The latter are then examined to determine whether they contain isotope derived from the genetically excluded phage. Similar experiments have been performed before, by Kozloff and others. We have repeated them because we were dissatisfied with the results.

Beta rays. A preparation of T2, labeled with P32, was inactivated by suspending it for 10 days in a 2 per cent solution of peptone containing one mc per ml of P32. This exposure to beta rays killed 99 per cent of the phage, as determined by plaque titration. The particles were then washed free of external P32 by centrifugation.

The inactivated phage was found to adsorb normally to bacteria, as measured by its radioactivity. When a suspension of cells with attached virus was spun in a high-speed homogenizer (Hershey and Chase), 60 to 80 per cent of the radioactivity could be stripped from the cells. Only 20 per cent of the radioactivity of a sample of the same phage that had not been exposed to beta rays was removed under the same conditions. This shows that most of the inactivated particles do not inject their DNA into the cells to which they attach.

When the inactive phage was used to produce mixed infections with live, unlabeled phage, and samples of the culture were lysed at successive times by the method of Maaløe and Watson, it was found that most of the radioactivity sedimented either with the bacterial debris or with the small-particle fraction containing the viral progeny. The distribution between these two fractions was nearly equal, and was the same for lysates prepared before and after viral offspring appeared. This shows that the genetically excluded phage does not contribute detectable amounts of phosphorus to the progeny of the mixed infection. We suspect that the behavior of X-ray–inactivated phage will prove similar to that described above, and that the conclusions drawn from the earlier experiments of this kind are incorrect. On the other hand, the correlation between genetic and metabolic impotence of phage inactivated by ionizing radiations must be regarded as trivial, since the exclusion occurs outside the cell wall. Doermann, and Bertani and Weigle, have shown by genetic experiments that a minority of the particles inactivated by ionizing radiations does contribute to the yield. Unfortunately, the limitations of the methods do not permit radiochemical analysis of this minority.

Ultraviolet light. When P_{32}-labeled T2 or T4 is exposed to 150 times the average lethal dose of ultraviolet light, the subsequent ability to inject is only moderately diminished. This has been tested both by the blender experiment and by measuring the sensitivity of the viral DNA to deoxyribonuclease after adsorption of irradiated phage to heat-killed bacteria (Hershey and Chase). Apparently, ultraviolet light damages almost exclusively the injected material. The situation is a promising one, therefore, for the attempt to correlate genetic and metabolic potency. This attempt is all the more timely because Doermann is currently analyzing the genetic effects of ultraviolet light in quantitative terms.

Garen is making intensive efforts to measure the corresponding metabolic effects. He finds that the transfer of atoms from irradiated phage is not, as previously supposed, independent of radiation dosage. Whether the genetic and metabolic effects are directly correlated, however, is not yet clear.

Superinfection. When bacteria are infected with T2, and then reinfected 2 to 10 minutes later with P_{32}-labeled T2, half the labeled DNA is quickly split into acid-soluble materials (Lesley, French, Graham, and van Rooyen). A negligible proportion of the DNA of the first particles to infect is split. Simultaneously with this effect, the superinfecting phage fails to contribute genetic markers to the viral yield produced by the first infecting particles (Dulbecco), and likewise fails to contribute labeled atoms to it (French *et al.*). Apparently the metabolic exclusion is primary, and the breakdown secondary, since streptomycin (which inhibits bacterial deoxyribonuclease) prevents the breakdown without affecting the exclusion. We have reexamined this situation, using a more specific method of inhibiting the bacterial nuclease, namely, low magnesium concentration.

We were interested in two questions. First, does the 50 per cent breakdown of superinfecting phage mean that viral DNA splits into two fractions having different properties? Second, does the viral DNA enter the cells when breakdown is prevented, and, if so, is it excluded from genetic and metabolic function?

The answer to the first question is no, because we find that only half the DNA of the superinfecting phage enters the cells, and practically all of this is broken down. If the blender experiment is performed using cells superinfected with P_{32}-labeled T2 in a medium containing 10^{-3} M magnesium chloride, 50 per cent of the P_{32} is excreted in acid-soluble form within 5 to 10 minutes, and 80 per cent of the remainder can be stripped away in the blender. If, instead, the medium contains 10^{-5} M magnesium, less than 5 per cent breakdown occurs, and only 40 per cent of the P_{32} is strippable. The low magnesium concentration does not itself affect injection, as is shown by similar experiments using P_{32}-labeled phage for the primary infection.

At the low magnesium concentration, the 50 per cent of the superinfecting DNA that is injected remains in the cells until lysis. Phage growth is normal, and so is the transfer of P_{32} from primary infecting phage to progeny. A repetition of Dulbecco's experiment under these conditions shows that the superinfecting phage makes no genetic contribution to the progeny. Similarly, P_{32} contained in the DNA of the superinfecting phage does not appear in the progeny. These experiments show clearly that part of the DNA of superinfecting phage, although it enters the cells and is accessible to at least one bacterial enzyme, can persist there without metabolic function in viral growth. The correlation between genetic and biochemical results in this instance is encouraging, but not of great significance until more can be learned about the mechanism of exclusion.

DISTRIBUTION OF TRANSFERRED ATOMS AMONG PROGENY

When a bacterium is mixedly infected with the related phages T2 and T4, progeny of both parental types are obtained. We were interested in determining whether, during mixed growth, the atoms contributed by one parent are located exclusively in its own progeny type, or in both types.

Garen examined this question in the following way. Stocks of T2 and T4 were first carried through several cycles of mixed growth in order to prepare "isogenic" strains, which could grow equally together under conditions of mixed infection. Bacteria were then mixedly infected with P32-labeled T4 and unlabeled T2 in buffer, and transferred for growth to a salt-free nutrient medium. The absence of salt prevented readsorption of phage after lysis. The total phage yield was 150 per cell and consisted of equal amounts of T2 and T4. The lysate was purified by treatment with deoxyribonuclease, followed by centrifugation at low and high speeds. In order to eliminate phenotypic mixing, the phage was put through a second growth cycle under conditions of single infection. Bacteria were infected in buffer with 0.3 particle per cell and then transferred to salt-free nutrient medium at a concentration of 1.5×10^8 bacteria per ml. After 40 minutes' growth, sufficient for lysis of the infected bacteria, the culture was filtered. The yield in the filtrate was 160 particles per infected cell, and again it consisted of equal amounts of T2 and T4. The phage was purified by two cycles of alternate low- and high-speed centrifugation.

The T2 and T4 progeny from the second cycle of growth were separated by successive additions, first of B/2,4 bacteria to remove nonspecifically adsorbable material, then of B/4 to adsorb T2, and finally of B to adsorb T4. The results are given in

table 2. They show that P32 originally present in the parental T4 phage appeared not only in T4 progeny, but to a large extent in T2 as well. The results are in agreement with those of Hershey, Kamen, Kennedy, and Gest, who showed by a different kind of experiment that parental phosphorus atoms must be distributed to at least two progeny particles. At present it is not known what mechanism is responsible for the distribution of parental DNA, but one or both of the following are likely to be involved: transfer by way of genetically nonspecific units; material interchange during genetic recombination.

TABLE 2
DISTRIBUTION OF P32 DERIVED FROM T4 AMONG PROGENY OF MIXED T2-T4 INFECTION

Adsorbing bacteria	Phage adsorbed (%)	P32 adsorbed (%)
B/2,4	....	5
B/4	T2: 92	22
B	T4: 85	36

THE ORIGIN OF HOST-RANGE MUTANTS OF T3

In work done by Fraser and Dulbecco at the California Institute of Technology, it was found that two host-range (*h*) mutants of the small bacteriophage T3 differed from the parent T3 in several loci, which showed segregation and recombination in crosses. Since the methods of isolation of these two *h* strains would normally be expected to provide single-factor mutants, it was of interest to discover why mutants differing from wild type in three or four factors were recovered instead. Fraser has pursued this question, with results suggesting that the mutants arise by interactions between phage and bacterium, possibly in the nature of genetic substitutions.

The physiological conditions controlling mutation. It has been found that a certain number of the host bacteria (*Escherichia coli,* strain B) infected with T3 fail to lyse at the end of the usual latent period, and that the progeny of such late-lysing bacteria contains host-range mutants, frequently comprising about 1 per cent of the yield, sometimes much more. Among the progeny issuing from cells that lyse promptly, no mutants are found within the limits of the method of detection (<1 in 30,000).

The proportion of such late-lysing bacteria, here referred to as complexes, depends on the physiological state of the bacteria before infection and on the treatment after infection. If the bacteria are starved by aeration in buffer for 2 hours before infection in broth, nearly all the bacteria form persistent complexes. If the culture is diluted in fresh broth at pH 6, the complexes begin to lyse after about 15 minutes. They are more stable in broth at pH 5.

Experiments have been carried out in which the infected bacteria are kept for periods of several hours to several days at 37° C in broth containing anti-T3 serum to prevent further infection. A slow increase in the number of infected bacteria occurs under these conditions, indicating that the complexes are able to reproduce as such. It has not been possible, however, to obtain stable lysogenic clones. On dilution and plating, most of the bacteria lyse.

It seems likely that the complexes are the same as the cells of the "thin" colonies frequently found when phage T3 and strain B are plated together for the isolation of resistants. Like the complexes, cells of thin colonies lyse on dilution, yielding phage that contains a high proportion of host-range mutants.

Mutant types. The mutants observed are identified by plating phage on agar seeded with a mixture of *E. coli* strain B (Delbrück) and one of its mutants, B/3b, resistant to T3. The plates are incubated at 38.5° for 3 hours and then allowed to stand at room temperature for several more hours, during which time they are observed at intervals and variant plaques are

marked. All particles form plaques, and variants are picked by inspection.

The phenotypes observed under these conditions may be roughly classified into six types (the wild type gives phenotype III): (I) uniformly turbid; (II) turbid with a slight ring of clearing at the periphery and occasional clear areas in the center; (III) turbid with a clearing ring at the periphery and irregular clear areas in the center of each plaque; (IV) turbid with a marked clearing ring and an irregular clear area in the center (these plaques soon become entirely clear, except for a fuzzy edge); (V) clear with a narrow halo; (VI) clear with a sharp edge. Each of these phenotypic classes includes phage strains differing in genetic constitution. The differences are heritable and, apart from the production of further mutants, do not depend on the bacterial strain in which the phages are grown.

Distribution. Single bursts produced by complexes of B and wild-type T3 diluted after 4 hours in antiserum were studied. It was found that mutants occurred in bursts which usually also contained a majority of wild-type particles. There seemed to be a correlation between the frequencies of phenotype I and phenotype IV or V in individual bursts. Early lysates of complexes were made by the cyanide technique at 2-minute intervals after "induction" by dilution in broth at pH 6. Variants were picked from plates of each of these lysates and replated. Fifty per cent of the variant plaques from the earliest yield contained approximately equal mixtures of phenotypes I and IV or V. The proportion of mixed plaques decreased in later lysates, reaching 15 per cent at the time of normal lysis. Thus it appears that the variants must be formed in pairs by a single event. This is hard to explain by a theory of mutational origin of the variants.

Irradiation of the host. Ultraviolet irradiation of the bacteria before infection increases the frequency of mutation. In one experiment, B cells were irradiated for various lengths of time, infected with T3 wild type, and plated 10 minutes after infection. The viral yield from nonirradiated cells contained no variants among 4280 plaques. With increasing ultraviolet dose the proportion of mutants of all types increased, to reach a maximum of about 1 per cent in the yield of cells that had been irradiated 2 minutes. (One minute gave a bacterial survival of 5×10^{-6}.) In another experiment it was found that the mutants occurred in bursts containing a majority of wild-type particles. The mutants appeared to be the same phenotypes as those resulting from complex formation with B. This is in agreement with Jacob's results on the production of viral mutants by irradiation of the host cell (K-12S) before infection with nonirradiated phage (λ). The mutants in such experiments may be produced by the action of mutagenic substances in the cytoplasm of the irradiated bacteria, or may reflect some more specific type of phage-bacterium interaction.

Possible origin of mutants. The initial observation, that mutants differing from wild type in several loci had presumably been formed by a single event, suggested that the event in question might be a recombination rather than a mutation.

If the mutants arise as a result of recombination between variants of identical phenotype already present in the wild-type stock, they should not be found in the progeny of singly infected complexes. This point was tested by single-burst experiments with complexes formed by single infection. Mutants occurred with the usual frequency in these bursts. Therefore, if they arise by a process of recombination, it cannot be by recombination between two different strains of phage in the wild-type stock.

During the past year, the idea has been discussed that there may be an actual homology between the genetic material of the phage and part of that of the host cell. To test this possibility, the interaction between B and T3 has been provisionally regarded as a genetic cross, and on that basis predictions have been set up to be tested experimentally.

If one of the parents in a cross is changed, the nature of the recombinants should be changed. Then hosts differing in the chromosome section homologous to the phage in question should produce different classes of mutant. A study was made of the mutant phenotypes formed by complexes of T3 wild type and each of a number of B strains that had been carried independently for some years. With six such strains, the total number of mutants found did not differ significantly from strain to strain, but with three of them the ratio of the numbers of variants IV, V, and VI to I and II diverged widely from the value 0.56 found with the original B. These values ranged from 0.08 in strain S (Hershey) to 2.0 in strain B' (Hershey). The efficiency of plating of each phenotype on the various strains did not give any indication of selective differences that might account for the variations found. The possibility of intracellular selection should be investigated by means of further experiments.

In a genetic cross the recombinants are closer in genetic constitution to either parent than the two parents are to each other. If a recombinant is backcrossed to one of the original parents, the new recombinants will be closer still to that parent, and the process of isolation of successive recombinants of backcrosses may be continued, to approach the genetic constitution of the parent more and more closely. By a series of "backcrosses" of phage to bacterium—that is, by isolation of a series of successive mutants on the same bacterial strain—it might be possible to produce a phage strain which would not give rise to mutants because the homologous portions of the genetic material of phage and bacterium would be identical. This principle might permit a critical test of the hypothesis under consideration.

Preliminary work has been done along these lines. Three phage strains, nos. 3, 14, and 18, all giving no further mutants on B, have been derived from wild type by successive isolation of mutants through the following steps (phenotypes shown in parentheses):

$$\text{No. 1 (III, wild type)} \xrightarrow{\times B} \text{no. 2 (IV)} \xrightarrow{\times B} \text{no. 3 (I);}$$

$$\text{No. 1 (III)} \xrightarrow{\times B} \text{no. 11 (IV)} \xrightarrow{\times B} \text{no. 14 (I);}$$

$$\text{No. 1 (III)} \xrightarrow{\times B} \text{no. 4 (I)} \xrightarrow{\times B} \text{no. 7 (V)} \xrightarrow{\times B} \text{no. 10 (IV)} \xrightarrow{\times B} \text{no. 18 (I).}$$

No. 3 and no. 14 appear to be identical, and produce no recombinants when crossed together by the standard phage-cross technique. The intermediates in the isolation series, no. 2 and no. 11, are not identical. No. 18 is clearly different from no. 3 and from no. 14.

The yields from complexes of these three strains on a different host strain, B', all include mutant types. The mutants produced by no. 3 and no. 14 on B' have identical phenotypes and appear in about equal numbers. No. 18 on B' gives mutants of another phenotype. Thus the stability of nos. 3, 14, and 18 is specific for the host in which they were obtained. In terms of the hypothesis, these strains, which give no recombinants when crossed with B, do give recombinants when crossed with B'.

Relationship of the stable strains. It was thought possible that strains no. 3 and no. 18 might not be completely homologous with respect to each other if they had picked up, from the host, genetic material not entirely homologous to T3, and different in each case. They were crossed together, and two recombinants were isolated. The most conspicuous recombinant, of phenotype VI, comprises about 12 per cent of the yield, a very high percentage for T3. The second recombinant gives a very small, ragged plaque of phenotype I. If no. 3 and no. 18 were not completely homologous, it would be expected that the recombinants might differ in length of chromosome (and hence in size) from each other and from each parent. High-speed centrifugations of mixtures of the

two recombinants, and of the clear recombinant (VI) with each parent, show that the clear recombinant sediments less rapidly than any of the other strains. The techniques so far employed do not permit a definite conclusion as to whether this result is due to an actual difference in size of the phage particles, or to some fortuitous factor. A more convincing demonstration of a difference in amount of genetic material per phage particle might be obtained from studies of ultraviolet inactivation of mixtures. Work on this possibility is in progress.

PREASSIMILATED BACTERIAL SOURCES OF VIRAL MATERIAL

Sources of DNA. When C14-labeled bacteria are infected with T2 in nonradioactive glucose medium, radioactivity rapidly disappears from the DNA cytosine and simultaneously appears in DNA hydroxymethyl cytosine (fig. 1). If, instead, unlabeled bacteria are infected in radioactive glucose, radioactivity appears in DNA hydroxymethyl cytosine but not in cytosine. As Kozloff found, and we have confirmed, bacteria specifically labeled in DNA thymine transfer their label to viral DNA. These results show that some of the materials used for synthesis of viral DNA come from preformed bacterial DNA, others come from the glucose assimilated after infection, and bacterial DNA is not an intermediate along the second route. This means that the infection alters DNA metabolism in three ways: it stops formation of bacterial DNA instantly, it quickly initiates the synthesis of viral DNA, and it causes the existing bacterial DNA to all but disappear.

These facts suggest that the rapid conversion of bacterial DNA into viral DNA is probably quite as unique as the accompanying genetic events certainly are. This line of thought led us to inquire whether the virus initiates the breakdown of bacterial DNA, or merely intercepts the products of a normal breakdown mechanism, and whether bacterial DNA is the only characteristic bacterial substance used to make virus.

It turns out that bacterial DNA is not the only source of preassimilated materials used to make viral DNA. If, in the experi-

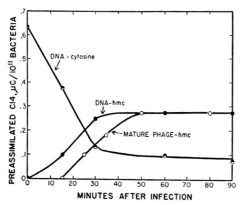

FIG. 1. Destruction of bacterial DNA and formation of viral DNA as observed by analysis of preassimilated C14 in cytosine and 5-hydroxymethyl cytosine (hmc) after infection.

ment in which one observes the entry of preassimilated C14 into viral DNA, one looks at any of the purines and pyrimidines other than the cytosines, one observes a slow, continuous increase in radioactivity. This is shown for thymine in figure 2. The rise usually amounts to 30 per cent or more 90 minutes after infection, and is not an analytical artifact because it does not occur when the cells have been specifically labeled by feeding radioactive thymidine.

The following experiment shows that the previously undetected source is probably ribose nucleic acid (RNA). Bacteria were labeled differentially by growing them in C14 glucose supplemented with nonradioactive uridine, cytosine, and orotic

acid. Such cells contain nonradioactive pyrimidines in RNA and DNA, but all or most of their other constituents should be radioactive. After labeling, the cells were transferred to a medium containing nonradioactive glucose and allowed to grow 30 minutes longer to use up any labeled intermediates they might contain. At this time the cells were infected with T2, and samples were withdrawn for analysis of the whole intrabacterial DNA immediately after infection and 90 minutes later.

Table 3 shows the results. (Radioactivi-

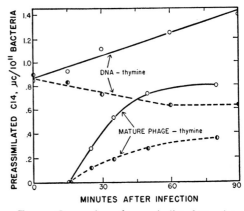

FIG. 2. Conversion of preassimilated C14 into DNA thymine and into phage thymine in infected bacteria. Continuous lines, culture without competitor; dashed lines, parallel culture with added thymidine. Phage growth was linear after 15 minutes, reaching 150 to 180 per bacterium at 80 minutes in both cultures.

ties shown for the pyrimidines have been corrected by subtracting assays of blanks cut from the rims of the spots, and are not very precise.) It can be seen that purine radioactivity increased significantly after infection, but no radioactivity entered the DNA pyrimidines. This result shows that the rise in pyrimidine radioactivity observed in similar experiments with uniformly labeled cells (fig. 2) comes from RNA or from some unknown source of preformed, metabolically stable pyrimidines.

The likelihood that the source is RNA is supported by the observation that labeled

DEPARTMENT OF GENETICS   219

RNA phosphorus and carbon decrease after infection, more than enough to account for the rise in labeled DNA. This decrease is slow and small, however, never exceeding 10 to 20 per cent of the total RNA, and does not lessen the special significance we wish to attach to the dramatic conversion of bacterial DNA into viral DNA illustrated in figure 1.

TABLE 3

DISTRIBUTION OF DNA C14 BEFORE AND AFTER INFECTION IN BACTERIA LABELED BY GROWTH IN C14 GLUCOSE (0.025 µc/µg C UNIFORM LABEL) PLUS NONRADIOACTIVE URIDINE

TIME AFTER INFECTION	µC C14 PER 10^{13} BACTERIA				
	Guanine	Hmc	Cytosine	Adenine	Thymine
0 minutes ..	57	0	0	61	10
90 minutes ..	75	0	0	79	8

Sources of protein. The following experiments were designed to measure the extent to which preassimilated bacterial sulfur is used to make viral protein in infected bacteria.

Bacteria were labeled with radioactive sulfur (S35), washed free from external S35, and allowed to grow for 30 minutes in nonradioactive medium (glucose, sulfate, phosphate, tris hydroxymethylamino methane buffer, salts). Such cells contain radioactive proteins and also relatively large amounts of labeled glutathione, which is a potential protein precursor but is not used up except in cells starved for sulfur (Abelson et al., Year Book No. 52, pp. 133–135).

Such labeled cells were infected with T2, and samples were lysed in the presence of cyanide and a large excess of lysing phage (inactivated by ultraviolet light) at various times during the period of viral growth. The lysed samples were treated with ribonuclease and deoxyribonuclease before fractionation, which reduced the radioactivity of the small-particle fraction from uninfected bacteria very appreciably.

Carrier bacteria (heat-killed B/2), which do not adsorb the phage, were then added, and the lysates were separated into a large-particle fraction and a small-particle fraction by successive low- and high-speed centrifugation. Other details of these methods are given in previous publications from this laboratory. S35 was assayed in the sedimentable fractions of the lysates, and also in trichloroacetic acid precipitates ("total protein") of other samples with-

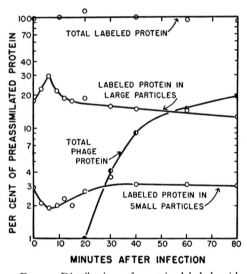

FIG. 3. Distribution of protein labeled with preassimilated S35 after lysis of infected cells. Preassimilated protein sulfur per bacterium measured 1060 phage equivalents (2.8×10^{-9} µg). Total phage protein is computed from the yields of infective particles, for example, 200 per bacterium at 80 minutes.

drawn from the infected culture at various times. Yields of phage were measured by plaque titrations of the lysates; 90 per cent of the infective particles were recovered in the small-particle fraction.

Figure 3 shows the results of a typical experiment of this kind. The data are expressed as percentages of the total labeled protein assimilated before infection. These can be converted into micrograms of protein sulfur per bacterium from conversion factors (given in the legend) computed from the specific activity of the preassimilated sulfur and corrected for the 30-minute period of growth in nonradioactive medium before infection. Total sulfur in phage protein is computed from the results of plaque titrations, multiplied by the sulfur content per phage particle (2.5×10^{-12} µg). The two basic constants, protein sulfur per bacterium and per phage, have been measured radiochemically in numerous independent experiments, some of which are described in a later part of this report.

The main facts shown in figure 3 are as follows. First, total labeled protein remains constant after infection. Second, labeled protein in small particles does not show a significant rise correlated with the formation of mature phage particles. Third, the initial fall and subsequent rise in S35 content of the small particles is inversely correlated with changes in the large particles. This is the effect of a temporary resistance to lysis characteristic of infected bacteria, first described by Visconti and also noted visually in the present experiments.

These facts show that there is no extensive conversion of preassimilated sulfur into phage protein. On the other hand, it is clear that centrifugal fractionation leaves about 3 per cent of the total bacterial protein, equivalent to 32 phage particles per bacterium or 16 per cent of the phage yield at 80 minutes, as a contaminant of the small-particle fraction. More specific methods of purification are required to detect smaller amounts of labeled sulfur that may be contained in the phage particles.

Adsorption to bacteria offers one such method. Two phage preparations, both isolated from the culture described in figure 3, were examined by this means. One consisted of the small-particle fraction obtained after lysis at 60 minutes, and contained 151 phage particles per bacterium. The other was obtained after spontaneous lysis yielding 500 phage per bacterium, and was purified by repeated fractional centrifugation, during which only 27 per cent of the phage was recovered. Samples of phage were mixed with suffi-

cient numbers of heat-killed bacteria (strain H) to adsorb all the phage (98 per cent by titration), and measurements were made of radioactivity sedimentable with the bacteria or precipitable from their supernatant fluid by trichloroacetic acid. The specificity of adsorption was controlled by identical measurements using similar numbers of heat-killed bacteria (strain B/2) that did not adsorb the phage. The sensitivity of the measurement was greatly increased by a preliminary treatment of each sample with B/2, which served to remove material adsorbed or sedimented nonspecifically. The assay methods were controlled by requiring that the radioactivity of sediment and supernatant fractions account for the total input (all measured under conditions of negligible self-absorption). Results are given in table 4.

The table shows, as concluded previously, that most of the radioactivity in the material from the 60-minute lysate is contained in contaminating bacterial protein. Only 12 per cent of the radioactivity, corresponding to 0.4 per cent of the labeled bacterial protein, or 2.6 per cent of the total phage protein, fails to be separated from the phage by this method. Of the radioactivity contained in the phage isolated from the terminal lysate, about half is specifically adsorbable. This half, after correction for the loss of phage during purification, corresponds to 1.1 per cent of the total labeled bacterial protein, or 2.2 per cent of the terminal yield of phage protein.

We conclude that preassimilated sources of phage sulfur, like preassimilated sources of phage DNA other than bacterial DNA, are drawn on slowly and inefficiently as compared with the utilization of the bacterial DNA itself.

These experiments do not prove that any bacterial protein is converted into phage protein. To test this, one would have to use labeled bacteria that had been starved to deplete their acid-soluble reserves (Abelson et al.).

Nucleic acid turnover in uninfected bac-

DEPARTMENT OF GENETICS   221

teria. Bacterial DNA is destroyed quickly, and RNA slowly, in infected bacteria. Does the virus initiate these decompositions, or merely block resynthesis?

If bacterial RNA or DNA or both were continuously being broken down to the level of common precursors of these two materials, one would expect, except under very special conditions, that labeled purines and pyrimidines would pass preferentially in one direction between them, and perhaps out of both, during the growth of labeled cells in unlabeled medium. More certainly, the addition of specific unlabeled

We conclude that if any breakdown of nucleic acid occurs during bacterial growth, it involves only a very small percentage per generation of the bacterial DNA, and still less of the more abundant RNA, or it yields unknown intermediates not interceptable by the available competitors. It follows that the rapid breakdown of DNA in infected bacteria is stimulated by the infection.

It should be noted that our experiments do not exclude time rates of turnover comparable with those observed in the classic experiments with resting mammalian tis-

TABLE 4

Test for constituent S35 in phage grown on S35–labeled bacteria

(The sediments contain S35 adsorbed to successive additions of heat-killed bacteria; the supernatant contains residual unadsorbed S35.)

	Radioactivity (cpm per ml)			
	60-minute yield		Terminal yield	
Fraction analyzed	(1)	(2)	(1)	(2)
First sediment (B/2)	1100	1220	129	140
Second sediment (B/2)	330	. . .	114	. . .
Second sediment (B)	. . .	659	. . .	785
Second supernatant	1045	825	915	320
Whole sample	. . .	2750	. . .	1420
Percentage specifically adsorbed	. . .	12	. . .	50

precursors to the medium would cause specific losses of labeled constituents in a manner predictable from the results of ordinary isotope-competition experiments.

Breakdown of nucleic acids in uninfected bacteria was tested along the lines indicated, using cells uniformly labeled by feeding radioactive glucose, and allowing these to grow for 7 generations (6 hours) in the presence of nonradioactive glucose supplemented with ribo- or deoxyribonucleosides. At the end of this period of growth, the bacterial DNA was analyzed for labeled purines and pyrimidines. No difference in amount or distribution of radioactivity was found between the DNA in the terminal and that in the starting cultures, for any of five competitors tested.

sues, in which the period of observation generally spanned several days. The point of our findings, which are in agreement with other results for microorganisms, is that rapid synthesis of cell constituents does not call for an acceleration of the slow processes of decay that could doubtless be observed also in microbial cultures by experiments of sufficient duration.

Biochemical route from bacterial to viral DNA. As already shown, the intermediates in the transfer of labeled viral DNA from parents to offspring cannot be intercepted by supplementing the culture medium with specific DNA precursors. Before concluding from this that the transfer is, in some sense, direct, it is necessary to study by the same methods a comparable situation in which the transfer is indirect. The transfer from bacterial DNA to virus recommended itself for the purpose. This simple aim has broadened considerably in the course of the work, leading us and our collaborators in Toronto to undertake some rather ambitious experiments. The present stage of the project can be summarized only briefly here.

When nonradioactive purines or pyrimidines, or their nucleosides or nucleotides, are put in competition with the conversion of labeled bacterial DNA to viral DNA, a decided but incomplete suppression of the transfer of thymine is observed in response to thymidine. This effect is illustrated in figure 2. Thymidylic acid also has a slight effect, and other competitors seem to have specific effects, but these are so small that it is difficult to be sure of them.

The competition observed with thymidine operates equally well when the bacterial DNA thymine is specifically labeled, which shows that the observed competition is not directed primarily against non-DNA sources.

These results partly confirm our expectations, but serve chiefly to raise new questions. On the one hand, we find that competition between bacterial DNA and external substrates as precursors of viral DNA can be demonstrated, and wish to contrast this evidence of indirect use with the failure of competition in the use of the parental viral DNA. On the other hand, the observed competition is severely limited, showing that the use of bacterial DNA by the virus does not involve primarily simple intermediates such as free bases, nucleosides, and perhaps nucleotides. This we wish to interpret as an indication of close biological and biochemical relationships between the two kinds of DNA, in spite of their obvious differences. Whether this confusion of ideas will lead any farther remains to be seen.

Both the biochemical and the general interpretation of these results depend on experiments in progress to locate the point of competition by thymidine, presumably some late step in the synthesis of DNA. It may be that this system will prove to be uniquely suited to such questions, the answers to which are essential for either genetical or biochemical thinking about DNA.

Synthesis of Viral Protein

Our immediate aim in studying protein synthesis in infected bacteria is to answer the question: Is protein synthesis part of, or otherwise necessary for, the replication of genetic determinants? As a start we have examined the assimilation of radioactive sulfur, and its incorporation into protein and virus, at various times after infection. The preliminary results already permit some interesting deductions.

As mentioned previously, bacterial protein formed before infection is not available in appreciable amounts for conversion into viral protein. This allows us to focus on proteins formed after infection.

If $S35$ is added to a bacterial culture immediately after infection, its incorporation into acid-insoluble materials occurs at nearly the rate characteristic of uninfected bacteria. This was shown by adding $S35$ to a growing bacterial culture, immediately dividing it into two parts, and infecting one part with T2. Labeled sulfur subsequently incorporated into acid-insoluble materials was then measured at intervals in both cultures. Results will be expressed here in multiples of the sulfur content of one phage particle (2.5×10^{-12} µg). In the uninfected culture, labeled sulfur was assimilated at the initial rate of 10.6 phage equivalents per bacterium per minute, and measured 820 equivalents at the end of one generation (51 minutes). In the infected culture, the rate was initially 8.3 phage equivalents per bacterium per minute, and gradually decreased after 30 minutes. In another experiment the rate for the infected culture was 90 per cent of that for the uninfected. The difference, and especially the decreasing rate after 30 minutes, must be due in part to early lysis of

DEPARTMENT OF GENETICS

some of the cells. We conclude, in confirmation of Cohen's early experiments, that protein synthesis continues without interruption during and after infection.

Figure 4 shows an infected culture analyzed more completely by this method. The acid-insoluble sulfur content per bacterium at the time of infection (measured radiochemically from a parallel culture uniformly labeled with S35) was 1040 phage equivalents. At the observed growth rate of 1.18 generations per hour, this must have been formed at the exponential rate

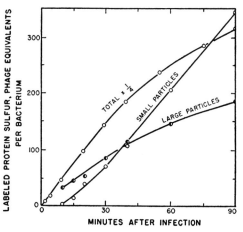

FIG. 4. Incorporation of S35 assimilated after infection into acid-insoluble protein and particulate fractions of lysates. S35 added at time zero. Multiply scale by 4 to read total assimilation.

of 1.35 per cent per minute. Converted to the equivalent linear rate for nonmultiplying cells, this amounts to 14 phage equivalents per bacterium per minute. The observed assimilation was linear during the first 30 minutes, and measured 19 equivalents per bacterium per minute, the excessive rate being due in part to the fact that an appreciable fraction of the sulfate in the medium had already been assimilated before the start of the experiment.

How much of this protein is a precursor of viral protein? Figure 4 shows that much of it is not, especially that portion that is incorporated into large particles.

Moreover, at 30 minutes after infection, labeled protein is being formed at the rate of 19 equivalents per bacterium per minute, whereas labeled protein in small particles is being formed at a rate of 4.8 equivalents per bacterium per minute, showing that only 25 per cent of the total protein being synthesized is going into phage.

Figure 5 shows the relation between the phage titer of lysates prepared at successive times and the labeled protein content of the small-particle fractions isolated from

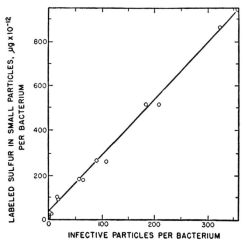

FIG. 5. Relation between phage titer and post-assimilated S35 content of small particles in lysates prepared at successive times after infection.

them, for the experiment just described. Such data justify the conclusion that most of the labeled protein is in phage particles, and this was, in fact, confirmed by tests of adsorption to bacteria. The slope of the straight line in figure 5 corresponds to 2.5×10^{-12} µg of sulfur per phage, in good agreement with other radiochemical analyses.

We have shown that the infected bacterium synthesizes considerably more protein than is incorporated into phage particles. From figure 4 it can be seen that the rate of synthesis of phage protein increases in relation to the rate of total protein synthesis as the infection progresses, an effect suggesting that the cells gradually change over from one kind of protein synthesis to another. The following type of experiment was designed to analyze this phenomenon in greater detail. It also permits an estimate of the amount of viral-precursor protein per infected bacterium.

S35 was added to infected cultures at various times after infection, and the assimilation was stopped 4 or 5 minutes later by adding sufficient neutral ammonium sulfate to reduce the specific activity of the medium 200-fold. The subsequent incorporation of S35 into protein and into phage

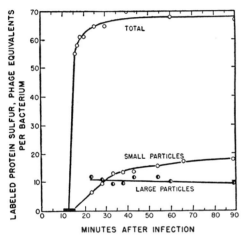

FIG. 6. Incorporation of S35 assimilated between 10 and 15 minutes after infection into acid-insoluble protein and particulate fractions of lysates. Phage growth was linear after 20 minutes, reaching 125 per bacterium at 89 minutes.

particles was then assayed periodically, as already described.

Figure 6 shows one of a series of experiments of this type that are consistent among themselves, in which the first virus particles begin to form at 20 minutes instead of the usual 15. (The slow phage growth was eventually traced to a deficiency of iron in the culture medium.) In this experiment, S35 was assimilated between 10 and 15 minutes after infection. The principal results may be summarized as follows. (1) The total amount of labeled protein synthesized corresponds to 13.6 phage equivalents of sulfur assimilated

per bacterium per minute during the 5-minute period, in close agreement with the rate of synthesis by uninfected bacteria. (2) Most of the labeled protein is formed during the period of assimilation of S35, and the maximum amount is reached within a few minutes afterward. The rate of incorporation is conveniently measured in terms of the interval between the midpoint of the assimilation period and the time when half the maximum labeled protein has been formed—about 2 minutes. (3) The corresponding time for the half-maximum rise in labeled phage protein is 13 minutes. If we assume that this protein was formed with the 2-minute half-time of the total protein, it must have spent about 11 minutes in the cell as phage-precursor protein. (4) Phage-precursor protein receiving its label from the sulfur assimilated between 10 and 15 minutes measures 22 per cent of the total protein similarly labeled.

Other experiments of this series showed that only about 5 per cent of the protein labeled by assimilation between 1 and 5 minutes after infection was phage-precursor protein, and that the proportion increased to 50 or 60 per cent for assimilation between 30 and 35 minutes. Atoms assimilated late persisted in phage-precursor protein for about 8 minutes. Similar experiments in which viral-precursor DNA was labeled with P32 under the same conditions showed that the time spent in DNA precursor was about 14 minutes. These results seem to show that the pool of viral-precursor DNA is nearly twice as large (in terms of phage equivalents) as the pool of viral-precursor protein, and therefore that the DNA that enters a given virus particle is formed earlier, on the average, than the bulk of its protein. The precision of our experiments leaves something to be desired, however, and we propose to seek confirmation of this conclusion by other methods.

The efficiency of conversion of labeled protein into phage as a function of the time of assimilation of S35 has been meas-

DEPARTMENT OF GENETICS

ured with satisfactory accuracy. Table 5 presents the results of a series of simultaneous measurements, made by the method just described except that total labeled protein was measured by a single analysis 30 minutes after the beginning of the interval

TABLE 5

RATE OF INCORPORATION OF S35 INTO TOTAL PROTEIN AND PHAGE PRECURSOR AT VARIOUS TIMES AFTER INFECTION

(S/B/min = phage equivalents of sulfur assimilated per bacterium per minute)

Interval tested (min)	Total protein (S/B/min)	Phage precursor (S/B/min)
Before infection	13.2	..
1–5	14.5	1.9
5–10	12.4	3.1
10–15	9.4	5.0
16–21	9.0	5.3
22–27	8.2	4.6

tested, and labeled phage protein by analysis after an additional 30 minutes. The culture medium was supplemented with 10^{-5} M $FeCl_3$. As was found before, the efficiency rises to a maximum of 53 to 59 per cent as the infection progresses, although the time scale is altered by the more rapid growth of phage. The limit is not imposed by early lysis of cells, because a control culture that assimilated $P32$ between 4 and 8 minutes after infection showed an efficiency of incorporation of labeled DNA into phage of 76 per cent.

In summary, we have preliminary evidence that the synthesis of viral DNA precedes the synthesis of viral protein, and very good evidence that infected cells form at least two classes of protein, one viral-precursor and one not. During the course of the infection the ratio of precursor to total protein being synthesized gradually increases, but never exceeds about 60 per cent. Whether or not the nonprecursor protein is virus specific in some other sense, and what role it may play in viral growth, our experiments do not tell. The fact that no sudden change in rate of synthesis follows infection, however, suggests that the infected cell continues to make most of the proteins it was making before. If so, the failure of infected cells to form typical bacterial enzymes must be regarded as an exceptional circumstance that itself deserves further investigation.

The report summarizes continued investigations on the relationships between intracellular phage DNA and protein and their counterparts in mature virus particles.

The effort to follow up a report that phage in which thymine has been replaced by 5-bromouracil are noninfectious was misguided. It was later shown that even those particles in which substitution is complete are, in fact, infectious (although exposure to fluorescent light commonly found in laboratories can inactivate them).

The radiobiology experiments reported are a fine example of the generally fruitless fun the Phage Group had with radiation-abused phages. Since the work was carried out in total ignorance of the variety of repair systems that operate on radiation-inactivated phages, there is no wonder that it never answered any questions.

GROWTH AND INHERITANCE IN BACTERIOPHAGE

A. D. Hershey, Elizabeth Burgi, Alan Garen, and Norman Melechen

During recent years the work with bacteriophage has been guided by the following questions: (1) Is nucleic acid the sole agent of genetic continuity in bacteriophage T2? (2) Is multiplying T2 a structure of molecular dimensions, or does it resemble the extracellular form of the virus, which measures 65 mμ in least diameter and is bounded by a protein membrane? (3) How can one investigate the

DEPARTMENT OF GENETICS

relation that must exist between structure and function in genetic material? The study of genetic interactions between dissimilar viruses, and between viruses and their host cells, has brought to light a bewildering variety of relations. These include interconversion (of nucleic acid), substitution, modification, incompatibility, synergism, functional coexistence, and, apparently, simple cross contamination. Two genetic principles are serving to bring order out of this confusion. They are structural homology and functional compatibility. Recent discoveries suggest possible ways to bridge the gap between these genetic principles and the chemistry of nucleic acids.

The work summarized below has a direct bearing on all these questions. It was supported in part by a grant (C-2158) from the National Cancer Institute of the Institutes of Health, U. S. Public Health Service, and by a fellowship to Garen from the National Foundation for Infantile Paralysis. We received valuable assistance from Miss Carole Lyons, in charge of laboratory services, and from Mr. Henry Jones, Department photographer, who prepared many radioautographs.

Protein Content of T2

The experiments of Hershey and Chase, previously reported, showed that deoxyribonucleic acid is the principal component of the virus to enter the cell at the start of infection. One way of defining genetic material, therefore, was to analyze the viral nucleic acid for associated substances.

We have recently completed an examination of the viral proteins. We find two components in addition to the membrane protein.

The membrane protein, defined as material sedimentable after osmotic shock, adsorbable to bacteria, and precipitable by antiphage, amounts to about 95 per cent of the total viral protein. The remainder can be separated into fractions soluble and insoluble in trichloroacetic acid.

The acid-insoluble fraction, comprising 3 per cent of the total protein, resembles membrane protein in amino acid composition, but differs from it antigenically. It does not seem to be combined with the nucleic acid.

The acid-soluble fraction, including all material of low molecular weight in the virus, yields on hydrolysis amino acids equivalent to about 1 per cent of the total viral protein. The distribution of amino acids does not suggest a basic protein, although it is different from the distribution characteristic of ghost protein. This fraction has not been examined further.

These results do not exclude a possible genetic function of viral protein, but strongly suggest such a function for the viral nucleic acid.

Effects of Chloramphenicol on Nucleic Acid Synthesis

Melechen has studied the effects of chloramphenicol on the synthesis of nucleic acid in infected bacteria, as another means of assessing the role of protein in viral growth. Chloramphenicol was chosen for the following reasons.

In uninfected bacteria, this antibiotic inhibits protein synthesis, but has little or no effect on ribo- or deoxyribonucleic acid synthesis. In infected bacteria, qualitatively similar results are observed when protein synthesis is inhibited by deprivation of amino acids, or by addition of 5-methyltryptophane or chloramphenicol. These facts suggest that the action of chloramphenicol described below is due to a primary and specific effect on synthesis of protein.

Infected bacteria form protein (measured radiochemically as acid-insoluble sulfur) at the rate of 7 to 12 phage-equivalent units per bacterium per minute. Chloramphenicol reduces this rate to 10 per cent at 10 µg per ml, 5 per cent at 20 µg, and 1 per cent at 100 µg. The inhibition is established within 2 or 3 minutes after addition of the antibiotic, and is independent of time of addition.

Infected bacteria form nucleic acid at a rate that becomes linear about 10 minutes after infection, and measures 2 to 3 phage-equivalent units per bacterium per minute. If chloramphenicol is added to the culture at the time of infection, synthesis of nucleic acid and synthesis of protein are suppressed about equally. If chloramphenicol is added 10 minutes after infection, nucleic acid synthesis is scarcely affected. Additions at intermediate times yield intermediate linear rates.

These findings may be summarized by saying that nucleic acid synthesis is independent of concurrent protein synthesis, but is dependent on prior protein synthesis. Similar conclusions have been reached independently by Burton and by Tomizawa. They suggest several lines of further investigation, one of which is pursued below.

State of Phage-Precursor Nucleic Acid in the Cell

Melechen's results provide a tool for investigating the question, Does or does not the replication of viral nucleic acid occur within a virus-specific membrane?

To answer this question Melechen studied cultures of the following kind. Bacteria were infected (at 0 minutes) with phage. At 7 minutes, 10 μg per ml of chloramphenicol was added. At 45 minutes, the culture was centrifuged to remove chloramphenicol. At 60 minutes, the infected cells were returned to a culture medium that lacked chloramphenicol, causing phage growth to start. Radiochemical analysis of viral growth under these conditions revealed the following.

During the period in chloramphenicol, about 100 phage equivalents of nucleic acid were formed. Nearly all of this was incorporated into phage particles after the removal of chloramphenicol.

During the first 60 minutes, about 60 phage equivalents of protein were formed, mostly before the addition of chloramphenicol. Only 10 equivalents of this protein were incorporated into phage particles after the removal of chloramphenicol.

The first phage particles to be formed after the removal of chloramphenicol each contained about 1 per cent of the labeled viral-precursor nucleic acid. This was true whether the precursor was labeled during its formation or by infecting the cells with previously labeled phage particles. The agreement showed that nucleic acid of parental origin, and all the nucleic acid formed in the presence of chloramphenicol, shared a common pool.

Melechen interprets these results to mean that nucleic acid formed in the presence of chloramphenicol is genuine viral precursor, and that its formation requires neither synthesis of appreciable amounts of viral-precursor protein nor concurrent synthesis of protein of any kind. It follows that viral nucleic acid is not formed inside the membranes in which the finished virus particles are enclosed. It also seems likely that the specific characteristics of a molecule of virus-precursor nucleic acid do not depend on associated protein. This last point, however, calls for confirmation by genetic methods.

Viral-Precursor Protein

A systematic study of viral-precursor protein is being made, chiefly to furnish a proper background for experiments of the kind already mentioned. Detailed results will not be reported here. The chief points to emerge are the following:

1. Antigenically specific viral protein that is precursor to phage particles can be demonstrated in infected cells. The amount and rate of turnover of this protein show that it is a major precursor.

2. The time required for atoms of sulfur to pass from the sulfate in the culture medium into antigenic precursor protein (about 2.5 minutes) is similar to the time required for entry into cellular proteins in general. This shows that there is little or no nonantigenic protein that is precursor to the antigenic protein.

DEPARTMENT OF GENETICS

3. Under the conditions studied, one infected cell contains about 25 phage equivalents of precursor nucleic acid, and 13 equivalents of precursor protein. The inequality is readily explained by Melechen's experiments showing that synthesis of precursor nucleic acid can precede synthesis of precursor protein.

An Unnatural Viral Nucleic Acid

Dunn and Smith reported recently that the thymine analogue 5-bromouracil can be incorporated into the nucleic acid of T2, and that such "substituted" phage particles are noninfectious. Burgi has confirmed their results, and is attempting to locate the point at which viral function is blocked. Her preliminary results suggest that the noninfective particles attach normally to bacteria, and inject their nucleic acid. She plans to investigate the chemical and genetic consequences of this unusual situation.

Genetic Homology between Virus and Host

Many aspects of the interaction between temperate phages and their hosts can best be interpreted in terms of structural homology between the genetic materials of the two organisms. Fraser's pioneer work (Year Book No. 53) extended this concept to the virulent phage T3, and suggested that actual incorporation of host material into virus can occur. Garen, in collaboration with Norton Zinder, has now obtained radiobiological confirmation of similar ideas. The facts are as follows:

1. As measured by inactivation caused by decay of preassimilated P^{32}, or exposure to X-rays, the nucleic acids of T2 and of a temperate *Salmonella* phage are equally radiosensitive.

2. When subjected to ultraviolet irradiation, however, the nucleic acid of the *Salmonella* phage is much more resistant than that of T2.

3. When measured in terms of ability of the irradiated viruses to grow in irradiated bacteria, the sensitivity to ultraviolet light of the nucleic acid of *Salmonella* phage approaches that of T2. The radiosensitive target in the bacterium is about equal to that in the phage.

Garen and Zinder interpret these facts in the following way. Since the nucleic acid of T2 is qualitatively different from that of its host, genetic homology is excluded, and radiation damage produces a lesion that is irreparable by substitution. The nucleic acid of *Salmonella* phage is equally sensitive to primary radiochemical damage, as indicated by points (1) and (3) above. In this phage, however, ultraviolet-damaged nucleic acid can be replaced by undamaged homologous material from the host, as indicated by point (2). For some reason this substitution is not possible after P^{32} decay or X-ray damage.

This interpretation is not intended as the only possible explanation of the facts. Final proof will have to come from genetic experiments. One useful prediction is already possible. According to the interpretation made, a fundamental division among phages should be sought on the basis of nucleic acid structure, not on the basis of temperate or virulent character. The radiobiological results tend to place T2, T4, and T6 in one class and all other known phages in another.

Good bookkeeping on the fate of isotopes that label parental phage particles allowed the Phage Group to let go of the idea that phages contain some DNA that is transferable to offspring and some that is not. The attention to detail that this work required epitomized Hershey's scientific style.

Radiobiological analysis suggested that 40% of the DNA of a virus particle is special and radiosensitive. This proposal seemed to agree with Levinthal's conclusion that phage particles contain a large piece of DNA that is transfered intact to offspring. The concept of a special piece of DNA in T-even phages faded from the scene within the next two or three years, thanks in large part to better methods, perfected by Hershey, for measuring molecular weight of DNA.

In genetically mixed infections, the transfer of radioactive atoms into progeny phage particles was correlated with the transfer of genetic markers. Experiments correlating physical and informational transfer in viral infections became convincing, however, only in 1961, when equilibrium density centrifugation was applied by other groups to the analysis of genetic recombination in phage λ.

GROWTH AND INHERITANCE IN BACTERIOPHAGE

A. D. Hershey, Elizabeth Burgi, Joseph D. Mandell, and Norman E. Melechen

Our major effort during the past year has again been aimed toward an understanding of nucleic acid (DNA) transfer from parental to offspring phage (see Year Book No. 53, 1953–1954, pp. 210–225). Recent work shows that all or much of the transfer is direct, in the sense that it does not involve breakdown below the polynucleotide level, and indeed that some of the intermediates are large, genetically specific fragments. Our experiments are concerned primarily with the functional significance of conservation during transfer. The independent studies of Levinthal and Stent and their collaborators measure the size of the conserved fragments of parental material among individual offspring phage particles. The several modes of attack complement one another in important respects, although a satisfactory interpretation of all the results is not yet possible.

Our work has again been partly supported by a grant (C-2158) from the National Cancer Institute of the National Institutes of Health, U. S. Public Health Service. Mandell is the recipient of a fellowship from the Carnegie Institution of Washington. Isotopes are supplied to us by the Oak Ridge National Laboratory, on allocation from the Atomic Energy Commission.

EFFICIENCY OF TRANSFER

One of the striking facts about the transfer of DNA (labeled with P^{32}, C^{14}, or N^{15}) from parental to offspring phage is its low efficiency. Usually the offspring particles, when recovered as completely as possible, contain only about 40 per cent of the labeled atoms that their parents possessed.

By minor improvements in technique, we have recently increased the efficiency of transfer to about 60 per cent. More important, we believe that we can account for the losses in terms of the following facts.

First, some of the parental particles are noninfective because they fail to inject their DNA into the bacterium to which they attach. The proportion of such noninfective particles varies greatly from preparation to preparation. At best it amounts to 10 to 15 per cent of the particles.

Second, some of the particles inject improperly, with the result that their DNA ends up not in bacterial cells but in the form of acid-soluble materials in the culture medium. The nature of the accident causing this effect is unknown, but such DNA is clearly unavailable for transfer to offspring. The loss does not exceed 5 per cent in properly designed experiments.

Third, some of the infected cells lyse spontaneously about 25 minutes after infection. They release DNA in two forms, in offspring phage particles and in the free state. Neither form contributes to the over-all transfer measurement. They can be separately detected as follows.

The labeled phage particles released by premature lysis of some cells are promptly adsorbed to others. As a consequence, about half the labeled DNA they contain is broken down to acid-soluble material. This loss can be estimated from the sharp rise in acid-soluble P^{32} in the culture observed after 25 minutes. The remaining half of their DNA remains attached to bacterial debris after lysis, and can be measured from the parallel rise in P^{32} content of that fraction. The total amount of parental DNA lost in this way varies with experimental conditions, but usually measures about 15 per cent.

The free DNA released into the culture medium by spontaneous lysis of some cells can be measured in various ways. The simplest way is to observe the difference in acid-soluble P^{32} in parallel cultures with and without added deoxyribonuclease. This measurement accounts for another 10 per cent of the parental DNA that fails to be incorporated into offspring.

The losses mentioned, together with the measured transfer, account for about 90 per cent of the parental DNA. We conclude that, if the transfer in a single productive bacterium infected with a single phage particle could be measured, a very efficient transfer would be observed. The losses encountered in actual measurements are due to trivial accidents that have nothing to do with basic mechanisms. Parental DNA shares common intracellular pools with newly synthesized DNA in infected bacteria. Both are efficient precursors of phage particles as long as they remain in the precursor pools.

This conclusion implies that the losses during transfer from parental to offspring phage are random losses. The same thing is suggested by the following facts. Maaløe and Watson showed that the transfer occurs with constant efficiency during successive cycles of growth, as it must if the losses are random. We found that the purine-pyrimidine composition of the transferred DNA is the same as that of the parental DNA (Year Book No. 53). Furthermore, the composition of the transferred DNA is independent of the presence of competing nucleosides added to the culture medium during transfer, suggesting direct transfer with random losses. We have repeated and extended the earlier experiments along these lines during the current year, with confirmatory results.

It seems clear that the bulk of the parental viral DNA is a potential precursor of offspring DNA. Probably little or none of the transfer occurs by way of intermediates smaller than polynucleotides. At least part of the transfer conserves large, genetically specific structures, as the following summary shows.

TRANSFER OF RADIATION DAMAGE

Kozloff found several years ago that if bacteria are infected simultaneously with P^{32}-labeled, ultraviolet-inactivated phage particles and live unlabeled particles, so that each bacterium is infected with one or more particles of each kind, the resulting phage yield contains P^{32} derived from the irradiated particles. This finding seemed to divorce material transfer from genetic function.

We questioned this conclusion on two grounds. First, it was not clear whether the transfer under these conditions is normally efficient (Year Book No. 53), although we now believe that it is, as Kozloff supposed. Second, it was possible that offspring particles receiving radiation-damaged DNA from their parents might themselves be defective, in which case the genetic inferences would be quite different.

Only recently have we succeeded in obtaining sufficiently precise transfer measurements to decide these questions. The technical improvements, which are minor, need not be described here. The principles of measurement, and the results, are summarized below.

Several people had looked, by more or less direct methods, for noninfective offspring particles from bacteria infected with irradiated virus. Their attempts, like our own, always failed. The failure merely showed that the number of noninfective offspring per irradiated parent is small. It could not exclude the possibility that irradiated DNA is preserved intact from generation to generation in mixedly infected bacteria.

To test this possibility we asked whether the P^{32} derived from irradiated parental particles ends up in infective or in noninfective particles. The techniques that could answer this question were suggested by the properties of the irradiated parents themselves.

If bacteria are infected singly with P^{32}-labeled, ultraviolet-inactivated phage particles, no viral progeny are produced, and of course no P^{32} transfer is recorded. This is why the irradiated particles are said to be noninfective. If, however, bacteria are infected with the labeled, inactive particles plus live, unlabeled particles, the mixed progeny contains P^{32}. This shows that the inactive particles inject their DNA, which participates somehow in the mixed infection. The results of the two transfer measurements, in single versus mixed infection, provide a means of detecting P^{32}-labeled, inactive particles in a population containing active unlabeled particles, no matter how small the fraction of inactive particles.

This test was applied to the offspring from irradiated, P^{32}-labeled and live, unlabeled T2. It readily showed that much of the P^{32} transferred under these conditions was contained in noninfective particles.

The actual proportion depended on two main variables. The effect of exposure to radiation was maximal at about 10 times the average lethal dose as measured by plaque counts. Higher doses neither increased nor diminished the proportion of transferred P^{32} contained in noninfective particles. Varying multiplicity of infection was studied by using a fixed exposure to ultraviolet light, approximately 13 average inactivating doses. The effects observed were not large, but showed a trend of the following kind.

The result was not detectably dependent on number of live, unlabeled phage particles per bacterium, except that the bacteria had to be infected with at least one live particle if transfer was to occur at all. For mixed infections, the proportion of transferred P^{32} contained in inactive offspring particles varied from 40 per cent to 70 per cent as the number of inactivated parents varied from one to ten per bacterium. At the higher ratios, so-called multiplicity reactivation occurred; that is, one could get viable offspring even if the unirradiated parent was omitted. This had no striking effect on the radiochemical result.

The first question to be asked was the following. Are the inactive offspring particles noninfective because they contain DNA derived from the irradiated parent, or because of spreading metabolic effects of the presence of irradiated DNA in the cells? This question was answered by introducing the label from the culture medium, or by infecting with irradiated, unlabeled parents and labeled, unirradiated parents. In either case the P^{32} in dead particles among the offspring amounted to only 0 to 30 per cent of the total P^{32} content as the number of irradiated parents per bacterium varied from one to ten.

The possible interpretations of this result can be illustrated by two alternatives. First, suppose that virus particles contain two kinds of DNA, one photosensitive and one not, as is suggested by the response to varying doses of ultraviolet light. All our results can be explained on the basis of this hypothesis if the photosensitive portion measures about 40 per cent of the transferred DNA and if the remainder is distributed at random between live and dead particles among the offspring. According to this interpretation, the label derived from the live parent and transferred to dead offspring measures the number of dead offspring, which in turn is dependent on the multiplicity of infection with the irradiated parent. The number of dead offspring is determined in addition by the efficiency of transfer and by the physical dispersion among the offspring of the irradiated parental DNA. The cause of death of the offspring is the incorporation into them of a piece of photosensitive parental DNA carrying one or more persistent lesions.

This interpretation is interesting because it suggests a functional differentiation of the viral DNA, presumably into a photosensitive chromosomal part and a photoresistant nonchromosomal part. It also suggests that the manner in which radiation effects are transmitted to offspring particles could provide information about the manner in which parental chromosomes are transmitted.

The second interpretation is less interesting. It can be assumed that a dead offspring particle contains the intact DNA from one irradiated parental particle or none. Dead particles containing irradiated DNA represent parental material that has been entirely excluded from participation in growth. Those not containing irradiated DNA must be produced by an independent mechanism. The live particles containing irradiated DNA reflect complete erasure of the effects of irradiation. This interpretation fails to explain the response to radiation dosage. It has not been formally excluded, however.

Associated Transfer of Parental Atoms and Parental Genes

If one infects bacteria with a mixture of two genetically marked phages, for example the host-range mutant h and non-mutant h^+, the two types reappear among the offspring in the same relative numbers with which the bacteria were infected. In theory, the two types of phage among the offspring can be readily separated from each other, though there are practical difficulties that need not be discussed here. It is feasible, therefore, to measure the exchange of atoms between the two types of phage by using additional isotopic markers. The appropriate experiment may be thought of as an $h \times P^{32}$ cross; bacteria are infected with P^{31} h and P^{32} h^+ parents, and the P^{32} content of h and h^+ offspring is measured.

Garen described a preliminary experiment of this type (Year Book No. 53, p. 214). He found that considerable amounts of P^{32} passed from particles of one genotype to those of another. We have studied this phenomenon more carefully. The results can be summarized as follows.

If bacteria are infected with equal numbers of the two parental phages (5 particles each per bacterium), approximately two-thirds of the conserved isotope is found in offspring particles having the genotype of the labeled parent. The result is independent of the genotype of the radioactive parent.

This result can be interpreted in at least three ways. First, one can suppose that the parental genetic material is fragmented by genetic recombination, in which case only a certain fraction will remain in association with a single genetic marker. Second, one can suppose that the parental nucleic acid is of two kinds, one kind inseparably associated with a single genetic marker, the second transmitted independently of the genetic marker. Third, one can suppose that the transfer involves two mechanisms, one that conserves the original association between atoms and genotype, and one that does not. These three hypotheses make different predictions concerning the experiments described below.

In the following experiment, we test the effect of altering the relative input of the parental phages. Bacteria are infected with an average of 0.5 particle of P^{32} h phage and 10 particles of P^{31} h^+ phage. This has the effect of greatly increasing the fragmentation of the genome of the h parent as tested by conventional genetic experiments. Nevertheless, we find about one-third of the conserved P^{32} atoms associated with the h particles among the offspring. This result tends to exclude genetic recombination as the sole cause of redistribution of label.

In another experiment, we test the effect of repeated cycles of mixed growth. Bacteria are infected with 5 particles each of P^{32} h and P^{31} h^+ phage. Two-thirds of the P^{32} in the offspring is found in particles of the h genotype. The mixed offspring particles are then used as parents during a second cycle of growth, following infection with 10 particles per bacterium. Of the P^{32} conserved among the second-cycle offspring, only 59 per cent is contained in particles of the h genotype. This result tends to exclude the second hypothesis stated above, unless it is assumed further that the two hypothetical classes of nucleic acid are conserved with different efficiencies.

A third experiment, employing two or more genetic markers and permitting analysis of recombinant genotypes for P^{32}, is feasible in theory but we have not yet succeeded in carrying it out.

It should be admitted that the experimental errors in these measurements may preclude any detailed inferences. Only one fact is clear: all the results show that an appreciable amount of parental DNA remains associated with the parental genotype during transfer.

DISCUSSION

Our experiments show clearly that DNA can pass in functionally intact form from parental to offspring phage. The experiments with irradiated parents suggest further that only a part of the parental DNA is conserved in this form, and that the specifically conserved pieces are large in size and few in number relative to the number and DNA content of irradiated parental particles.

The experiments of Stent and his collaborators, employing entirely different methods, also suggest conservation of parental DNA partly in the form of large pieces, and further suggest that these pieces are not subject to progressive fragmentation during viral growth. The function of these pieces is unspecified except that phage particles containing them are subject to inactivation by decay of incorporated radiophosphorus.

Levinthal has introduced a very powerful method of studying these questions. In effect, he can measure the size of a single piece of DNA down to an equivalent molecular weight of about 10 million. His experiments show that phage particles contain large pieces of DNA, and that these are conserved more or less intact among the offspring. These facts raise all the questions discussed in this report. Levinthal's method, radioautography of P^{32}-labeled DNA, supplements other physical methods in an indispensable manner.

It is to be hoped that results obtained by the various methods mentioned above can be fitted into a single scheme. Such a scheme would necessarily say something about possible mechanisms of genetic replication and recombination. For this reason the need for further experiments along current lines is acute.

The demonstration by George Streisinger and Victor Bruce (published years later) that T-even phage have but one genetic linkage group helped focus studies on the physical nature of phage DNA.

The demonstration that the negative charges of phage DNA are typically neutralized by polyamines was foreshadowed by the study of minor proteinaceous components of phage particles. The lack of genetic significance of these components was demonstrated, leading Al to the view, which apparently surprised him, that the phage chromosome was essentially just DNA. This view was strengthened by the demonstration that the DNA formed while protein synthesis was inhibited was equivalent to the DNA in phage particles.

GROWTH AND INHERITANCE IN BACTERIOPHAGE

A. D. Hershey, Elizabeth Burgi, Joseph D. Mandell, and Jun-ichi Tomizawa

Phage T2 behaves in genetic experiments as though each particle contains a single set of linear chromosomes. According to recent experiments by Streisinger, there is in fact only one such chromosome. There is little doubt that this invisible but thoroughly familiar chromosome contains nucleic acid (DNA). In current work we are asking three questions about the chromosomal substance. Is it DNA exclusively? Is all the DNA in phage particles chromosomal DNA? Is the formation of chromosomal DNA in infected bacteria independent of the formation of phage protein?

Somewhat to our surprise, the answer to the first question is probably yes. The other questions are by no means answered, but we, and others, seem to be coming to grips with them.

The work summarized below is partly supported by a grant (C-2158) from the National Cancer Institute of the National Institutes of Health, U. S. Public Health Service. Mandell is a Fellow of the Carnegie Institution of Washington.

Composition of Chromosomes

Phage-precursor DNA can be formed in the presence of chloramphenicol, which almost completely inhibits protein synthesis as measured by incorporation of radiosulfur into acid-insoluble materials. In the course of work with various systems, Melechen found that chloramphenicol seldom permitted a normal rate of DNA synthesis, though no inhibition had been noticed in earlier experiments by him or by others. His observation, perhaps not interesting in itself, prompted a search for particular phage proteins whose synthesis might be poorly suppressed by chloramphenicol and essential to synthesis of DNA, a search otherwise necessary in any event.

As a preliminary to this work, a systematic examination of particles of phage T2 was made in an attempt to identify as many minor components as possible, especially those that might associate with DNA. Only two new components were found: an acid-soluble peptide, composed chiefly of aspartic and glutamic acids and lysine; and "substance A," probably representing one or two unidentified free amino acids. Substance A can be regarded as a single substance in metabolic experiments, and is chiefly characterized by the fact that arginine is a major precursor. The peptide and substance A each contain about 1 per cent of the total carbon of phage particles.

When C^{14} arginine is fed to infected bacteria and the yield of labeled phage particles is isolated, only substance A, among acid-soluble constituents, is labeled. Similarly, from phage particles labeled with C^{14} lysine it is easy to separate the labeled acid-soluble peptide from other labeled substances. These facts have facilitated experiments yielding the following information.

DNA and substance A, but not lysine- or arginine-labeled protein or acid-soluble peptide, are transmitted from parental to offspring phage. Substance A is not a constituent of DNA, however, and its transfer from parental to offspring phage is of no genetic significance. Substance A is readily separable from DNA by dialysis if the phage particles are first disrupted by osmotic shock, and it is efficiently incorporated into phage particles if supplied to infected bacteria as a constituent of the culture medium.

These and other findings show that phage DNA, and hence chromosomal DNA, are not permanently associated with more than about 1 per cent by weight of phage protein, if with any. Since chloramphenicol suppresses formation of total phage protein and of acid-soluble peptide about equally, it seems unlikely that any particular fraction of the protein is spared by the antibiotic. Our results are consistent with earlier work of Hahn, Wisseman, and Hopps in suggesting that chloramphenicol inhibits peptide synthesis generally and specifically.

Autonomy of Synthesis of Chromosomal DNA

In earlier experiments with Melechen we found that phage-precursor nucleic acid (DNA) can be formed in infected bacteria whose capacity to form protein is blocked by chloramphenicol. By adding chloramphenicol 9 minutes after infection and removing it at 60 minutes, for instance, one can obtain phage particles containing almost exclusively the phosphorus of DNA synthesized before the sixtieth minute and protein synthesized after the sixtieth minute. A detailed analysis of this result by kinetic tracer experiments seemed to prove that synthesis of phage DNA and synthesis of phage protein are sequential processes (see Year Book 54, 1954–1955, pp. 216–219). This conclusion evidently calls for genetic tests for the accumulation of chromosomal DNA in the presence of chloramphenicol. Several such tests have been designed and are being applied. Only one of them has so far yielded interpretable results.

Tomizawa finds that if infected bacteria are irradiated with five phage-lethal doses of ultraviolet light at various times during the period of treatment with chloramphenicol, and phage particles are subsequently isolated after removal of the chloramphenicol, a constant number of phage particles is obtained of which a variable fraction is noninfective. The number of noninfective particles is roughly proportional to the amount of DNA in the cells at the time of irradiation. If the amount is large, most of the particles formed are noninfective. The noninfective particles have all the properties, as far as these have been tested, of irradiated phage particles. They attach to bacteria, inject their DNA, and make genetic contributions to the offspring of mixed infection with live particles. These facts strongly suggest that phage-precursor DNA formed in the presence of chloramphenicol is functionally equivalent to the DNA in phage particles.

Fractionation of Phage DNA

A number of people are attempting to fractionate phage DNA by various physical and biological means, in the hope of recognizing functional diversity if it exists. Burgi and Mandell have made appreciable further progress along these lines.

On the physical side, they have modified the methylated serum albumin column of Lerman so that it now yields authentic fractions. So far these fractions have been identified only in terms of their behavior on the column itself. At least three fractions, on repeated test, elute from the column at slightly different concentrations of sodium chloride. The relative amounts of these three fractions are different in whole phage particles and in the P^{32}-labeled DNA transmitted from parental to offspring phage. The transfer seems to enrich the fractions that elute from the column at lower salt concentrations. Unfortunately, this result is incomprehensible in terms of any of the existing ideas about mechanism of transfer. At any rate it cannot be interpreted until something has been learned about the chemical basis of the fractionation and the metabolic origin of the fractions.

Conclusion

The theory that the chromosomes of future phage particles multiply in the form of naked molecules of DNA in infected bacteria has been substantiated.

Preliminary work on the physical fractionation of phage DNA shows that labeled DNA transferred from parental to offspring phage consists of a characteristic fraction. At the moment this result merely suggests that existing ideas about the mechanism of transfer are inadequate.

Al reported analyses of infections in which DNA is permitted to replicate and recombine without concurrent protein synthesis. These analyses further reduced doubts regarding the sufficiency of DNA as the carrier of replicating and recombining genetic information.

Experiments on the effects of UV-irradiation on genetic recombination seemed to yield to a simple interpretation. The thoughts hinted at here were developed in a daunting contribution to the 1958 Cold Spring Harbor Symposium, Volume 23, pages 19–46. If you examine that article, you may wonder how Al could have delivered such a paper to a Symposium audience. He didn't. His oral presentation at the 1958 Symposium was on a different subject.

Koch's experiments on kinetic properties of phage precursor protein synthesis were heroic. Such attempts to understand phage particle assembly became moot when Dick Epstein and Bob Edgar and their colleagues demonstrated the power of conditional lethal mutants for analyzing phage development (Symposium Volume 28, pages 375–394).

GROWTH AND INHERITANCE IN BACTERIOPHAGE

A. D. Hershey, Gebhard Koch, André Kozinsky, Joseph D. Mandell, René Thomas, and Jun-ichi Tomizawa

This report deals with five topics: (1) an attempt to recognize genetic consequences of the synthesis of phage-precursor nucleic acid (DNA) in the presence of chloramphenicol, (2) an attempt to determine whether or not DNA synthesis involves transfer of information to protein or protein-containing substances as an intermediate step, (3) an attempt to separate and identify physically and biologically different fractions of the DNA of phage T2, (4) an attempt to characterize the damages produced by irradiating phage-precursor DNA with ultraviolet light, and (5) an attempt to analyze kinetically the synthesis of phage-precursor proteins in infected bacteria.

Our work is supported in part by a grant (C-2158) from the National Cancer Institute of the National Institutes of Health, U. S. Public Health Service. Thomas is a Fellow of the Rockefeller Foundation, and *chargé de recherches* at the Fonds National de la Recherche Scientifique (Belgium). Tomizawa is on leave from the National Institute of Health of Japan, and is the recipient of a Fulbright Travel Grant.

Relation between Chromosomal Replication and Protein Synthesis in Phage T2

In bacteria infected with phage T2, DNA synthesis proceeds when protein synthesis is stopped, some 5 or 10 minutes after infection, by the addition of chloramphenicol. Under these conditions DNA synthesis continues for about 60 minutes, at which time the cells contain about 130 phage-equivalent units of DNA per bacterium. If chloramphenicol is left in the culture after this time, there is a progressive leakage of cell constituents into the culture fluid, and the cells soon lose the capacity to produce phage when transferred to media lacking chloramphenicol. If the chloramphenicol is removed from the culture 40 to 60 minutes after infection, however, protein synthesis resumes, and phage particles are formed which contain DNA synthesized before and protein synthesized after the removal of the antibiotic. Given these facts, one immediately asks whether or not the phage-precursor DNA that accumulates in the presence of chloramphenicol represents the finished chromosomes of future phage particles.

Tomizawa's experiments have demonstrated that phage-precursor DNA synthesized in the presence of chloramphenicol is subject to damage by ultraviolet light which resembles, qualitatively and quantitatively, the damage produced by irradiation of phage particles themselves (Year Book 56, p. 363). Such damage is unmistakably localized in chromosomes. Considered alone, this evidence shows quite satisfactorily that phage-precursor DNA already contains, at the time of synthesis, the genetic specifications of phage particles formed subsequently. The following experiments are also consistent with that conclusion.

Genetic Recombination in the Presence of Chloramphenicol

It has generally been assumed that genetic replication and genetic recombination are concurrent processes during phage growth. This assumption has never been proved. The following experiments tend to show that it is correct and to confirm our previous conclusion that DNA synthesized in the presence of chloramphenicol is in fact finished genetic material.

At various times, in our laboratory and elsewhere, people have tried to determine whether or not genetic recombinants accumulate in the presence of chloramphenicol when phage crosses are performed with the appropriate antibiotic treatment. Significant effects were never observed, but the interpretation of the experiments was complicated both by the inaccuracy of the genetic methods and by theoretical uncertainty as to how big an effect should be expected. Tomizawa has now obtained significant results by employing in the crosses *rII* mutants of phage T4, kindly supplied by Chase and Doermann. With these mutants the recombination frequencies can be measured by precise selective methods because the wild-type recombinant, but not the mutant phages, can form plaques on certain strains of *Escherichia coli* (Benzer).

The cross $r59 \times r61$ was performed in peptone broth in the usual manner except

recombination frequency than was observed at comparable phage yields obtained without chloramphenicol treatment. These results are reasonably consistent with the hypothesis that recombination frequency is proportional to number of generations of DNA replication in a pool of replicating and mating chromosomes, and that neither replication nor recombination is suppressed by chloramphenicol.

Frequency of genetic recombination can be greatly increased by ultraviolet irradiation of the phages before infection (Jacob and Wollman) or after infection (Burgi). This fact makes possible a somewhat dif-

TABLE 1. Effect of Chloramphenicol on Frequency of Genetic Recombination

	No Chloramphenicol, Lysis at Minute t_2		Chloramphenicol Present from Minute 7 to t_1, Lysis at t_2			
t_2	Phage per Bacterium	Recombination Frequency, %	t_1	t_2	Phage per Bacterium	Recombination Frequency, %
10	0.5	1.0	15	22	5	2.6
12.5	7	1.5	15	25	20	2.6
16	52	2.6	20	27	4	3.0
21	126	3.5	20	32	21	3.0
26	194	4.1	30	39	6	2.9
31	263	4.5	30	49	28	3.2
41	260	4.8	40	53	6	3.6
			40	60	19	3.9

that, to a portion of the culture, 30 μg/ml of chloramphenicol was added at 7 minutes after infection. Samples of the culture without chloramphenicol were lysed at intervals, to measure the normal recombination frequency and its dependence on time of lysis. Samples of the culture containing chloramphenicol were diluted at various times to release the inhibition by the antibiotic; and then samples of these cultures were lysed at appropriate times to permit measurement of recombination frequency. The results are shown in table 1. They reveal a small progressive rise in recombination frequency as the period of exposure to chloramphenicol was increased. Even a short period of exposure to the antibiotic yielded phages showing a higher

ferent test of the hypothesis that genetic recombination can occur in the presence of chloramphenicol. The test, suggested to us by Burgi and Streisinger, can be described as follows.

Bacteria are infected, in a glucose-ammonia medium, with two mutant phages between which genetic recombination normally occurs with low frequency. Chloramphenicol is added to the culture 9 minutes after infection. At 12 minutes after infection, the culture is irradiated with a dose of ultraviolet light (7 phage-lethal hits) sufficient to increase recombination frequency about fivefold in the absence of chloramphenicol. The irradiation does not appreciably affect DNA synthesis or phage growth under these conditions. Now, if genetic replication occurs in the presence of chloramphenicol during the period following irradiation, it can be asked whether the products of replication will consist of chromosomes of the parental genotype only or will include recombinants as well. To determine the answer, chloramphenicol is removed from the culture 60 minutes after infection, phage particles are allowed to form, and recombination frequency among them is scored. The result shows that phage particles produced under these conditions include recombinants whose frequency depends on the dose of ultraviolet light in the same way as that of phage particles produced in cultures not subjected to chloramphenicol. It confirms in a quantitatively satisfactory way that genetic recombination occurs during the chloramphenicol period, and not exclusively after removal of chloramphenicol.

Superinfection Experiments with Phage Lambda

The most straightforward method of measuring genetic replication in phage-infected bacteria makes use of the following principle. If a bacterium is infected at time zero with a genetically marked phage (say h^+), and is then superinfected at various times with a second genetically marked phage (in this case h), the second phage makes a progressively smaller genetic contribution to the eventual phage yield as the time of superinfection is delayed. In the absence of complicating factors, the rise in the ratio of h^+ to h among the phages produced should measure the multiplication of h^+ chromosomes up to the time of superinfection. Such experiments with phage T2 were reported by Visconti and Garen (Year Book 52, p. 221). These investigators noted, however, that quantitative interpretation was difficult owing to the resistance of T2-infected bacteria to superinfection. Hershey, Burgi, and Melechen attempted to apply this method, in experiments with T2, to the demonstration of genetic replica-

tion in the presence of chloramphenicol, but failed to surmount the technical difficulties.

In the meantime, similar methods have been employed by Whitfield and Appleyard, Jacob and Wollman, and others, in experiments with phage lambda. With this phage no resistance to secondary infection develops. Thomas has therefore undertaken a study of the effects of chloramphenicol in this system. He uses virulent derivatives (c mutants) of phage lambda exclusively, making primary infections with h^+ phage and superinfections with h. His results may be summarized as follows.

1. In the absence of chloramphenicol, the genotype ratio h^+/h in the final phage yield increases rapidly as the time of superinfection is delayed. This finding is taken to mean that h^+ chromosomes begin to multiply in the cells soon after infection.

2. If chloramphenicol is added at the time of primary infection, and the cells are superinfected at various times in the presence of chloramphenicol, which is then removed to permit phage particles to form, the ratio h^+/h remains constant and equal to that found by simultaneous infection with the two phages in the absence of chloramphenicol. This observation is interpreted to mean that infection in the presence of chloramphenicol is not followed by multiplication of phage chromosomes. Such a result was to be expected since, with phage T2, DNA is not synthesized under these conditions.

3. If chloramphenicol is added some minutes after primary infection, and the cells are then superinfected, the ratio h^+/h remains constant independently of the time of removal of chloramphenicol, and is close to the ratio found by superinfection at the same time in the absence of chloramphenicol. This result is taken to mean that, if chromosomes multiply in the presence of chloramphenicol under these conditions, the primary infecting phage, which prepared the scene for multiplication, and

the superinfecting phage, which is not permitted to direct protein synthesis, participate equally in the multiplication.

4. If chloramphenicol is added some minutes after infection, and the cells are then superinfected *at various times,* chloramphenicol being removed at still later times, the ratio h^+/h increases about twofold after the addition of chloramphenicol, and then remains constant. This result seems to show that only a limited multiplication of chromosomes occurs after the addition of chloramphenicol to cultures infected with phage lambda. Thomas does not yet have information about the effects of chloramphenicol on DNA synthesis in this system, with which the genetic results can be compared.

It should be added that the experiment described in item 3 can also be performed satisfactorily with T2, and yields an identical result, as Burgi and Melechen discovered. The results of the other experiments cited above when performed with T2 cannot be interpreted, owing to partial exclusion of the superinfecting phage.

Item 3 is the focus of interest in these experiments, because it suggests that, with either phage lambda or T2, no information transfer involving protein synthesis is necessary as a preliminary to genetic replication. With phage lambda, the result is not yet quite satisfying because chloramphenicol seems to limit rather severely the rate of genetic replication (item 4). With phage T2 there is independent evidence of genetic replication in the presence of chloramphenicol, but the superinfection method is in general technically unsatisfactory with this phage. Therefore our results tend to support the idea of autonomous replication of DNA but fall considerably short of an elegant demonstration.

It is worth recalling here the principal findings that suggested the idea of information transfer as a prerequisite to DNA synthesis. When bacteria are infected with phage and are then exposed to radiation (ultraviolet light or decay of assimilated radiophosphorus) at various times, the phage-producing capacity of the infected bacteria proves to be sensitive to radiation damage during the first few minutes of viral growth, but soon becomes resistant (Luria and Latarjet; Stent). Chloramphenicol prevents the progress of the stabilizing reactions even under conditions in which DNA synthesis proceeds (Tomizawa and Sunakawa; Stent). In explanation of these facts, both Stent and Tomizawa suggested transfer of information to non-DNA sites of DNA synthesis by a process dependent on protein synthesis. Our results do not confirm their hypothesis, but neither do they suggest an alternative explanation of the radiobiological findings.

Fractionation of T2 DNA

A particle of phage T2 contains an amount of DNA equivalent to an aggregate molecular weight of about 100 million. Various physical measurements suggest that this represents 5 to 20 molecules of DNA. Genetic experiments show that the phage particle contains only one chromosome. How is the relation between these two numbers to be understood? Some smaller phages contain much less DNA (Sinsheimer, Tessman). Does this mean that the chromosome of T2 is a more complex structure than the chromosome of these smaller phages, or does T2 contain both chromosomal and nonchromosomal DNA? We are exploring the second alternative, which is suggested by the following facts: (1) The DNA in a particle of T2 (analyzed by autoradiography of P^{32}-labeled material) seems to consist of a single large piece and several smaller ones (Levinthal; C. Thomas). (2) The DNA of T2 can be separated into two chemically distinct fractions by adsorption to and fractional elution from a basic protein (Brown and Martin). (3) In experiments in which phage-precursor DNA is irradiated with ultraviolet light and subsequently analyzed by biological methods for the presence of radiochemical damages, evidence can be obtained that the effective damages are localized in about 40 per cent of the irradiated DNA (Tomizawa). Up to now, however, it has not been possible to correlate or interpret these and other facts satisfactorily.

In order to pursue these questions further, Mandell is developing an additional method for fractionating DNA. His method is very similar to that of Brown and Martin except that he uses methylated

More recently, Mandell has investigated the dependence of observed properties of T2 DNA on methods of preparation. He has compared DNA liberated from phage particles by osmotic shock, DNA prepared by the phenol method (Gierer and Schramm; Kirby), and DNA prepared by the chloroform method. As tested by behavior on the fractionating column, the first two methods yield a very similar product, which is quite different from DNA subjected to chloroform treatment. This difference is illustrated in figure 1, which

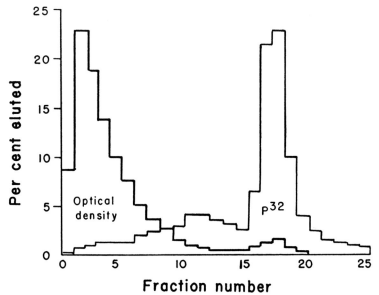

Fig. 1. Separation of P^{32}-labeled, untreated DNA from unlabeled, chloroform-treated DNA by elution with increasing salt concentrations from a column of methylated serum albumin. Concentrations of NaCl range from 0.7 M to 0.8 M.

serum albumin supported on a column of diatomaceous earth (Lerman) instead of a natural basic protein. In earlier experiments Mandell and Burgi found that valid fractions could be obtained in this way, and that the DNA of T2, for example, could be separated from that of T4; but they did not obtain any evidence of a biologically interesting fractionation of the DNA of T2. In this work they used preparations of DNA that had been deproteinized in the usual way by shaking with chloroform and octanol.

shows the elution diagram of a mixture containing a small amount of P^{32}-labeled DNA, prepared from T2 by osmotic shock, and a relatively large amount of chloroform-treated DNA. It will be seen that the two are almost completely separable. Nothing has been learned yet about the chemistry of the separation.

Effect of Ultraviolet Light on Genetic Recombination

Evidently, much of our recent work depends for its interpretation on an understanding of the action of ultraviolet light on DNA. Owing chiefly to the work of Doermann and his students, it is virtually certain that the effects of ultraviolet light on phage particles are due, at least in part, to localized damage to phage chromosomes. Our results further illustrate the importance of this general conclusion. Therefore we have been impelled to look in other directions for additional clues to the nature of the genetic damage.

According to one hypothesis, radiation damages produce local blocks to the replication of DNA. This hypothesis provides a ready explanation of two phenomena: "multiplicity reactivation" of irradiated phage particles in bacteria infected with two or more such particles, and the increased recombination frequency that accompanies multiplicity reactivation in crosses between irradiated phages. The hypothesis is interesting also because it seems to call for a copy-choice mechanism of recombination, as opposed to a mechanism involving breakage and reunion of finished chromosomes.

We have therefore tested the following model for the effects of ultraviolet light on recombination frequency in phage crosses. First, we assume that only radiation damage to the chromosomal region lying between or near the markers can affect recombination frequency. Second, we assume that multiplicity reactivation occurs by a copy-choice mechanism that selects undamaged chromosomal segments for replication; damage points force "switches" from one source of information to another during chromosomal replication. Third, we assume that each switch of information source involves a new random choice of partners (as opposed to switches confined to two members of a pair). These assumptions predict the following relation:

$$L = \frac{nx}{\log\left[(B - R_0)/(B - R)\right]}$$

where L is the length of the phage chromosome, x is the length of the chromosome segment in which radiation damages force recombination between a given pair of markers, n is the number of radiation damages per chromosome, R_0 is the recombination frequency between the specified markers in the absence of irradiation, R is the dose-dependent recombination frequency after irradiation, and B (0.43 in our experiments) is the recombination frequency at genetic equilibrium. If it is assumed that the same damages effective in the inactivation of phage particles are responsible for the increased recombination frequency, all the quantities entering into the equation can be measured independently.

We have tested this theory in several crosses with T2 and T4, with considerable success, as indicated in figure 2. Needless to say, the success can scarcely be taken as strong support of all the assumptions of our theory. The results nevertheless encourage further exploration of the stated lines of thought about mechanisms of genetic recombination and effects of radiation. In particular, they suggest a "group mating" model for genetic recombination that is susceptible of independent test (Bresch; Steinberg and Stahl).

Phage-Precursor Proteins

Work in several laboratories has shown that bacteria infected with phage T2 contain, besides phage particles, several structures morphologically and serologically related to the protein coat of the particles. The status of these materials as precursors or by-products of phage growth has never been clarified. Tracer methods applicable to this question have been in the course of development in our laboratory for several years (Year Book 53, p. 210). Koch has now improved and applied them.

In his experiments he labels phage-precursor proteins with radioactive sulfur (S^{35}), and expresses his results in terms of a phage-equivalent unit of protein, which in these experiments amounts to 1.5×10^{-12} μg of sulfur, measured as S^{35} insoluble in trichloroacetic acid after appropriate fractionation of lysates of phage-infected bacterial cultures.

Two constituents of the lysates can be directly assayed by radiochemical methods: mature phage particles, isolated by fractional centrifugation, and "surplus antigen," precipitated from supernatant fluids by specific antiserum after removal of the phage particles. Koch finds that surplus antigen starts to form in the cells a few

and leaves it somewhat less rapidly afterward. The maximum labeling of surplus antigen is observed about 30 seconds after the end of the pulse. During the next 15 minutes, S^{35} contained in surplus antigen falls to a low minimum value concomitantly with the incorporation of S^{35} into phage particles.

A balance of S^{35}-labeled protein in all fractions of the lysates shows, however, that only about 60 per cent of the phage-

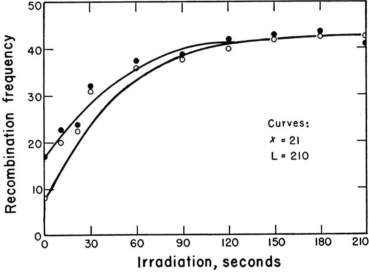

Fig. 2. Dependence of recombination frequency on dose of ultraviolet light given to infected bacteria 5 minutes after infection. Open circles: premature lysates yielding about 1 phage particle per bacterium. Filled circles: spontaneous lysates yielding about 100 particles per bacterium.

minutes before phage particles appear, and soon reaches a level of about 20 units per bacterium. This amount remains constant from the 20th to the 70th minute after infection, and then tends to increase somewhat. During the same time-interval the number of intrabacterial phage particles increases continuously, numbering about 120 per bacterium by the 60th minute.

In experiments in which a "pulse" of S^{35} is fed to the culture, for instance between the 10th and 16th minutes after infection, the phage-precursor status of the surplus antigen is readily demonstrated. It is found that S^{35} enters surplus antigen very rapidly during the feeding period,

precursor protein is precipitated by antiphage serum. One can thus speak of "antigenic" and "nonantigenic" proteins, both of which include phage precursors in the sense that S^{35} flows out of them simultaneously with the incorporation of equal amounts of S^{35} into phage particles. In this sense, 87 per cent of the surplus antigen is phage precursor, and 50 per cent of the nonantigenic water-soluble protein in the lysates is also phage precursor.

Once S^{35} is incorporated into phage particles, all of it, of course, is precipitated by antiserum. One might expect that the incorporation of precursor proteins into precipitable structures would be a stepwise process, more or less completed for a given phage particle in advance of the moment at which the particle achieves its finished, infective form. If this were so, the kinetic experiments should show clearly that S^{35} passes successively through a pool of non-antigenic precursor into a pool of antigenic precursor and finally into phage particles. There is in fact a suggestion of this in the experiments described, but the bulk of the nonantigenic phage precursor seems to enter phage particles as a terminal step. This strongly suggests that phage particles contain two types of protein subunits, each comprising a major fraction of the total phage protein. It may be possible to verify this suggestion by direct analytical methods.

The demonstration of protein in phage heads that is injected with the DNA generated no second thoughts regarding the view that the genetic material of the phage is DNA. One such protein (the product of gene 2) is now recognized as a "cap" that protects injected DNA from attack by the *E. coli* enzyme exonuclease V.

The vision of a sugar coat on DNA that does not interfere with DNA functions broadened our view of nature.

The demonstration, using improved methods of column chromatography, that T2 DNA molecules are uniform in size (and glucose content) was a milestone on the journey to demonstrating that each phage particle has just one DNA duplex, corresponding to its one genetic linkage map.

GROWTH AND INHERITANCE IN BACTERIOPHAGE

A. D. Hershey, Carlo Cocito, and Teiichi Minagawa

The guiding motive of our work continues to be the hope of learning something about the molecular organization of the chromosome of phage T2, which is clearly a structure composed of DNA. We believe that our experiments exclude the possibility that phage proteins play any direct role in the stabilization of this structure. In connection with other questions about structure, we are still preoccupied with technical problems that may or may not prove relevant.

The work discussed below was supported in part by a grant (C-2158) from the National Institutes of Health, U. S. Public Health Service.

Some Properties of the Internal Protein of Phage T2

Particles of phage T2 contain an internal protein fraction amounting to 5 to 7 per cent of the total phage protein. The fraction is defined as acid-insoluble protein that is free from materials serologically related to the surface proteins of the phage particles. It resembles whole phage particles in amino acid composition, but contains more lysine. To isolate such protein we subject phage particles to osmotic shock, centrifuge the viscous solution in 0.15 M NaCl for 5 hours at 38,000 rpm to throw down phage membranes and DNA, and precipitate the protein from the supernatant fluid with trichloroacetic acid (or simply dialyze the supernatant solution). Dr. Lawrence Levine and Dr. Helen Van Vunakis have kindly verified for us that the principal, though not the only, component of this fraction is the internal antigen discovered by them. The protein is conveniently assayed when labeled with radiosulfur.

Minagawa has established the following facts about the internal protein.

1. The internal protein that is eventually incorporated into phage particles begins to form immediately after infection of bacteria with T2.

2. The protein is not synthesized in the presence of chloramphenicol.

3. The internal protein of the infecting phage particles is not reincorporated into the offspring particles.

4. The internal protein synthesized during the first few minutes after infection is incorporated into the first few phage particles subsequently formed, independently of the amount of phage-precursor DNA in the cells (which can be made large or small by treatment of the culture with chloramphenicol).

Items 2, 3, and 4 show that the internal protein is not a part of the phage chromosome essential to the preservation of its specific structure during replication. Item 1 is interesting because it shows that the 6-minute delay after infection before phage-head protein starts to form (Year Book 57, p. 385) is peculiar to that protein. The mechanisms that control the timing of synthesis of different proteins may be amenable to future study in phage-infected bacteria.

Conservation of DNA Glucose during Replication

Cocito finds that the glucose and adenine contained in the DNA of labeled phage particles are reincorporated with equal efficiency (about 50 per cent) into the offspring particles. The glucose must lie in one of the grooves of the DNA helix, and is absent from the DNA of other organisms. One might have guessed that the glucose would have to be removed in the course of replication. Cocito's result, however, shows that the glucose does not introduce any steric hindrance to essential DNA functions.

Homogeneity of T2 DNA

Last year (Year Book 57, p. 382) we discussed the possibility that the DNA of T2 consists of chromosomal and nonchromosomal fractions differing in physical properties. We have by no means abandoned this hypothesis. We now believe, however, that *isolated* DNA is homogeneous in physical properties—a conclusion reached after further work with a column of basic protein capable of fractionating DNA.

We found that a major technical problem arises in this kind of work owing to the strong tendency of DNA to cause channeling in the column. To avoid this we now form our columns of two layers, an upper one consisting of a mixture of inert and DNA-binding materials, and a lower one containing most of the basic protein. Thus the DNA is fed gradually into the bottom layer, where sharpening of the band occurs without mechanical distortion. To avoid breakage of DNA during the preparation (Year Book 57, p. 383), we extract phage particles at a concentration of 10^{12} per ml by gentle shaking with an equal volume of water-saturated phenol.

DNA prepared in this way from T2 proves to be almost perfectly homogeneous, as is indicated by the fact that samples taken from the leading or trailing edges of the elution band are not separable when rerun on the column. The meaning of this result depends, of course, on the sensitivity of our method for discriminating between different types of DNA. Two kinds of evidence show that it is extremely sensitive. First, the DNA of T2 is readily separated from the DNA of T4 on our column, presumably owing to the slight difference in glucose content between the two nucleic acids. Second, we can separate DNA molecules of different chain lengths, artificially prepared by stirring at different speeds. The fragments resulting from single breaks not only can be separated from intact molecules but also can be fractionated into several classes of different length.

T2 DNA isolated as described above therefore appears to consist of molecules that are identical with respect to glucose content and chain length. It is possible that the isolation procedures themselves break bonds. If so, they must be weak bonds of a special type, holding together nearly identical subunits.

Classes of phage proteins were discriminated according to the time during infection that they were synthesized. Al succinctly recognized that "Identification of the mechanism controlling the time of expression of different DNA functions may be considered a central problem of biological development."

Pursuant to an understanding of that mechanism, preliminary analyses of RNA made during infection were compatible with the view that "Synthesis of a protein must be preceded by the synthesis of one or more ribonucleic acids of corresponding specificity." In 1961, Brenner, Jacob, and Meselson demonstrated messenger RNA by analysis of phage-infected cells.

Continued work with chromatography on columns of basic protein confirmed the ability of such columns to distinguish DNA molecules of slightly different molecular weights and led to the proposal, later rejected, that each T2 particle contained *two* DNA molecules, of equal size.

Following the onset of infection, the ability of an infected cell to produce progeny phage particles becomes progressively radiation-resistant. Experiments that monitored this effect were consistent with the view that the increasing resistance measures primarily increase in amount of phage DNA in the cell. The advent of conditional lethal mutants for T-even phages soon ended these indirect methods for studying phage development.

GROWTH AND INHERITANCE IN BACTERIOPHAGE

A. D. Hershey, Elizabeth Burgi, Carlo Cocito, Laura Ingraham, Edward H. Simon,
and Teiichi Minagawa

According to views widely held at the present time, the sole carrier of hereditary information in most organisms is deoxyribonucleic acid (DNA). If so, the species character is a manifold message written in at least two codes. One code is never broken, but serves to ensure that DNA molecules are reproduced without change. Other codes are translated into action through irreversible processes by which the varied potentialities of the species are expressed. In protein synthesis, for example, the translator is thought to be ribonucleic acid (RNA), which, guided by a certain sequence of bases in DNA, responds by directing a certain sequence of amino acids in protein.

At present, actual knowledge of this scheme goes little beyond the generalities of the preceding paragraph. The scheme itself is not merely verbal, however; it suggests numerous ways in which it might be invalidated. For some tests of the scheme, experimental study of phage infection is particularly appropriate. When a bacterium is infected with a particle of bacteriophage T2, the consequences of the literal substitution of one DNA for another in an otherwise nearly constant metabolic system can be observed.

The goal of our work is sufficiently indicated by the foregoing remarks. In the long run, however, no scheme can be tested short of learning the details of its operation. This is evidently a task for generations of biologists. We report below what we hope is perceptible progress.

Our work is supported in part by a grant (C-2158) from the National Institutes of Health, U. S. Public Health Service. Simon holds a postdoctoral fellowship from the National Institutes of Health.

Metabolic Categories of Phage-Specific Protein

Several phage-specific proteins have been recognized in bacteria infected with T2: an enzyme that hydroxymethylates deoxycytidilic acid; a lysozyme operative in bacterial lysis; the phage-coat proteins; an internal protein fraction found in phage particles; and a major constituent of this fraction, called internal antigen, identifiable by serologic tests. These proteins have been studied by various investigators,

step in phage growth. No clue to its function has been found, however.

Proteins of class 2, exemplified by the phage-coat materials, do not start to form until several minutes after infection, at about the time synthesis of DNA begins. They are not formed in cells infected with irradiated phage particles. Irradiation of cells some minutes after infection, however, especially cells in which phage-precursor DNA has been allowed to accumulate in the presence of chloramphenicol, does not prevent subsequent synthesis of class 2 proteins. Evidently the synthesis of these proteins depends on a DNA func-

TABLE 1. Phage-Specific Proteins in T2-Infected Bacteria

Protein	Synthesis Begins, min.	Phage-Precursor?	Transfer from Parental to Offspring Phage?	Synthesis Initiated by Irradiated Phage?
Class 1				
Internal protein	<2	Yes	No	?
Internal antigen	<2	?	?	Yes
Hydroxymethylating enzyme	<2	No	. . .	Yes
Class 2				
Coat proteins	7	Yes	No	No
Lysozyme	10	?	?	No

necessarily by different methods, but some degree of correlation is now possible. An attempt at classification is presented in table 1.

The table reveals two metabolic classes of phage-specific proteins synthesized in infected bacteria. Proteins of class 1, exemplified by the internal protein, are synthesized without delay after infection, and can be synthesized in bacteria "infected" with phage particles killed by ultraviolet light, under conditions in which nucleic acid synthesis fails. Presumably the synthesis of these proteins is initiated by the DNA of the infecting phage particles. The biological reason for early synthesis is obvious in the case of the hydroxymethylating enzyme, which is needed before DNA synthesis can start. Perhaps the internal protein also is prerequisite to some early

tion that is not itself sensitive to ultraviolet light but can be performed only after the cell contains relatively large amounts of DNA. Either cooperative action among DNA molecules is necessary for measurable class 2 protein synthesis or, more likely, these proteins are synthesized during a particular developmental stage, which itself depends on one or more prior events. With respect to the coat proteins, it is reasonable to assume that the critical stage would be the transition from the extended to the condensed configuration of DNA accompanying the formation of phage particles. Minagawa's recent finding that the phage lysozyme also begins to form late is not intelligible under this hypothesis, however. Identification of the mechanism controlling the time of expression of different DNA functions may be considered a central problem of biological development. Concerning it, one must tolerate ignorance and cultivate receptivity to ideas.

Analysis of Intrabacterial RNA

It is now recognized that bacteria infected with phage T2 synthesize RNA, though little is known about the rates or kinds. Volkin and Astrachan obtained evidence suggesting that the RNA formed after infection differs from the whole intrabacterial RNA. In general, this might mean that the RNA formed is characteristic of the infection, or that the infection merely alters the proportions of the several types of RNA formed in uninfected bacteria. With these and related questions in mind, Cocito has explored the possibility of applying to them chromatographic methods originally developed in this laboratory for analysis of deoxyribonucleic acids. His preliminary results may be summarized as follows.

Chromatography, employing a fractionating column in which the active principle is basic protein, separates two classes of RNA from either infected or uninfected bacteria. They correspond to the "soluble" RNA and "ribosomal" RNA previously identified in uninfected bacteria. So far there is no indication that ribonucleic acids synthesized after infection can be separated by chromatography from their counterparts extracted from uninfected bacteria. In infected bacteria, both types of RNA are formed, apparently about as fast as they are formed in uninfected bacteria, but do not accumulate because they are rapidly destroyed again. Ribosomal RNA in infected bacteria (identified by chromatography) differs metabolically from ribosomal RNA (actually in ribosomal particles) in uninfected bacteria. Only the former exchanges phosphorus atoms rapidly with other intrabacterial constituents.

Evidently these results do not contradict the idea that synthesis of a protein must be preceded by the synthesis of one or more ribonucleic acids of corresponding specificity.

The DNA in Phage Particles

Last year we reported the extraction of DNA from phage particles in a form that proved to be chromatographically homogeneous. By applying the same methods of analysis (chromatography on a column of basic protein), Burgi and Hershey have now shown that the column is capable of resolving DNA, artificially rendered heterogeneous by stirring, into numerous molecular-weight classes. Half molecules and quarter molecules can be identified and isolated without resort to molecular-weight measurements, because stirring a homogeneous collection of DNA molecules at a critical minimum speed breaks most of them near their centers. The majority class can then be purified by chromatographic fractionation. Measurements of sedimentation coefficient show that the column can resolve 10 or more classes between whole molecules ($s=63$) and half molecules ($s=43$), and a similar number in the range between half molecules and quarter molecules ($s=30$). The sedimentation coefficient and intrinsic viscosity of quarter molecules correspond to a molecular weight of about 13 million according to the calibration curve, based on light-scattering measurements, obtained by Doty and his collaborators. If this estimate is correct, a particle of phage T2 contains two molecules of identical molecular weight, 50 million, one or both of which must make up the phage chromosome. Some other types of measurement do not yet yield results consistent with the interpretation stated, and work along these lines is being continued.

Phage-Precursor DNA

Simon has compared the DNA in phage particles with DNA (synthesized in the presence of chloramphenicol and labeled with radiophosphorus) extracted from infected bacteria. He finds that an appreciable part of the phage-precursor DNA is identical in chromatographic properties to that contained in phage particles. Difficulties with the methods of extraction have so far prevented a complete account of the intrabacterial DNA.

DNA Synthesis and Evolution of Radiation Resistance

When bacterial cultures infected with phage T2 are irradiated with ultraviolet light at various times, to measure the resistance to irradiation of the phage-producing ability of the cells, a remarkable increase in resistance is observed during the first 10 minutes or so after infection (Luria and Latarjet). Two of the proposals offered in explanation of this phenomenon will be mentioned here. Genetic information carried initially by DNA may be transferred to radiation-resistant structures other than DNA, for example, protein. Alternatively, DNA synthesis itself, by producing multiple and interacting sources of identical information, may produce a radiation-resistant complex, as it is known to do under other circumstances. The first explanation is inconsistent with the general scheme outlined in the introduction of this report, but some evidence for it has been brought forward in the past. The following results obtained by Simon, however, tend to correlate the evolution of radiation resistance with DNA synthesis.

Under the conditions employed, resistance to ultraviolet light begins to increase within 2 minutes after infection of bacteria with T2 and reaches a maximum at 9 minutes. If chloramphenicol is added to the culture at any time, protein synthesis stops within 30 seconds. What effect does the inhibitor have on the evolution of radiation resistance? Simon finds that, when chloramphenicol is added to a culture, resistance to irradiation continues to increase in qualitatively the same way as in the absence of the inhibitor. The later the addition of chloramphenicol, the higher is the rate of evolution. If chloramphenicol is added 2 minutes after infection, no resistance develops. The effect of the inhibitor on rate of development of resistance is thus very similar to its (previously studied) effect on rate of DNA synthesis, not to its effect on protein synthesis. Simon's results are contrary to those of previously published experiments of the same type, performed under different conditions.

Radiation and Genetic Recombination

A rather different index of the effect of radiation on intrabacterial DNA is gained by measurements of recombination frequency in phage crosses: ultraviolet irradiation greatly increases the frequency of recombination. Simon has inquired how the results of this measurement vary with the time of irradiation of the culture. He finds that a given dose of radiation has the same effect whether applied immediately after infection, at 6 minutes, or at 9 minutes. This finding suggests that the radiation-sensitive target at all these times is DNA, and that the eventual yield of phage particles among which recombination frequency is scored is always the progeny of the irradiated DNA. Simon's results are in agreement with earlier results of the same type obtained by Burgi, but take advantage of the more precise measurements now possible. They are consistent with the idea that the information transfer involved in reproduction is directly from DNA to DNA.

With the column chromatography of DNA under tight control, Al's lab characterized chromosomes from various phages and then expoited those characterizations. These studies were exciting in two regards — they led to reliable determinations of molecular weights of phage chromosomes and, hence, of DNA in general, and they revealed unanticipated differentiation along the length of phage chromosomes. In reading these reports, one gets the feeling that Al has reached his goal — he has seen, and played a leading role in, the establishment of satisfying connections between chemistry and genetics.

For phage λ, different segments were shown to have different average base compositions. Skalka exploited this differentiation to demonstrate that different regions of the λ chromosome were transcribed at different times. When amber (or *sus*) mutations became available, Skalka could demonstrate the genetic control of this differential transcription and determine the segmental location of particular genes.

The ends of the λ chromosome stick to each other and to the ends of other chromosomes. This discovery played a crucial role in our understanding of the λ life cycle and is currently exploited when cosmids are used as gene-cloning vectors.

The discovery of T4 particles carrying DNA molecules of abnormal length raised questions regarding the mechanism of chromosome length determination. Mosig soon made excellent use of these particles in testing ideas regarding genetic recombination and DNA replication.

The chromosome of T5 has fragile regions. Later work in other laboratories showed that these are the result of single-strand nicks at specific sites. T5 mutants that are unable to nick or be nicked seem to be perfectly viable. Oh, well.

T4 DNA extracted from infected cells had about twice the average length of chromosomes extracted from phage particles. These observations supported the proposal of Streisinger and Meselson that T4 chromosomes were circularly permuted and terminally redundant.

Delbrück, the Pope of the Phage Church, was attracted to phage because he saw them as the best material for understanding biological replication, which he identified as the fundamental question of biology. He hoped that this pursuit would uncover new principles of physics. That's one way of saying that Delbrück hoped that the laws of the physical sciences, as we then understood them, would be inadequate to explain the most basic biological phenomena. In these research reports, Al lets us know, with characteristic economy, how he feels about such vitalist thinking. *En passant*, he remarks on the publishing habits of some dear friends:

> Streisinger, Edgar, and Denhardt, in the first of a series of three papers written between 1961 and 1963, and published between 1964 and 1967, predicted faithfully on the basis of genetic experiments the structure of T4 DNA verified by

Thomas and Rubenstein (1964) and McHattie, Ritchie, Thomas, and Richardson (1967). So much for the complementarity principle according to which genetics and chemistry were to provide immiscible information.

And earlier in the same report, Al footnotes:

Stent cites Schrödinger and Bohr as prophets of "other laws of physics" presumed to reside in the hereditary substance. The curious reader should be able to find a textbook of physiological chemistry, probably out of date when I read it around 1930, whose author expressed in his preface the opinion that atoms composing living matter would turn out to express unique attributes.

These pot shots tell us that Al is now satisfied that secure connections have been made between the physical and the life sciences.

GROWTH AND INHERITANCE IN BACTERIOPHAGE

A. D. Hershey, Elizabeth Burgi, H. J. Cairns, Fred Frankel, and Laura Ingraham

Our efforts are directed at present toward an elucidation of the gross structure of DNA molecules isolated from phage particles. We hope that methods may be developed suitable for testing the hypothesis of collinearity between the molecular structure of DNA and the linkage map of the chromosome. The results reported below show that the phages provide favorable materials for this purpose. In short, we find that DNA can be extracted from T2 to yield a single molecule per phage particle. The molecular weight is 130 million, and the molecules measure about 50 μ in length, in agreement with expectations for a double helical structure. Other phages also appear to contain a single molecule of DNA, in keeping with the genetic results calling for a single chromosome.

Dr. Cairns, on leave from the Australian National University, joined our group for a year as a fellow of the U. S. Public Health Service. Dr. Frankel holds a post-doctoral fellowship from the same agency, which also contributes a grant (C-2158) in support of our research.

The DNA of Phage T2

Last year we reported that DNA could be extracted from particles of phage T2 in the form of uniform molecules of high molecular weight, and that the molecules could be broken successively into halves and quarters by subjection to critical rates of hydrodynamic shear. In collaboration with Dr. I. Rubenstein and Dr. C. A. Thomas, Jr., of Johns Hopkins University, we have now measured the molecular weight of these materials.

The measurements were made by autoradiography of DNA labeled with radiophosphorus of high specific activity (Levinthal). When labeled phage particles or DNA molecules are imbedded in a "nuclear" emulsion, the emitted β particles produce tracks which, radiating from a point source, give rise to characteristic star-like figures. A count of the average number of rays per star measures the number of radioactive atoms per particle, which, if the specific radioactivity of labeling is known, is simply related to the mass of DNA in the particles. In our experiments, the specific radioactivity of the phosphorus fed to the phage-bacterial culture was measured directly, and then checked in the phage particles themselves by observation of their rate of "suicide."

The molecular weight of isolated T2 DNA, calculated for the sodium salt, proved to be 130 million, and the phage particles were shown to contain a like amount of DNA per particle. Therefore, all the DNA in a particle is contained in a single piece. This conclusion was reached simultaneously with our work by Davison, Levinthal, and their colleagues.

It would be good to know something about the organization of the piece of DNA in T2. Since it shows a relatively small sedimentation coefficient (63 svedbergs, as compared with 750 to 1000 for phage particles with only twice the mass) and a high intrinsic viscosity (250 dl/g), it is an extended structure, as is also indicated by its extreme sensitivity to mechanical breakage. The molecules are rapidly broken in two at stirring speeds of about 500 rpm in ordinary laboratory stirrers. Moreover, they do not exhibit preferential breakage points, but can be broken anywhere at a stirring speed inversely proportional to the 1.4 power of the length of the resulting fragments. The molecules are stable on heating, however, approximately as would be expected for DNA of conventional structure.

What is the actual length of the molecules? Cairns succeeded in obtaining fine-resolution autoradiographs of DNA molecules labeled with tritiated thymine. In spite of many difficulties in getting the molecules properly stretched out, he could recognize a characteristic length of about 52 μ. This agrees with the length (44 to 62 μ, depending on water content) expected for a molecular weight of 130 million according to the accepted double helical structure. An autograph of the labeled molecules is reproduced in plate 1.

Michael Beer at Johns Hopkins University has developed a method for visualizing long molecules in the electron microscope. Obtaining statistically significant lengths in this way is very difficult, but our quarter molecules, at least, show in his hands about the proper length-to-mass ratio for a two-stranded helical structure.

From the evidence cited above and from previously known facts, we conclude that the particle of T2 contains a single molecule of DNA whose molecular weight is 130 million (± 15 per cent or so), and that the molecule can be described as a two-stranded polynucleotide helix. If the molecule is composed of subunits joined by bonds of unknown kind, these are at least as strong as the phosphodiester bonds.

Physicochemical Measurements of Molecular Weight

Apart from their biological interest, the phages provide unique materials for the study of nucleic acids, for several reasons, the first being the ease of extraction of the DNA. Existing methods for the isolation of DNA call for stirring or shaking, which can be done without breaking the DNA only when the concentration of DNA in the solution is high, a fact that we have signalized by the term self-protection. On the other hand, deproteinization without loss of DNA requires a solution of low protein content. Owing to the high DNA-to-protein ratio of the particles, the phages meet the practical requirements for successful extraction of DNA very well.

A second advantage lies in the diversity of phage species available, each containing its unique DNA. Furthermore, if we may anticipate the generalization that all phages contain a single molecule of nucleic acid, the molecular weight of the nucleic acid can be determined independently of other methods by measuring the DNA content per particle. We stress this point because, in the past, molecular weights of DNA could be measured only by methods largely insusceptible of check. The high molecular weight of the phage DNA's, which at first seemed an obstacle because the conventional methods of measurement were not applicable to them, now appears to be a boon.

Finally, the DNA molecules from T2 are structurally and functionally uniform, and the same is likely to be true of other phages but not, as far as is known, of DNA's from other sources.

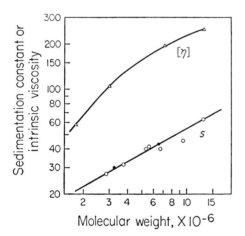

Fig. 1. Relation of sedimentation coefficient, *s*, and intrinsic viscosity, [η], to molecular weight for T2 DNA and its fragments. Open circles, autoradiographically measured molecular weights. Solid circles, relative molecular weights deduced from the kinetics of breakage under shear. Triangles, molecular weights measured from sedimentation coefficients.

In figure 1 we show the relation between molecular weight and physical properties of T2 DNA and its breakage products. The sedimentation coefficient (*s*) proves to be a useful index of molecular weight and is convenient to measure. It is related to molecular weight (*M*) according to

$$s = 0.0022M^{0.55} \qquad (1)$$

In effect, the coefficient in this equation was measured independently by molecular autoradiography and from the DNA content per phage particle. The exponent was measured independently by autoradiography and by analysis of the kinetics of breakage.

According to equation 1, molecular weights of DNA reported in the past are too low. For example, the historically important DNA from the pneumococcus, exhibiting a sedimentation coefficient of 17 to 24 S, has been assigned a molecular weight of about 4 million, on the basis of light-scattering measurements by several investigators. According to equation 1 it should have a molecular weight of 12 to 22 million.

It is not our intention to assert that the light-scattering method yields erroneous results. Even if equation 1 is correct for T2 DNA, its applicability to other DNA's remains to be shown. Also, equation 1 is derived from data covering only the range of 63 to 27 S. Our results merely indicate that the light-scattering measurements should be questioned until they have been checked by some of the alternative methods now available.

The DNA of Phage T5

Phage T5 is smaller than T2 and contains somewhat less DNA per plaque-forming particle. The isolated DNA (*s* = 48.6) has a molecular weight of 81 million if equation 1 is applicable. Fragments of half length, isolated from the products of single breaks per molecule, exhibit *s* = 33, indicating a half-molecular weight of 40 million. When the DNA's from phages T2 and T5 are stirred at a series of high speeds, they yield products showing an identical relation between sedimentation coefficient and speed of stirring. Finally, the maximum speed of stirring that just fails to break the DNA of T5 rapidly breaks the longer molecules of T2 DNA but does not break fragments of T2 DNA of *s* about 50 or less. These results indicate that equation 1 is applicable to the DNA's of both T2 and T5, and also that the two DNA's are equally sensitive per unit length to breakage under hydrodynamic shear. We believe that these findings indicate a close structural similarity between the two DNA's, apart from the difference in molecular weight.

The DNA of Phage λ

Some years ago, in collaboration with Dr. George Streisinger, we began a study of phage λ in the hope of obtaining some clue to the molecular basis for genetic deletions, of which interesting examples had been demonstrated by the work of Weigle, Arber, and others. The only immediate result of our work was the discovery that the DNA of phage λ shows two striking differences from the DNA's of several other phages so far examined. First, λ DNA fails to form a band on passage through our fractionating column, but instead streaks badly and produces multiple peaks. Second, the DNA (prepared by any of three methods) exhibits an extremely diffuse boundary in the ultracentrifuge. Both anomalies disappear when the DNA is broken by stirring. These results are unexplained, but suggest the possibility of molecular aggregation.

Burgi has found that heating dilute solutions of λ DNA at 75°C for 30 minutes, or even at 37°C for several hours, modifies part of the DNA so that it bands properly on the column and forms a sharp boundary in the centrifuge. The sedimentation coefficient is reduced from 40 to 36 S by this treatment, which suggests that the modified DNA has a molecular weight of about 46 million, in good agreement with that estimated from the DNA content per infective phage particle.

Kellenberger, Zichini, and Weigle isolated a mutant phage, called λb_2, selected for its decreased buoyant density in cesium chloride solutions. The decrease in density accompanying the b_2 mutation, if due solely to a reduction in DNA content per particle, calls for a reduction of 17 per cent.

Burgi finds that a mixture of the DNA's from these two phages, subjected to heat treatment as described above, can be resolved on our fractionating column. The DNA from phage λb_2 elutes at the lower salt concentration, a result consistent with the hypothesis that the mutation is accompanied by a reduction in molecular weight of the DNA. Direct support for this hypothesis is still lacking, however.

Related Work in Other Laboratories

The first indication that a phage particle might contain a single molecule of DNA was obtained by Sinsheimer with the small-particle phage φX-174. He found that the phage particles contain 25.5 per cent DNA, weigh 6.2×10^6 daltons (measured by light scattering), and yield DNA with a molecular weight of 1.6 million (also measured by light scattering). The DNA itself is unique in being composed of only one polynucleotide chain. The only other small-particle phage so far examined contains RNA, very likely a single molecule having a molecular weight of about 1 million (Loeb and Zinder). These two phages can be regarded as special cases, both because of the types of nucleic acids they contain and because of their small size. In fact, the suggestion has been made several times that all natural nucleic acids have a molecular weight of about a million, in which event the phages mentioned above would be the first members of a series in which particle size was directly related to number of molecules.

Our results for phages T2, T5, and λ suggest, on the contrary, that all phages contain a single molecule of nucleic acid, with molecular weights ranging upward to at least 130 million, according to species. This conclusion modifies several problems in phage genetics.

In the first place, it is now clear that the genetic properties of the phage particle must reflect exclusively the intramolecular structure of its DNA, and of all its DNA. The idea frequently suggested in the past, by some of our work as well as that of others, that phages contain nonchromosomal as well as chromosomal DNA, can be abandoned. Likewise, the possibility, which seemed very likely on the grounds of comparative biology, that the phage chromosome is a hyperstructure composed of several DNA molecules can be rejected. (Our own work on DNA structure was intended to resolve these questions in quite a different way; we meant to ask whether the several DNA molecules per phage particle were alike or different.) A perennial genetic problem can now be presented in new guise: why does phage T2 require a molecule of 180 thousand nucleotide pairs to carry out roughly the same functions for which phage φX-174 employs a polynucleotide composed of only 4.5 thousand residues? A plausible if not quite satisfying explanation would be simply that T2 carries a complement of some 180 genes or gene vestiges, whereas historical accident has eliminated all but four or five from φX-174. It is evidently imperative to learn something about the functional complexity of the smallest viruses. T2 undoubtedly does employ a multiplicity of genes. It is also imperative to look for regions in DNA molecules that may have functions not detectable by the usual methods of genetic analysis.

A new question about DNA structure has been raised by the finding (Streisinger) that the genetic map of phage T4, like that of certain bacteria, is circular. According to our results, the DNA molecules in phage particles are not circular: ring-shaped molecules should not break into halves at critical rates of shear, and they should be identifiable microscopically. Streisinger and Stahl point out that the circular map can be interpreted in two ways: in terms, that is, of circularity existing at the time when genetic exchanges occur, or of circularity existing only at some other stage in the developmental cycle. In the second case, but not necessarily in the first, the transition from cir-

cular to linear shape would have to occur by a randomly placed cut. The necessity arises, therefore, of determining whether or not the ends of DNA molecules correspond to fixed positions on the linkage map, which is an interesting question quite apart from the idea of circularity.

It was shown some years ago, in our laboratory and elsewhere, that the DNA originating in a single particle of phage T2 is dispersed during growth into numerous rather large pieces among the phage progeny. As long as the DNA complement was thought to include several molecules, it seemed hopeless to try to separate molecular dispersion from various types of fragmentation that might be involved.

A new approach became possible with the development by Meselson, Stahl, and Vinograd of the cesium chloride density gradient method for separation of DNA's differing in density. By the application of this method Roller found that parental T2 DNA labeled with heavy isotopes is fragmented during growth into pieces so small that their incorporation into progeny DNA molecules has little effect on density. This finding is in contrast to the result seen in *Escherichia coli* and other cells, in which progeny DNA molecules (if they contain isotope at all) tend to retain a density midway between that of the labeled parents and that of the unlabeled offspring. The results with T2, and other evidence, were variously interpreted as suggesting a unique mode of replication in viruses, or as bringing into question all evidence supporting the mechanism of replication proposed by Watson and Crick.

These matters have been largely cleared up by Kozinski and by Roller, who now find that, if the progeny DNA descending from heavy parents of T2 or T4 is mechanically broken into fragments, all the labeled fragments show the correct hybrid density. Moreover, if the mass-to-length ratio of T2 DNA fragments as measured by us and by Beer is accepted, it seems clear that the conserved fragments are segments of single polynucleotide chains. These results add considerably to the evidence that the two strands of the helix really separate during replication.

The replication of T2 DNA therefore involves two separate processes: strand separation and reunion, seen in all organisms so far examined; and fragmentation and re-formation of the strands, to a considerable extent unique in T2. Since T2 is also unique in showing a very high genetic recombination frequency, it is natural to relate recombination to fragmentation. At any rate, if we are to believe anything about the genetic role of DNA, we must believe that fragmentation and re-formation of polynucleotide chains provides a mechanism of genetic recombination.

The considerations outlined above led Meselson and Weigle and their colleagues to turn to other phages in which the frequency of genetic recombination is lower. In experiments with phage T7, Meselson observed semiconservative replication not only of DNA molecules but also at the level of the whole phage particle. That is, when the progeny of a cross between heavy (isotope-labeled) and light phages was analyzed by density-gradient centrifugation, phage particles of two densities were found, namely, those whose DNA was virtually free of isotope, and those containing only DNA molecules of hybrid density. Since the DNA of T7 exhibits a sedimentation coefficient of 29, which was supposed to indicate a molecular weight of 17 million, or half the DNA content of the phage particle, Meselson was forced to postulate a bimolecular chromosome that is somehow subject to the semiconservative principle of replication understandable only for single DNA molecules.

According to our equation 1, the molecular weight of T7 DNA is 32 million, the phage particle contains a chromosome composed of one such molecule, and Meselson's result is understandable. Thus, in T7, replication of DNA molecules is accompanied by strand separation and reunion without

the necessity for fragmentation of strands. Fragmentation does occur as an accessory phenomenon, but it is much less frequent than in T2.

Meselson and Weigle also studied crosses with phage λ by heavy-isotope labeling and density-gradient separation of the intact phage progeny. They could discern, as in phage T7, semiconservative replication on which was superimposed considerable fragmentation of strands. Moreover, some of the genetic recombinants exhibited a density intermediate between the hybrid density and the lowest density, the exact density being correlated with genotype, showing that a strand breaking between the two genetic markers often survived without additional breaks.

Kellenberger, Zichini, and Weigle demonstrated similar principles in beautiful experiments employing natural density markers in phage λ. In their crosses two mutants, characterized by densities denoted here by 2 and 3, gave rise to the lighter double mutant of density 1 and the heavier wild type of density 4. In a cross in which the parental phage of density 2 was labeled with radiophosphorus, most of the radioactivity transferred to the progeny was found in phages of density 2, smaller amounts in the recombinant classes of densities 1 and 4, but very little in the initially unlabeled parental class of density 3. Thus replication in phage λ, like replication in general, involves strand separation and reunion, usually without breakage of polynucleotide chains, which in this phage are about 24 μ long. When a strand is broken, it tends to break only once; and if the break occurs between two markers it results, with a high probability, in genetic recombination.

Analogous experiments performed in our laboratory several years ago with phage T2 yielded no such elegant results, for reasons that are now understandable. In T2 every polynucleotide chain of parental origin is broken several times during the cycle of phage growth.

Our discussion above suggests that breakage and rejoining of strands is the sole mechanism of genetic recombination, and the contrast between results with T2 and with the two other phages so far studied supports this interpretation. Meselson and Weigle point out, however, that their experiments do not rule out additional mechanisms. The phenomena involved in negative interference in phage crosses, in fact, require special explanations and seem to require different mechanisms. If additional mechanisms do exist, it may be anticipated that a third class of phages will be found in which recombination is not accompanied by fragmentation of polynucleotide chains.

It will be seen that the evidence for equating the single chromosome and the single DNA molecule in phages ranges from the results of genetic linkage tests to the measurements of length of DNA molecules, with identification of specific fragments and specific genes seemingly all but realized. The most direct mode of attack on the purely topological aspects of the chromosome remains to be mentioned.

Kaiser and Hogness developed a system in which bacteria can be infected simultaneously with "helper" phage particles and with solutions of DNA extracted from phage λ. The contribution from the DNA solution is recognized by appropriate genetic markers. They found that, when multiple markers were used, the contribution was all or none. Thus they could conclude either that the phage particle contains a single DNA molecule or that one of several is multipotent. The corresponding experiments in which the genetic characters of bacterial cells are modified by "transformation" with DNA solutions have always given the contrary result, sometimes (incorrectly) described as indicating a single gene per DNA molecule. It will be interesting to see how the results of linkage tests with bacterial markers are modified when precautions are taken to prepare the transforming DNA with a minimum of mechanical breakage of the molecules.

Summary

The work from our own and several other laboratories discussed above establishes the following principles:

1. Phage particles may contain DNA of the usual type, single-stranded DNA, or RNA. Most of the species so far examined contain conventional DNA.

2. In T2, T7, and λ (containing conventional DNA), the chromosome is single and is composed of a single molecule of DNA, ranging in molecular weight from 30 to 130 million, depending on the species. In T2 and T5 DNA, at least, the length-to-mass ratio is close to that demanded by the Watson-Crick two-stranded helical structure.

3. Replication of DNA molecules is accompanied by separation and reunion of strands.

4. Fragmentation and rejoining of individual strands occurs as a separate phenomenon not essential to replication. It may be frequent, as in T2, or infrequent, as in λ. In λ it appears that the two-stranded chromosome can break and rejoin without separation of strands.

5. Fragmentation and rejoining of strands provides a mechanism of genetic recombination. In phage λ, preliminary indications can be seen that individual genetic markers are disposed along the DNA molecule in accordance with their positions on the genetic map.

6. In phage T2, the frequency of strand cleavage is high in two senses: each parental strand is broken during replication, and it is broken in several or many places. In phage λ, the frequency is low in both senses.

7. In phage T2, the high frequency of breakage is correlated with a high frequency of genetic recombination. In phage λ, both frequencies are low. Nevertheless, additional mechanisms of recombination not involving breakage of strands cannot be excluded at present and should, perhaps, be anticipated.

GROWTH AND INHERITANCE IN BACTERIOPHAGE

A. D. Hershey, Elizabeth Burgi, Fred Frankel, Edward Goldberg, and Laura Ingraham

Several methods applicable to the characterization of DNA molecules have been developed in recent years. Examples are the optical analysis of thermal denaturation, chromatographic analysis, measurement of fragility, measurement of buoyant density, and specific enzymatic tests. Such methods make it possible to distinguish DNA's from different sources and to detect alterations produced experimentally. They do not, however, yield direct information about molecular structure. Structure is a more or less plausible inference that serves to unify diverse measurements, as did, for example, the notably successful model of Watson and Crick.

These remarks sufficiently explain why our work, though directed toward rather specific biological goals, is for the moment devoted to the exploration of physical techniques.

The Molecular Weight of T5 DNA

The above generalities are illustrated in experiments of the following type, by which we arrive at an estimate of the molecular weight of the DNA of T5 and, more important, evaluate a novel technique.

As a preliminary, a sample of DNA isolated from phage T2 is stirred under conditions that produce single, clean, transverse breaks near the centers of molecular length. The DNA is labeled with radiophosphorus, because the tracer permits analysis of extremely dilute solutions (less than 0.1 μg of DNA per ml) in which molecular interactions can be neglected. The stirred solution now consists of fragments of DNA molecules ranging in length from about $\frac{1}{3}$ to $\frac{2}{3}$ of the length (50 μ) of unbroken T2 DNA molecules. The fragments are next sorted out into length classes by chromatography on a column of methylated bovine serum albumin. The separation is possible because the basic protein in the column acts as an ion exchanger from which the acidic DNA fragments are removed at different salt concentrations, depending on their length. The resulting fractions are listed in table 1, where they are characterized by their sedimentation coefficients and corresponding molecular weights (*Year Book 60*).

The DNA of phage T5 exhibits a sedimentation coefficient of 48.5 S. From table 1 we see that fragments of T2 DNA sedimenting at the same rate have a

TABLE 1. Equal Fragility under Hydrodynamic Shear of T5 DNA and Fragments of T2 DNA of Similar Sedimentation Coefficient

Sample	Chromatographic Interval, %	Sedimentation Coefficient, S	Molecular Weight, $\times 10^{-6}$	Breakage at 630 rpm, %
T2 DNA fragments				
tube 30	10.6	31.8	37	--
tube 31	23.4	36.2	47	--
tube 32	36.6	38.6	55	--
tube 33	49.4	42.4	64	0
tube 34	60.7	44.1	71	15
tube 35	69.8	44.9	73	27
tube 36	77.9	47.2	79	51
tube 37	83.8	48.5	82	60
tube 38	88.5	49.5	86	73
tube 39	92.3	51.0	89	95
T5 DNA	----	48.5	?	65

molecular weight of 82 million. It does not immediately follow that this is the molecular weight of T5 DNA, however, because sedimentation constants depend on molecular shape as well as molecular weight, and are also influenced by molecular interactions that may differ from one DNA to another.

To circumvent the difficulties last mentioned, we apply a very different criterion to the same materials. The last column in table 1 gives the results of fragility tests, in which we measure the fraction of DNA broken when very dilute solutions are stirred for 30 minutes at 630 rpm. From the results with the T2 fragments, it is clear that molecular fragility is a sensitive index of molecular weight. T5 DNA exhibits a fragility corresponding to that of T2 fragments of molecular weight lying between 82 and 86 million. In other words, T5 DNA matches practically the same fragments of T2 DNA either in terms of sedimentation coefficient or in terms of fragility. The agreement permits the following conclusions.

1. A single relation between molecular weight and sedimentation velocity applies to both T2 DNA and T5 DNA.

2. A single relation between molecular weight and fragility under hydrodynamic shear applies to both DNA's.

3. The bonds broken by stirring must be of equal strength in both DNA's.

4. The molecular weight of T5 DNA is about 84 million, as compared with 130 million for the DNA of T2.

5. Since T2 DNA shows about the proper mass per unit length for a double helical structure, the DNA of T5 must also have this structure.

6. Since particles of phage T2 contain a single molecule of DNA, the same must be true of T5 particles, which are somewhat smaller.

These conclusions are interrelated in such a way that all must be correct or all or most of them incorrect. That they are correct follows as a plausible inference from the data presented, although no

single measurement forces this conclusion. The alternative can be rejected on the basis that it would require an improbable set of coincidences.

Finally, independent checks of any of the stated conclusions reinforce them all. Such checks will be reported in another publication.

The DNA of Phage Lambda

Comparisons of the sort illustrated in table 1 can also bring to light DNA's having unusual properties. The DNA of phage lambda is an example. It is unusual in the following respects.

1. It does not emerge from our fractionating column in a single band, but trails over a wide range of salt concentrations and fails to elute completely.

2. It shows a broad range of denaturation temperatures, similar to that of the bacterial DNA's and in contrast to the exceedingly narrow range characteristic of other phage DNA's. A comparison with the DNA of phage T1, chosen for its similar composition and molecular size, is made in figure 1. The flatter curve for

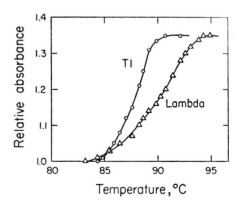

Fig. 1. Thermal denaturation of DNA from phages T1 and lambda.

lambda DNA probably reflects the character of the individual molecules, though the possibility that the DNA is composed of a mixture of species with different melting temperatures has not been excluded.

3. Lambda DNA forms a broad or double boundary in the optical ultracentrifuge. Zone centrifugation according to methods developed by Britten and Roberts at the Department of Terrestrial Magnetism confirms that even in very dilute solutions the DNA exists in two or more differently sedimenting forms.

4. On stirring at low speeds, too low to break DNA molecules of the maximum molecular weight (50 million) represented by the DNA content of the phage particle, lambda DNA is converted into a form homogeneous by ultracentrifugal criteria and exhibiting the sedimentation coefficient (32 S) corresponding to the slowest form present in unstirred DNA. The behavior on heating and chromatography is not affected by this treatment.

These properties suggest a diversity of molecular shapes or aggregation products, together with and perhaps related to other structural peculiarities responsible for the unusual response to heating. It is evident that different DNA's do exhibit idiosyncrasies not yet accounted for in structural terms, and it is interesting that an example should be found in phage lambda, itself a biological curiosity in several respects.

A cursory examination of DNA from phages T1 and P22, as well as more thorough studies of T2 and T5, has not revealed comparable unexpected properties. Since lambda is a phage of well known and complex genetic properties, we propose to continue our physical studies in the hope of reaching conclusions of biological pertinence.

The Nature of Self-Protection

Some years ago we noted that the breakage of DNA by stirring depends strongly on the concentration of the solution. In more recent work we find that, as more and more dilute solutions of T2 DNA are subjected to stirring, susceptibility to breakage increases to a maximum at a concentration lying between 0.2 and 0.1 μg/ml. Further dilution is without effect. At 0.2 μg/ml, the molecules are separated by an average distance of about 28 μ, or half the actual length of the molecules. Self-protection is therefore a remarkably sensitive index of molecular interactions.

Such interactions could involve specific intermolecular forces of biological interest. We have looked for evidence of such forces by testing the ability of T2 DNA to protect that of T5, and vice versa. We find that the protection is not specific and therefore does not involve specific intermolecular bonding.

The protective effect greatly depends on the length of the molecules. We suspect that it results simply from their viscous drag, which may reduce the maximum local shear rate in the stirring vessel without, of course, affecting the average rate of shear in the vessel as a whole.

Local Denaturation by Hydrodynamic Shear

For a number of experimental purposes it is desirable to be able to fragment DNA molecules without producing unwanted side effects. In the course of our work with T5 DNA we found, however, that stirring can produce denaturation as well as breakage. In order to understand, and the better to avoid, such denaturation, we studied it in some detail. The principal facts are the following.

1. When T5 DNA is stirred in salt solution at 25°C at speeds just sufficient to initiate breakage of the molecules, subsequent chromatography reveals that a considerable fraction of the DNA has been altered in such a way that it attaches irreversibly to the basic protein in the column.

2. Such denaturation can be demonstrated also by the action of Lehman's phosphodiesterase, a bacterial enzyme that acts on denatured but not native DNA.

3. The extent of denaturation, measured by either of these methods, increases

as the salt concentration at the time of stirring is lowered or the temperature raised. Salt concentration and temperature do not affect breakage, however. At 5°C and 0.6 M NaCl, or at 25°C and 2.6 M NaCl, DNA can be broken with very little denaturation.

4. The speed of stirring is also critical, depending, of course, on the other variables mentioned. At 25° in 0.6 M NaCl, little denaturation is produced at speeds too slow to cause breakage. Remarkably, rapid stirring also fails to produce denaturation, which requires a critical stirring speed just sufficient to break the molecules slowly. At higher temperatures and lower salt concentrations, denaturation can be produced either without breakage, by stirring at low speeds, or in spite of rapid breakage caused by stirring at high speeds.

5. The existence of a critical stirring speed for denaturation is explained by the fact that DNA fragments produced without denaturation by stirring at a low temperature are relatively resistant to denaturation on restirring at a higher temperature.

6. The critical speed at which denaturation occurs varies with DNA concentration in the same way as the speed required to produce breakage: high concentrations protect against both denaturation and breakage.

7. If a sample of DNA is stirred under conditions that produce both partial denaturation and partial breakage, the denaturation can be demonstrated in both broken and unbroken molecules. Similarly, molecules previously denatured without breakage can be shown to exhibit about the same susceptibility to breakage on restirring as their native counterparts. Breakage and denaturation are independent events.

8. The denaturation produced by stirring does not alter the characteristic melting curve of the DNA.

9. When DNA denatured by stirring is restirred at low temperature to reduce the size of the fragments, the fraction that will subsequently pass a column is greatly increased but the susceptibility to phosphodiesterase is not affected.

10. This type of denaturation cannot be repaired by a regime of gentle heating and slow cooling.

11. Denaturation showing all the above characteristics can also be produced by heating DNA in the absence of shear to temperatures (84° to 88°) near the midpoint of the melting curve.

These results can be interpreted in the following way.

When a DNA solution is heated to a characteristic range of temperatures, the weak bonds responsible for maintaining the two-stranded configuration are loosened and local separation of strands is permitted, even though a sufficient number of bonds persists to hold the two chains in apposition. This effect can be measured by the increased absorption of ultraviolet light as the temperature is raised. A major factor in the process is the purely mechanical effect of Brownian motion. When the solution is cooled at this stage, the original structure is regained as far as can be determined by optical measurements. By chromatographic or enzymatic tests, however, local molecular lesions can be shown to persist. If the molecules are now fragmented without further denaturation, the denatured and undenatured regions of individual molecules are separated and the weight fraction of the DNA that can pass a column is increased.

Subjecting the molecules to shearing forces can produce the same result at low temperatures: in effect such forces merely lower the temperature of thermal denaturation, as does decreasing the salt concentration. The stresses generated by shear are of a special character, however, in that they depend not only on the speed of stirring but also on the length of the molecules. They are localized, moreover, near the centers of molecular length. Thus if a collection of fragments of different lengths is subjected to stirring at a given speed, the shortest ones will survive because subjected to little stress, the longest will quickly be broken to fragments that are likewise resistant to denaturation, and only those of a critical length will resist breakage and at the same time undergo the repeated mechanical distortions that eventually produce permanent local denaturation. Only if the original population consists of molecules of uniform length can a majority of them be denatured in this way at low temperatures, and then only at the critical speed of stirring appropriate to that length, namely, the maximum speed that just fails to cause rapid breakage. If the temperature is sufficiently low and the salt concentration sufficiently high, the stresses required for denaturation exceed those required for breakage, and denaturation cannot be observed.

As far as we can surmise, permanent denaturation of this type, whether the consequence of gentle heating or of stirring, must result when locally separated strands rejoin out of register, producing unpaired loops or similar structural irregularities. As can be demonstrated with fully denatured DNA, such structures have a strong affinity for the basic protein of our fractionating columns and are the natural substrate for Lehman's phosphodiesterase.

In short, local denaturation of DNA molecules can be produced by either heating or stirring, the effects of which are identical except in one respect. When hydrodynamic shear is a dominant cause of denaturation, the molecular lesions tend to be central with respect to molecular length. Heating in the absence of shear, on the contrary, tends to produce terminal defects. These differences are readily accounted for theoretically and can be demonstrated experimentally.

The Reversibility of Thermal Denaturation

When a solution of T5 DNA is heated to 100°C for 10 minutes and then quickly cooled in ice water, relatively little denaturation is evident by optical criteria

and then only if the measurement is made at about 45°C. By chromatography or by tests with phosphodiesterase, however, complete denaturation can be demonstrated. Thus denaturation produced at 100°C is irreversible on rapid cooling, but optical criteria are poor measures of it.

As is well known, measurements of the absorption of ultraviolet light serve admirably in following the progress of denaturation during the actual heating (fig. 1). According to this criterion, T5 DNA is half denatured at 85°C. However, if the solution is heated to 85° and then cooled, either rapidly or slowly, little denaturation is detectable by enzymatic tests. What little there is is due to small imperfections in many of the molecules, which prevent them from passing a fractionating column.

Thus denaturation on heating occurs simultaneously in all the molecules but is largely reversible unless a temperature of about 90° is exceeded. The irreversible event occurring at this temperature is not reflected in the optical density measurements; it undoubtedly consists in the unwinding of the polynucleotide chains of the helix, which have already largely separated at lower temperatures.

The behavior described above is presumably characteristic of a molecularly homogeneous DNA and explains a puzzling feature of experiments performed in the past with inhomogeneous DNA preparations. In the earlier experiments it was found that progressive denaturation involved the irreversible "collapse" of individual molecules at different temperatures, indicating that the denaturation of a given molecule was an "all or none" event.

It is now clear that it is not the denaturation measured by optical means (i.e., the loss of regular structure) that constitutes the all-or-none event, but the subsequent separation of chains. And the failure to observe partly denatured molecules after heating to intermediate temperatures and subsequent cooling (by electron microscopy, for example) is due to the fact that partial denaturation of individual molecules is reversible on cooling. Our experiments now show that the recovery is often imperfect, but the resulting local imperfections that we see were not detected by the means previously employed.

It may be added that the reversibility of denaturation of a sample of DNA heated to the middle range of its melting curve should provide a new criterion of molecular homogeneity applicable, for instance, to the DNA of phage lambda.

Replicating DNA of Phage T2

DNA undergoing replication during growth of phage T2 can be studied in two ways: first, by infecting bacteria with isotopically labeled phage particles and then examining the labeled DNA subsequently isolated from the infected cells; second, by labeling DNA synthesized after infection, which is exclusively viral, and examining it. In either method chloramphenicol can be added to prolong the period during which DNA synthesis can be observed without complications due to the re-formation of phage particles.

Experiments of this type reveal at least two forms of DNA that differ from the finished molecules finally incorporated into phage particles. These are as yet poorly characterized, and little can be said about them except that they do not include any appreciable fraction of low-molecular-weight DNA, either single or double stranded.

On the other hand, the cells always contain a considerable fraction of their total DNA in a form indistinguishable from that found in finished phage particles. This finding shows that a mechanism for the preservation and determination of molecular length operates continuously during replication, and not only at some terminal stage in the formation of a phage particle—a conclusion pertinent to several hypotheses concerning genetic mechanisms.

Conclusion

The phage DNA's provide favorable materials for investigation, for three reasons. First, they can be isolated in a molecularly homogeneous state, a circumstance that permits for the first time satisfactory correlations between gross structure and biological function. Second, the synthesis of phage DNA can be studied in infected cells that have proved amenable to metabolic experimentation in the past. Third, intensive genetic study of a few phage species is yielding results to which physical and chemical findings can be constantly referred.

In the present state of knowledge concerning both genetic mechanisms and molecular structure, work of a purely exploratory nature is called for as well as efforts directed toward well defined goals.

SOME IDIOSYNCRASIES OF PHAGE DNA STRUCTURE

A. D. Hershey, Elizabeth Burgi, Fred R. Frankel, Edward Goldberg,
Laura Ingraham, and Gisela Mosig

Our work for the past five years, though often proceeding by indirection, has been guided by specific questions about the biology of phage. What is the structure corresponding to the single chromosome revealed by genetic experiments? What aspects of that structure can account for synapsis, heterozygosis, genetic deletions, and other phenomena familiar to geneticists but obscure to structural chemists? Since the phage chromosome is composed of DNA, our questions called for investi-

481

gation of DNA structure along lines for which, as it turned out, the tools had to be invented.

The first question was given the simplest possible answer by the finding that a particle of phage T2, T5, or lambda contains a single molecule of typical, double-helical DNA, whose length is characteristic of the species (*Year Book 61*). Other questions remain unanswered, but some of the tools are ready, as the following report will suggest.

The work was aided by a grant (CA-02158) from the National Cancer Institute, National Institutes of Health, and by postdoctoral fellowships from the U.S. Public Health Service, for Mosig and Frankel, and from The National Foundation, for Goldberg.

Molecular-Weight Measurements

Burgi and Hershey

A minimum requirement for the study of DNA structure is a means of determining size, shape, and uniformity of the molecules. In the absence of information about either size or shape, the analysis is a formidable task (see *Year Book 60*). Fortunately, most phage DNA's prove to be typical double-helical molecules. For these, length is the principal variable, and is conveniently measured from the rate of sedimentation in a density gradient of sucrose. The general relationship between lengths (L) and distances sedimented (D) of two DNA's proves to be $D_2/D_1 = (L_2/L_1)^{0.35}$. In addition, the pattern formed by the sedimenting bands says a good deal about the homogeneity of the preparation. Several applications of the method are mentioned in this report. Other details are being published in the *Biophysical Journal*.

Genetic Deletions in Phage Lambda

Burgi

Two mutations have been described in phage lambda that were recognized as hereditary changes in the density of the particles, apparently as a result of reduc-

tions in their DNA content (Kellenberger, Zichichi, and Weigle). One, called *b*2, lies in a segment of the DNA molecule that directly interacts with the bacterial chromosome in lysogeny. Another, *b*5, lies in the region controlling the immunity to lytic infection that characterizes bacteria in the lysogenic state.

Burgi has shown that the mutations were accompanied by literal deletions of two different segments of the lambda DNA molecule, measurable because the mutant molecules sediment more slowly, and resist higher rates of shear before breaking, than those of the wild-type phage. (Chromatographic differences support the other measurements.) The double mutant carrying both deletions has lost 20 per cent of the original molecular length. The remaining DNA contains about 48 per cent guanine-plus-cytosine, as compared with 46 per cent in the DNA as a whole. The two deletions therefore occurred in regions containing only 38 per cent of the bases named. Further work along these lines could lead to the assignment of specific phage functions to specific molecular segments.

Cohesive Sites in Lambda DNA

Hershey and Burgi

Analysis of molecular deletions in phage lambda was hampered for a long time by exceptional properties of its DNA. The difficulties were finally traced to the tendency of the DNA to form complexes. The phenomenon proved interesting in its own right, and was studied in some detail.

When freshly prepared by extraction from phage particles at very low concentration (2 μg/ml), the DNA consists of linear molecules about 15 microns long (molecular weight 31 million). These molecules are of typical double-helical structure, as shown by the normal relation between sedimentation rate and fragility under shear, as well as by radioautographic measurements of length (Cairns, *Year Book 60*).

The unusual properties of the DNA are best described in terms of the model needed to explain them. Each molecule carries two cohesive sites, one situated near each end, which have the property of adhering to each other but not to other parts of the molecule or to DNA in general.

At DNA concentrations below 5 μg/ml, only cohesive sites belonging to the same molecule can join, giving rise to folded molecules (presumably rings or hairpins). These structures are best recognized by the fact that they sediment 1.13 times faster than linear molecules in a centrifugal field. The transition from linear to folded molecules is rapid and complete at 60°C in 0.6 M salt solution. It occurs very slowly at low temperature in 0.1 M NaCl. At 75°C, the reverse change occurs, and one gets linear molecules by heating the solution briefly at that temperature and cooling it quickly.

At higher DNA concentrations, cohesive sites belonging to two or three molecules can join to form dimers sedimenting 1.25 times faster and trimers sedimenting 1.43 times faster than linear molecules. According to their sedimentation rates, these must be open, more or less end-to-end structures. Side-by-side aggregates do not form in appreciable numbers.

Aggregation and folding occur under identical conditions, except for the dependence on DNA concentration, and are reversed by the same treatments. Moreover, they are mutually exclusive: folded molecules do not aggregate, nor do aggregates fold. Therefore the same cohesive sites can engage in either process, and one site is satisfied by a single partner.

The several forms of lambda DNA behave differently under shear. Trimers are more fragile than dimers. Folded molecules are much more resistant than either, but can be opened up at a shear rate just insufficient to break the skeletal bonds. Thus linear molecules can be prepared by shearing as well as by heating

and, after either treatment, can be converted back to the alternative forms.

The identification of cohesive sites may furnish a clue to some of the ways in which chromosomes function. We shall try next to locate and analyze molecular segments having adhesive properties.

Intramolecular Heterogeneity
Hershey and Burgi

Systematic analysis of bacterial DNA's showed some years ago that their base compositions are species specific and vary among different species from about 25 to 75 per cent in guanine-plus-cytosine content. Fragments derived from any one species, on the other hand, are relatively uniform in composition. The variations can have little to do with the functions of genes, which must be generally similar in many different bacterial species as well as individually distinct in any one species. The conclusion was reached that all DNA's are subject to some constraint (presumably the coding requirement) tending toward a composition of 50 per cent, and that other evolutionary pressures (presumably part of the process of speciation) produce the deviants. According to these results, the base composition of DNA is a clue to its phylogenetic origin, and the internal homogeneity of the DNA in a chromosome is one measure of its "age," that is, of the time during which it has persisted as an isolated structural entity.

The meager information about the DNA's of higher organisms is consistent with these ideas. Those analyzed show about 50 per cent guanine-plus-cytosine but prove to be heterogeneous, perhaps reflecting the likelihood that individual chromosomes are older than their assortment in a given species. (If so, a sample of animal DNA, fractionated to obtain a uniform intermolecular composition, should prove to be intramolecularly homogeneous as well.)

Lambda DNA is exceptional in the

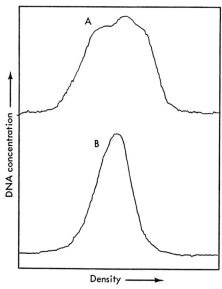

Fig. 1. *A*. Band formed in CsCl by fragments of lambda DNA, molecular weight 5 million. *B*. Band formed in CsCl by fragments derived from the DNA of *Escherichia coli*. The two bands are shown at the same magnification.

respects discussed, as shown in several ways.

1. It melts through a broad range of temperatures (*Year Book 61*), suggesting heterogeneity of composition that is not seen in other phage DNA's.

2. After breakage into 4 to 6 pieces, the molecules yield fragments that are readily separable into chromatographic fractions ranging in guanine-plus-cytosine content from 41 to 51 per cent as measured either by nucleotide analysis or in terms of density.

3. The *b2* deletion, already mentioned, occurred in a segment having only 38 per cent guanine-plus-cytosine.

4. The results cannot be explained in terms of a random assortment of small segments varying continuously in base composition, but call for a few (probably three) internally homogeneous segments. This interpretation is supported by the density distribution of fragments of molecular weight 5 million (still large compared with the dimensions of genes), which reveals a few classes of roughly equal size. By contrast, fragments of the same molecular weight derived from *Escherichia coli* DNA show a density distribution strongly centered about its average (fig. 1).

In view of these results, the similar over-all base compositions of *E. coli* and lambda DNA's, though conforming to a proposed rule for temperate phages and their hosts, appear to be an accident having neither evolutionary nor functional significance. Perhaps lambda phage contains a "young" chromosome of diverse origins. Note that this interpretation by no means excludes the possibility, which we hope to pursue, that the internal heterogeneity of lambda DNA reflects an effective functional differentiation among its segments. A clue pointing in that direction comes from the work of Cowie, which demonstrates for the first time that there are homologous parts in the DNA's of *E. coli* and lambda (see report of the Biophysics Section, Department of Terrestrial Magnetism, this *Year Book*). These like segments in virus and host must have a common origin and related functions, and they deserve study by all available means.

Variation in T4 Phage Particles
Mosig

Mosig had found, in Doermann's laboratory, that she could separate T4 phage particles on the basis of buoyant density into classes of different biological properties. Since joining our group, she has begun an analysis of the DNA content of the particles. In so doing she has turned up materials favorable for studying several different questions.

Most of the phage particles are of average density and contain a single DNA molecule of unit length (about 50 microns, molecular weight 130 million).

Another class, which may be quite numerous among phage progeny produced late after infection, consists of particles of lower than average density, which contain a DNA fragment measuring 0.67 of the unit length. This class is homogeneous, and intermediate lengths of DNA are not found. The particles are individually noninfective, but two or more attaching to a single bacterium can infect successfully. Genetic analysis shows that different particles lack different parts of the phage genome. According to an idea familiar in phage genetics, two particles can infect a bacterium when every gene is represented in one or the other of the particles. When the infection succeeds, viable progeny are produced that contain normal DNA, into which atoms from the parent fragments are incorporated. The preliminary results are consistent with the idea that the fragments represent chromosomal segments of fixed length cut at random with respect to positions on the circular genetic map. The ability of two segments of analyzable structure to cooperate in the production of phage particles with complete genomes provides a novel means for studying mechanisms of replication and genetic recombination. For example, the results so far obtained with genetically marked phages afford evidence for replication in one direction only through the marked region of the chromosome.

Other exceptional particles, denser than average, are formed particularly at early times after infection. These particles, when crossed, produce few recombinants. They contain DNA molecules of unit length and two classes of fragments averaging about 0.05 and 0.2 of a unit long, respectively. The dense particles are also potentially heterozygous, as shown by Doermann and Boehner. A simple interpretation would be that the small fragments represent the redundant region of a partially diploid chromosome. This interpretation is not entirely satisfactory, however, and a more attractive alternative is now being investigated.

All these findings raise interesting questions about the mechanism of determination of length of DNA molecules which, in this material, evidently operates at several lengths. The 0.67-unit fragments, for instance, are apparently determined neither by self-reproduction nor by recognition of specified end points, but by a device that cuts to measure.

Breakage Points in T5 DNA
Hershey and Burgi

When subjected to critical rates of shear, DNA molecules from T2 and lambda phage break near their centers. T5 DNA, on the other hand, often breaks first into a 40 per cent and a 60 per cent fragment, and the larger of these often breaks in turn into a 40 per cent and a 20 per cent fragment. (The weak points are very slightly weaker than other points, so that the over-all fragility is little affected.) We tried to show that the weak points lie at specific locations in the nucleotide sequence. The initial attempts failed because significant differences in density could not be detected among the resulting fragments. Thus molecules of T5 DNA, in contrast to those of phage lambda, are extremely homogeneous internally. This finding is consistent with the narrow range of temperatures through which T5 DNA melts.

Intrabacterial T2 DNA

Frankel and Ingraham

DNA synthesized in the presence of chloramphenicol after infection of bacteria with phage T2 can be recovered by phenol extraction and chromatographic purification in 5 to 25 per cent yield. Such material, when isotopically labeled, is chemically recognizable as phage-specific DNA. No small fragments are found. No DNA molecules of the length (50 microns) found in phage particles are recovered either, owing to the use of the antibiotic to prevent late stages in phage growth. The properties of the material suggest variable structures about 100 microns long. Apparently, therefore, a mechanism exists for joining ends of DNA molecules, presumably in the specific way required for genetic function. Such a mechanism is likely to be related to the circular form of the genetic map. Comparative experiments with a phage characterized by a two-ended map may illuminate that relation.

SOME IDIOSYNCRASIES OF PHAGE DNA STRUCTURE

A. D. Hershey

In the decade since Watson and Crick proposed a double-helical, base-paired polynucleotide structure for DNA, that structure has been confirmed in many ways, and no alternative has been proposed. A unique feature of the model, which its authors were quick to notice, is the built-in redundancy implicit in the pairing rules. One ideal DNA molecule differs from another only by its nucleotide sequence, which therefore defines the information content of the molecule. And since the sequence in one strand dictates the sequence in its complement, the information is carried in duplicate, and either strand, given an appropriate mechanism, can generate the other without gaining information. This scheme of replication has also been amply confirmed.

Given a principle of structural redundancy, probably unique to the nucleic acids, living things could put it to use in several ways: to ensure genetic stability while facilitating replication and repair; to gain some of the advantages of potential heterozygosity, even at the molecular level; and, though it is not clear how, probably to achieve something like molecular synapsis.

Some of the uses of redundancy call for minor modifications of the base-paired, double-helical structure. One of these we found by seeking to understand some unusual physical properties of DNA extracted from phage lambda. It turns out that that phage, in effect by transferring a small segment of one DNA strand from one end of the molecule to the other, created a structure capable of making highly specific end-to-end attachments. This device permits, in principle, the reconstitution of a chromosome from its fragments. What we found is not precisely what we were looking for, namely, a mechanism of synapsis, but something that may in fact prove to be its molecular equivalent.

The group whose joint and individual efforts are summarized in the following paragraphs includes, besides the writer, Elizabeth Burgi, Edward Goldberg, who is a Carnegie Institution Fellow, Laura Ingraham, Nada Ledinko, who is a Special Fellow of the National Institute of General Medical Sciences, U. S. Public Health Service, and Gisela Mosig. The work of the group was aided by grant CA 02158 from the National Cancer Institute, U. S. Public Health Service.

Nucleotide Distribution in Lambda DNA Molecules
Hershey

We first suspected an unusual intramolecular distribution of nucleotides in lambda DNA because of the very broad range of temperatures through which melting occurs. Our suspicion was confirmed when we found that fragments of the molecules are widely divergent in density (*Year Books 61* and *62*). Hogness and Simmons independently found a similar differentiation in DNA molecules extracted from a defective, transducing line of the same phage species. They separated the two molecular halves and identified genes in each one, showing that the left half, so called with reference to the genetic map, is the denser. Work along these lines in our laboratory has been directed toward separation and analysis of additional molecular fragments.

Our method of fractionation is based on three principles. First, DNA fragments of any desired length can be prepared by subjecting the molecules to the appropriate rate of shear in a stirring vessel. Second, lambda DNA molecules have terminal cohesive sites permitting, as we shall show, the end pieces only of fragmented DNA molecules to be joined together. (Methods of joining and disjoining the fragments will be described in connection with Burgi's work in the next section of this report.) The joined ends can then be separated from the remaining unjoined fragments by virtue of the faster sedimentation of the complexes in a density gradient of sucrose. Third, when the isolated end pieces are disjoined and passed through a methylated albumin column, the fractions eluting first contain mostly left ends (rich in guanine plus cytosine) and the fractions eluting last contain mostly right ends (rich in adenine plus thymine). Fractions of the two kinds are then purified by a second induction of pairwise joining followed by centrifugation through a density gradient of sucrose. This time the unjoined fragments, representing the numerical excess of left or right ends achieved by the preceding chromatographic fractionation, are recovered. The purity of such materials, measured in terms of density analysis, we

266 ❖ *Science*

582

80 87657

estimate at 90 to 95 per cent. Otherwise the method is inefficient, and we are trying to improve on it. In the meantime considerable information about lambda DNA structure has been gained through its use.

The unequal densities of the two molecular halves are clearly visible when the appropriately fragmented but unfractionated molecules are centrifuged to equilibrium in a density gradient of cesium chloride. Two bands containing equal amounts of DNA are formed (fig. 1A). To see this result, it is essential to cool the rotor to 5°C or below, or to add formaldehyde to the solution, to prevent spontaneous rejoining of the molecular fragments.

If the two halves are deliberately rejoined before centrifugation, only a

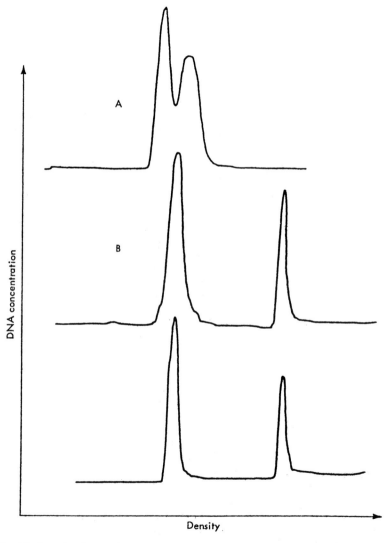

Fig. 1. Equilibrium density-gradient centrifugation of lambda DNA. *A.* DNA stirred to produce half-length fragments. *B.* Same after the fragments are rejoined by heating the solution to 70°C and cooling it slowly. The band at the right is a density marker. *C.* Unstirred DNA; density marker at the right.

single band is seen, which appears at the same position on the density scale as the band formed by the original unbroken molecules (fig. 1B and C). Measurement of sedimentation rates shows further that unbroken molecules and rejoined halves are of equal length. The rejoined structures must consist of two halves joined at the original molecular ends. Thus each half-length fragment carries a single cohesive site, as would be expected from our earlier conclusion that each unbroken molecule carries only two. Furthermore, the two cohesive sites on a single molecule are not equivalent, but are differentiated into left and right kinds, each of which joins specifically to its opposite number. If this were not so, the densities of the rejoined halves would span the range from that characteristic of right halves to that characteristic of left. Figure 1 shows, on the contrary, that the densities of rejoined halves vary only to the moderate extent expected from the fact that the half-length fragments of each kind themselves vary in density because they vary in length.

Direct analysis of the nucleotide composition of isolated half-length fragments yielded 55 per cent guanine-plus-cytosine for the left half and 45 per cent for the right, in excellent agreement with the measured densities of the same preparations. The densities of the different parts of the molecule so far measured, or calculable by difference, are summarized in figure 2.

In collaboration with Dean Cowie, we have made use of our materials to investigate some questions of biological interest. Some of the results are described by him in the report of the Department of Terrestrial Magnetism in this volume. The idea was to combine our methods for the separation of DNA fragments with the technique of Bolton and McCarthy for the recognition of matching nucleotide sequences in DNA. We were able to show that each of the halves of the lambda DNA molecule contains sequences absent from the other, that the DNA of wild-type lambda phage contains sequences missing in a line carrying a genetic deletion, and that the sequences in the lambda DNA molecule that interact with the DNA of the bacterial host are not confined to any one molecular segment but occur in all parts of the molecule that we could test. We also obtained other potentially interesting but still controversial results, which we hope to pursue.

Cohesive Sites in Lambda DNA

Burgi

The cohesive sites in lambda DNA were first identified in our work as small regions, two per molecule, capable of

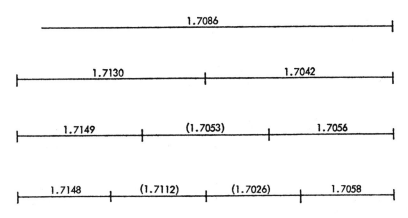

Fig. 2. Density map of lambda DNA. Densities measured with isolated fragments or (in parentheses) by difference.

joining reversibly to one another. When the cohesive sites are disjoined, a preparation of lambda DNA can be recognized for what it basically is, a collection of identical molecules with a characteristic rate of sedimentation. When the cohesive sites are joined, we see, depending on the conditions under which joining occurred, uniform monomers of altered, more compact shape, or a mixture of these with dimers and trimers formed by the joining of two or three molecules more or less end to end. Each complex is identifiable by its rate of sedimentation. Independently of our work, Chandler and Ris found rings and multimolecular threads in the electron microscope, and it is now clear that the cohesive sites lie at either end of the molecule.

Cohesive sites disjoin when a DNA solution is heated briefly to 75°C and cooled in ice water. They join when a solution is heated to 75° and cooled slowly, or is heated for a long time at 45° to 60°. The joining is reversible, subject to a temperature-dependent equilibrium in which the closed state is favored at low temperatures and the open state at high, the net rate of closure being optimal at around 60°C in 0.6 M NaCl. Joining is accelerated by high salt concentrations, and occurs rather rapidly at 25°C in concentrated cesium chloride solutions but very slowly at 5°C.

The conditions required for joining and disjoining cohesive sites recall those producing denaturation and renaturation of DNA, which means, basically, separation and rejoining of complementary polynucleotide chains. The analogy immediately suggests that lambda DNA molecules have single-stranded ends with complementary nucleotide sequences. According to this hypothesis, the rather low temperature at which cohesive sites disjoin (about 55° in 0.15 M salt) signifies that the joined region is short, the local guanine-plus-cytosine content low, or the complementarity inexact.

Two additional facts consistent with the stated hypothesis have already been mentioned. The cohesive sites are complementary in structure in the formal sense that they can join only in left-right pairs. Also, formaldehyde at 1.8 per cent prevents the joining of cohesive sites (at 25°C in concentrated cesium chloride solution) but does not open cohesive sites already joined. If, after action for 24 hours, the formaldehyde is dialyzed out of the solution, the cohesive sites are found still to be functional. The formaldehyde presumably interferes with base pairing by forming anhydrides with amino groups.

Burgi undertook some months ago to test the hypothesis that the cohesive sites in lambda DNA are single-stranded ends of complementary structure. She made use of enzymes having specific action on DNA, notably one isolated from lamb brain by Lawrence Levine that cleaves single-stranded but not double-stranded DNA. A typical experiment with this enzyme proceeds as follows.

A solution of P^{32}-labeled lambda DNA in the form of linear molecules (cohesive sites open) is exposed to the action of the enzyme. After various periods of time, samples of the solution are diluted tenfold in 0.6 M NaCl, heated to 75°C, and allowed to cool slowly. The concentration of DNA in the solution is now only about 1 μg/ml, at which concentration rings but not aggregates can readily form as a result of the closure of cohesive sites.

After the solution has cooled, a small amount of tritiated marker DNA is added, and the mixture is spun through a gradient of sucrose to observe the rate of sedimentation. With an appropriate amount of enzyme, Burgi finds that the P^{32}-labeled DNA taken from the digestion mixture during the first few minutes sediments as rings under the conditions stated, whereas samples taken after 30 minutes or more sediment as linear molecules. Samples taken at intermediate times sediment as a mixture of the two. That no fragmentation of the molecules is seen even after prolonged digestion confirms the specificity of the enzyme.

When the experiment is repeated starting with DNA molecules already in the form of rings, no action of the enzyme can be detected. Thus open cohesive sites (one or both per molecule) are destroyed by the enzyme, but closed sites are resistant, as expected if the interacting structures are complementary, single-stranded polynucleotides. It remains to be proved, however, that the enzyme acts hydrolytically to destroy, and not mechanically to block, the cohesive sites.

Burgi is pursuing these experiments with enzymes of diverse specificity, and hopes with their aid to define the structures of left and right cohesive sites individually.

Another experiment supports the idea that cohesive sites join by pairing of complementary nucleotides. When P^{32}-labeled DNA is heated and cooled slowly in the presence of unlabeled DNA previously fragmented by sonic treatment and denatured by boiling, labeled rings do not form. Evidently some of the denatured fragments can themselves react with the cohesive sites, blocking the end-to-end joining of the molecules. This result shows, at least, that one or both cohesive sites survive sonication and boiling.

If there are single-stranded regions in lambda DNA they are relatively short,

because enzymes that efficiently hydrolyze the DNA after it is denatured liberate only 0.5 per cent of the nucleotides from native lambda DNA, an amount not significantly greater than that liberated from other DNA's not suspected to have exposed single strands.

At present the simplest assumption about the structure of lambda DNA is that illustrated in figure 3. It is consistent with Burgi's results, and with the fact, reported by MacHattie and Thomas, that no discontinuity of structure at the molecular junction in dimers can be seen in the electron microscope.

Infectivity of T4 DNA

Goldberg

The discovery that cells can be genetically modified by artificial exposure to DNA molecules or fragments (Avery, MacLeod, and McCarty) provided a powerful means for correlating genetic properties with physical structure of DNA. The principle has been exploited most successfully, perhaps, by Kaiser and Hogness, who showed that the genetic map in phage lambda is colinear with the molecular structure of its DNA.

Application of the same principle to T4 is urgently needed because of some striking peculiarities of the genetics of

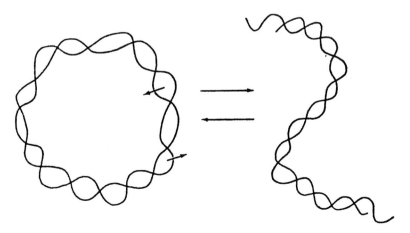

Fig. 3.   Probable structure of closed and open forms of lambda DNA.

that phage. For one thing, the genetic map is circular, whereas the DNA molecules are not. Van de Pol, Veldhuisen, and Cohen have described a system in which genetic transformation brought about by T4 DNA can be detected. The observed transformation, so far limited to a single genetic marker, is inefficient and difficult to reproduce.

With the help of advice from Cohen, Goldberg has succeeded in duplicating the results of the Dutch workers, and has obtained genetic transfer of additional markers, including examples of the general class of temperature-sensitive mutants. He is investigating the method systematically in the hope of being able to apply it in genetic studies, not only in connection with problems peculiar to T4 but also as a means of testing linkage in single DNA strands.

Methylcytosine in Lambda DNA

Ledinko

Some years ago Weigle and Bertani found that phage lambda exhibits different biological properties, depending on the host bacteria in which it is grown. Other phage species show similar host-determined, nonheritable variations. Recently Arber proved that the host alters the phage DNA itself and, indeed, that the phenomenon works both ways since a prophage carried by a bacterium can modify the bacterial DNA.

Host-induced modification of phages reveals a curious mechanism of speciation, in which chromosomal structure is determined in part by physiological means. The classical example is glucosylation in the DNA's of phages T2 and T4, where enzymes controlled presumably by one or two phage genes interacting with metabolic systems in the bacterium modify entire DNA molecules. It is now clear that analogous phenomena are widespread and diverse, suggesting not only a device for establishing genetic barriers but also a possible means for controlling gene action in a single species.

When phage λ is grown on three different strains of *Escherichia coli*, called W3110, C, and K12(P1), three different versions of the phage are produced. Each can infect the host on which it grew, but can infect only certain of the others according to a characteristic pattern. Ledinko examined DNA's from phages of each type in an attempt to determine the nature of the modifications.

DNA's from phage particles grown on W3110 and C did not differ in any of the following characteristics: nucleotide composition of unbroken molecules and of half-length fragments, sedimentation rate, density, fragility under shear, melting temperature, and possession of cohesive sites.

DNA's from phage particles grown on W3110, C, and K12(P1) were also examined for the presence of unusual nucleotides. For this purpose P^{32}-labeled DNA of each kind was hydrolyzed enzymatically, the nucleotide fractions were separated on ion-exchange columns, and individual fractions were rerun on additional columns, together with unlabeled marker nucleotides. One previously unidentified nucleotide, 5-methyldeoxycytidylic acid, was found in this way. It comprised 0.07 to 0.08 per cent of the recovered nucleotides, but did not differ in amount among the three kinds of DNA analyzed.

These results add to the growing evidence that rare bases occur in many nucleic acids, suggesting a general function that remains to be determined. Ledinko's findings do not support (nor do they necessarily disprove) the current notion that differences in methylation of DNA underlie host modification of phage lambda.

Diversity among T4 Phage Particles

Mosig

When a genetically homogeneous population of phage T4 is fractionated with respect to buoyant density in cesium chloride solution, particles showing correlated differences in density, DNA content, and biological properties are found (*Year Book 62*). Study of the several types of particles separated in this way should contribute to an understanding of the relation between chromosome structure and genetic behavior.

The analysis is conducted as follows. A purified stock of phage T4 is prepared, in which the proteins are labeled with radiosulfur and the DNA with radiophosphorus. A suspension of the particles, in a solution brought to density 1.5 g/ml with cesium chloride, is spun to equilibrium in an ultracentrifuge. After the tube is removed from the rotor, a hole is pierced in its bottom and individual drops are collected separately. The successive drops correspond to a series of known densities, owing to the fact that cesium chloride itself sediments to form a characteristic density gradient. Except for minor disturbances, the phage particles recovered from a given drop have a buoyant density equal to the density of the solution of which the drop is composed. The contents of the separate drops are analyzed for radiosulfur, radiophosphorus, and viable phage particles. The phage particles themselves are further characterized by physical and biological methods, and their DNA is extracted for separate examination. A reasonably complete examination of a given class of phage particles isolated in this way calls for numerous experiments, because the yields are small, the materials are unstable, and further purification is usually necessary. The classes so far recognized may be described as follows.

The majority class (class 1). In a typical cesium chloride run, about 96 per cent of the viable phage particles are recovered in a single band falling between 1.485 and 1.515 g/ml on the density scale. (The density at the center of the band is nominal, and the width of the band depends somewhat on the concentration of phage.) Different drops collected within these limits show a constant ratio of plaque count to radiosulfur (a measure of

the viability of the particles) and of radiophosphorus to radiosulfur (a measure of the DNA content of the particles or, more strictly, of the DNA:protein ratio). Particles of class 1 are therefore reasonably homogeneous in these respects, and serve as the reference with which exceptional particles may be compared. In practice, reference particles are taken from a narrow region in the center of the band.

Density 1.516 (class 2). Particles taken from the edge of the band corresponding to the higher solution density, chosen to contain about 2 per cent of the viable population in a typical lysate, do not differ outspokenly from the majority in DNA:protein ratio or in viability. They do differ in density, as can be seen when they are recentrifuged in mixture with suitably labeled reference particles. They contain, in fact, distinctive DNA, as shown so far only by chromatographic methods. (See, however, description of class 3.)

Particles of class 2 are more numerous in cultures lysed early after infection than in cultures lysing spontaneously; that is to say, the average density of the majority of the particles is higher in early lysates.

Particles of class 2, when propagated as a single clone, yield progeny with the normal density distribution.

Density 1.484 (class 3). On primary fractionation in cesium chloride, we recover from the edge of the band corresponding to the lower solution density a mixture of particles belonging to two distinct classes, which separate completely from each other on centrifugation in cesium chloride for a second time. The denser of these particles we call class 3; the less dense, class 5.

Particles of class 3 comprise only about 2 per cent of the viable population, more in late lysates than in early ones. They do not differ appreciably from the reference class in DNA:protein ratio or in viability.

DNA extracted from particles of class 3 is often compared with that from particles of class 2 rather than with that from reference particles. The molecules from class 3 particles sediment more slowly, and elute earlier from a column of methylated albumin, than molecules from class 2. The differences are clear but small. Together with the density difference of the particles, they suggest a difference in molecular weight of DNA of about 1 per cent.

Perhaps the simplest interpretation of classes 1, 2, and 3 is that the majority of phage particles contain DNA molecules varying continuously in length over a very narrow range compatible with viability. That the variable is length is strongly supported by the facts stated, but is not certain, particularly since DNA extracted from particles of class 3 is unusually prone to breakage during extraction, which suggests a structural difference unrelated to length.

Particles of class 3, propagated as single clones, yield offspring that again show a lower than average density, but the difference soon disappears on continued multiplication. The behavior described, together with results of the corresponding experiments with particles of class 2, suggests that both gains and losses of DNA are possible during clonal growth, the losses occurring more rapidly than the gains.

Density 1.520 (class 4). The densest region of the cesium chloride gradient from which viable phage particles can be recovered is of special interest because, when present in the population, heterozygous particles tend to collect there, as was first shown by Doermann and Boehner.

Particles of class 4 represent only about 0.1 per cent of the viable phage particles. Labeled protein and DNA are relatively abundant, however, indicating that only about 1 particle in 10 of the specified density is viable. The viable and inviable particles in the mixture can be separated fairly well by an additional centrifugation in a density gradient of sucrose. The inviable particles sediment faster than the viable ones.

The dense, noninfective particles recovered in this way attach very poorly or not at all to bacteria, and yield on extraction rather homogeneous DNA fragments sedimenting at the rate of quarter-length pieces of reference DNA. To explain the physical and biological properties of the particles, we may suppose that they contain several quarter-length DNA fragments per particle, and lack tails. That the particles should show two seemingly unrelated structural defects seems odd, and calls for further study.

The dense, infective particles of class 4 are themselves heterogeneous in sedimentation rate. Considered as a single class, they yield on extraction normal DNA molecules, half-length fragments, and quarter-length fragments (not a continuous distribution of lengths). The individual particles must contain more than one piece of DNA of the specified lengths in any combination adding up to more than the normal complement of DNA per particle. This description may apply to the class of dense, heterozygous particles, whose structure is being analyzed by genetic methods in Doermann's laboratory.

Density 1.474 (class 5). The separation of particles of class 5 from those of class 3 requires a second centrifugation in cesium chloride, as already described. Those of class 5 are found only in cultures lysed late after infection, when they may comprise 5 per cent of the population.

Particles of class 5 have been rather thoroughly studied (see Mosig, 1963). In brief, they are noninfective and of low density because they contain a single DNA fragment measuring only 0.67 of the normal molecular length, and consequently have incomplete genomes. Apparently, different particles lack different parts of the genome, and the particles are otherwise functionally intact, because two or more can successfully infect when they attach to a single bacterium. With genetically marked stocks, it can be seen that such bacteria yield very large numbers of recombinants when the end of one of the chromosome fragments lies in or near the genetically marked region. (This circumstance is recognized when the phage yield from a single bacterium contains some but not all markers from both parents.) Two chromosome fragments, each representing two-thirds of the genome and together representing the entire genome, possess one-third of the genome in common. The high recombination frequency indicates that the common regions pair directly or indirectly early in the course of DNA replication; indeed, we may surmise that pairing is prerequisite to replication under these circumstances.

The phage progeny produced by multiple infection with particles of class 5 contain DNA molecules of normal length, into which at least half of the atoms (and presumably polynucleotides) of the parental DNA fragments are incorporated.

Density <1.4 (class 6). In the usual cesium chloride fractionation, some phage particles float to the top. They sediment much more slowly in sucrose than normal particles, are noninfective, and contain little DNA. In early lysates, particles of class 6 are much more numerous (in terms of protein content) than the viable particles. They are phage particles (or degradation products of phage particles) that were not yet quite finished at the time of lysis, as shown some years ago by Maaløe and Symonds and by Koch and Hershey. Surprisingly, they prove to contain DNA fragments of characteristic size, sedimenting in sucrose as a compact band at a rate corresponding to 7 per cent of the normal molecular weight. It is possible, though, that the size merely reflects the extent to which the DNA originally present in the particles is degraded under the given conditions of lysis.

Discussion

The structure of lambda DNA is remarkable in two respects. First, guanine-cytosine pairs are noticeably concentrated in the left third of the molecule. Since the distribution is relatively uniform in other microbial DNA's, the local differentiation in lambda DNA suggests diverse historical origins for different parts of the molecule. Presumably such differences could have been preserved only if there was a corresponding functional differentiation among molecular parts. The functional differentiation could be physiologic if there are mechanisms of control of gene action that depend on recognition of base composition of DNA. According to current ideas about the nature of the genetic code, such mechanisms are conceivable. Or, as suggested by Dove, the functional differentiation could be genetic only, if different parts of the lambda DNA molecule are undergoing genetic exchange with different phage (or other episomal) species. Either version of the hypothesis of functional differentiation is consistent with genetic evidence that genes with related functions tend to occur in clusters in the chromosome, and both are subject to experimental test.

The second remarkable feature of lambda DNA structure is evident in the mutually cohesive properties of the molecular ends. The probable structure is best described as that which would result if a circular DNA molecule were opened by cutting the two strands at specified but slightly different levels, so as to leave short, single-stranded ends with complementary base sequences.

One strong prediction follows from the stated structural model: that the molecular ends in general, not just the cohesive sites, must have a common origin and should have the same guanine-plus-cytosine content. We have not yet tested that prediction. A more general prediction is that the chromosome of lambda should be potentially circular, although the genetic map certainly is not.

The idea of circularity has cropped up repeatedly since Jacob and Wollman first found that the genetic map of *Escherichia coli* is circular. The idea was introduced into lambda genetics when Campbell suggested that the gene order of the lambda chromosome in lysogenic bacteria might be a permutation of the genetic map deduced from phage crosses made in the usual way. It now appears that his suggestion was correct, and that the vegetative and prophage chromosomes correspond, respectively, to the lambda DNA molecule as isolated from phage particles and to its permutation created in the laboratory by breaking the molecule in the center and joining the original ends through their cohesive sites.

If the analogy just drawn is valid, there must be two genetically determined cutting points in each strand of the hypothetical circular DNA molecule. Such points would have to be marked, presumably by unique base sequences. It may be that the rare base 5-methylcytosine found by Ledinko in lambda DNA plays a role in the marking system.

In T4, Mosig has chosen to study a phage different from lambda in many ways but no less complicated, as her work shows. In short, exceptional T4 phage particles may contain, instead of or in addition to the DNA molecules characteristic of the majority of particles, molecular fragments of a few discrete lengths: about 25, 50, and 67 per cent of the normal length. (Some much shorter fragments, mentioned in last year's report, were probably accidental products of enzymic hydrolysis.) Among the molecules of normal length, variations in sedimentation rate corresponding to length differences of only 1 or 2 per cent have been detected. The other length classes are also relatively homogeneous.

The origin and role of the molecular fragments are obscure, but some of the fragments at least have demonstrable genetic functions. In studying the exceptional particles, Mosig hopes to gain clues to the molecular behavior of DNA in general.

For example, by infecting bacteria with particles of class 5, containing DNA fragments of the two-thirds length only, she finds that two incomplete fragments can cooperate to start off normal DNA replication, the products of which are necessarily biparental. Genetic recombi-

nants evoked under these circumstances show that the replication is accompanied or preceded by pairing of homologous parts of the fragments, particularly involving an end of one of them. A hypothetical structure generated by such pairing is illustrated in figure 4. The structure presents certain obligatory features. Since every part of the genome must be represented in the biparental structure, and if Mosig's inference is correct that any genetic marker can be contributed by both parents, there will usually be two regions of homology, comprising together one-third of the length of the reconstructed chromosome. Given appropriate genetic markers, one-third of the genome is recombinant with respect to a second third, and the remainder is heterozygous.

The crosses so far studied permit detection of only one heterozygous region per reconstructed chromosome. It shows three features: a high frequency of recombination in single-burst clones containing some but not all markers from both parents; polarized segregation, in the sense that both parents tend to contribute markers more frequently near one end of the heterozygous region than the other; and unidirectional polarization, meaning that segregation is strongly biased in favor of counterclockwise polarity in relation to the genetic map. Taking into account that in Mosig's material the heterozygous regions are exceptionally long for understandable reasons, the first two features (but not the third, concerning which see her paper) are exactly those seen by Doermann and Boehner in their study of natural heterozygotes. Therefore reconstruction of phage chromosomes from their fragments must depend on mechanisms closely related to those eliciting normal heterozygosis and genetic recombination. To this extent the diagrammatic representation of rejoined fragments in figure 4, though fanciful in detail, probably illustrates a basic mechanism exploited in different ways by both T4 and lambda phages.

Mosig's data support not only Doer-

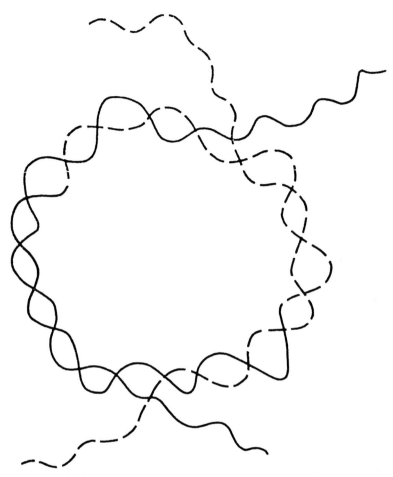

Fig. 4. Hypothetical paired structure generated by two complementing DNA fragments from phage particles of class 5.

mann and Boehner's inferences about the special role of molecular ends in the functioning of heterozygotes but also several postulates about genetic circularity put forward by Streisinger and his colleagues and sometimes called bizarre by them. They assume: (1) that T4 DNA molecules carry terminal redundancies of base sequence; (2) that the sequences are "circularly permuted" in the sense that the terminal redundancies represent different parts of the genome in different molecules; and (3) that the length of the molecule is not determined by the length of its unique base sequence but must be controlled by a separate mechanism, ultimately by certain genes.

Mosig's finding of a class of viable phage particles with slightly variable DNA content implies some form of redundancy. Circular permutation best explains the existence of T4 particles of class 5 that apparently contain DNA fragments two-thirds the normal size, each representing a different part of the genome. Moreover, when two of the fragments are able to complement each other, the products of replication are typical T4 DNA molecules—proof that the molecular length is a species character under independent control, as Streisinger predicted, and is not determined merely by the length of the parental molecules. (Indeed, Mosig's materials reveal a number of distinctive molecular lengths, each of which must be a species character.) The same conclusion follows from the fact that single viable phage particles, selected to contain a DNA molecule slightly longer or shorter than the average, produce clones of descendants among which the average is eventually reestablished.

The natural history of lambda phage shows that circular permutation, though perhaps a consequence of terminal redundancy in T4, is not a necessary consequence, and suggests that terminal redundancy plays a general role unconnected with the evolution of the circular map. As a more general role we propose the literal formation of rings, more or less as illustrated in figure 3, in keeping with growing evidence that DNA replicates in the circular form.

The recent achievements of our group studying DNA molecules can be summarized as follows. We have elucidated a mechanism for joining the ends of DNA molecules that was previously difficult to imagine and that apparently serves multiple biological functions. We have developed a technique applicable to lambda DNA for the isolation of specified molecular fragments, and have shown that the fragments will be useful in the investigation of several questions of biological interest. We have found, among exceptional T4 phage particles, materials offering clues to mechanisms of chromosomal replication, recombination, and length determination, clues that may or may not prove individually decipherable but will in any event have to be weighed with the final evidence.

STRUCTURE AND FUNCTION OF PHAGE DNAs

A. D. Hershey

The DNAs isolated from phage particles of different species, though possessing a basic structure in common, show a surprising number of individual differences. A few years ago my colleagues and I began studying some of them, pursuing the thought that elucidation of the differences in structure should contribute to an understanding of the common features.

Different phage DNAs must also perform common as well as individually distinct functions and, indeed, the structural features are mainly interesting as clues to function. DNA molecules in phage particles are inert products of

519

replication, and can furnish only indirect clues. For this reason several people in our laboratory and elsewhere have inquired into the structure of phage DNAs in the replicating state by extracting them from infected bacteria. Here too it appears that the different phages show both common features and individual differences, though further work will be needed before these can be understood or even adequately described.

A rather different approach to DNA function has been initiated by Anna Marie Skalka, who is taking advantage of the internal differentiation in lambda DNA to look for signs of order in the functioning of genes. She finds that after infection of bacteria with phage lambda, transcription of genetic information begins somewhere in the right half of the DNA molecule, and only later on extends to both halves.

The group studying DNA molecules this year includes Ruth Ehring and Edward Goldberg (Carnegie Institution Fellows), Mervyn G. Smith (Damon Runyon Fellow), Anna Marie Skalka (American Cancer Society Fellow), Elizabeth Burgi, Laura Ingraham, Gisela Mosig, and A. D. Hershey. The work is partly supported by grant HD 01228 from the National Institute of Child Health and Human Development, U.S. Public Health Service.

Lambda DNA
Burgi and Hershey

Burgi is continuing her analysis of the structure of the ends of lambda DNA molecules, and has arrived at a specific but still hypothetical model that should be testable in due course.

One of the requirements in this and other work is a practical method for the separation of right and left halves of lambda DNA molecules. In collaboration with C. I. Davern, we found that equilibrium density-gradient centrifugation in an angle rotor is a serviceable method. It is based on the principle that bands formed at an oblique angle to the tube axis in the spinning rotor spread apart when the rotor is stopped and the tube is brought to a vertical position. Although there is no theoretical gain in resolution under these circumstances, resolution is in fact improved because the expanded bands are less subject to mechanical mixing when the fractions are collected. Details of the method have been published by Hershey *et al.* in *Biochemical and Biophysical Research Communications.*

Another promising method, which is likely to facilitate our work, has been introduced by Nandi, Wang, and Davidson. In this method the density difference between molecular halves is increased by addition of mercury ions, which combine preferentially with parts of the molecule rich in adenine and thymine.

Weak Spots in T5 DNA
Ingraham and Hershey

When a solution of T5 DNA is stirred to generate the minimum rate of shear that will break the molecules, fragments 0.4 and 0.6 of the original molecular length are produced. These can be separated by chromatographic or centrifugal fractionation. When the fragments of 0.6-unit length are broken in turn, fragments 0.4 and 0.2 unit long appear. This behavior must be contrasted with that of some other phage DNAs, which do not show preferred breakage points but pull apart near their centers for purely mechanical reasons.

The breakage pattern of T5 DNA can be explained by several models, at least three of which are potentially distinguishable.

1. Two weak spots may lie at positions 0.4 and 0.6 of the molecular length from either end. According to this model, the 0.4-unit fragments resulting from the first and second breaks would be identical mixtures of pieces from both ends of the molecule.

2. Two weak spots may lie at positions

0.4 and 0.8 of the molecular length from a specified end. In this model, the 0.4-unit fragments resulting from the first and second breaks would represent different and nonoverlapping segments of the molecule.

3. The molecule may contain four weak spots dividing it into five segments of equal length. In this model, the 0.4-unit fragments resulting from the first break would be an equal mixture of the two ends, and those resulting from the second break would consist partly of ends and partly of subterminal sections. Thus the second breakage product would contain parts absent from the first, but not vice versa.

We explored these possibilities by application of the DNA homology test of Bolton and McCarthy. The experiments showed that the 0.4-unit fragments resulting from the first and second breaks each contain sequences absent from or less frequent in the other, but also contain many common sequences. These results exclude models 1 and 2 but not 3, and suggest additional possibilities.

Besides model 3, two principal alternatives remain. The structure may resemble that of model 1 if the two weak spots are of unequal strength or are situated at slightly unequal distances from the molecular center. Or model 2 may be correct if different parts of the molecule contain both unique and common base sequences. The interesting possibility that different molecular parts contain similar sequences has already arisen in our work with lambda DNA and will have to be pursued with that DNA, whose parts can be isolated without ambiguity.

The structure of the weak spots is unknown. One obvious possibility is that they are points at which one of the two polynucleotide chains of the molecule is already broken. The hypothesis is plausible inasmuch as several people have found evidence of single-strand breaks in T5 DNA but not in some other phage DNAs. Such breaks are detected by

measuring molecular weights in terms of sedimentation rates of the DNA in single-stranded form. In our experience the results do not support the hypothesis, because the sedimentation rates do not indicate sufficient uniformity of size of the polynucleotide chains to explain the rather precise locations of the weak spots. Both the method and its application to T5 DNA require further study.

The foregoing description refers to the DNA of a heat-stable mutant of T5. Irwin Rubenstein at Yale University has found that DNA of the wild-type phage differs from that of the mutant both in length and in location of weak spots. His results make the structure of T5 DNA more interesting but even more puzzling than it was before.

T4 Phage Particles With Incomplete Genomes

Mosig

Exceptional particles of phage T4 can be isolated that prove to contain a single DNA fragment two thirds of the standard length (*Year Book 63*, pp. 588–589). The particles are individually noninfective, but have the interesting property that two or more of them can cooperate, in a single bacterium, to regenerate complete chromosomes. Apparently the particles have incomplete genomes, and the missing parts are different in different particles, so that one particle can supply what is absent in another. That idea underlies the following experiments.

When a single bacterium is simultaneously infected with a normal phage particle and with a defective particle that is genetically marked, the marker from the defective particle may appear in the phage progeny. In this type of experiment different markers are "rescued" from the defective particles with identical frequencies, independently of their map position. The actual frequency of rescue of a single marker is about two thirds, as expected from the length of DNA in the particles. These results show that the rescue is highly efficient, that the DNA molecule and the genome are colinear, and that the two-thirds length is cut at random from an effectively circular genome. The results are consistent with other evidence that the genetic map of phage T4 is circular, and that the DNA molecules are "circularly permuted."

If, in the experiments just described, the defective phage particles are doubly marked, both markers can be rescued from the same defective particle even when they are widely separated in the genetic map. The frequency of joint rescue depends on the marker pair, providing a test of linkage. The linkage measures obtained in this way agree roughly but not precisely with distances on the genetic map. The discrepancies are unexplained but the following considerations apply. The rescue of a single marker from a DNA fragment appears to depend only on the presence or absence of the marker in the fragment. If so, the frequency of joint rescue of two markers from a continuous fragment of fixed length is a direct measure of the physical distance between them. The genetic map, on the other hand, is based on recombination frequencies that may be influenced by factors other than distances between markers.

In another type of experiment, bacteria are infected with defective phage particles only, to measure the efficiency of cooperation among them. This kind of cooperation may be called mutual complementation, to distinguish it from marker rescue by viable phage particles. If a bacterium is infected with one defective particle, phage growth fails, by hypothesis because phage genes are missing. If it is infected with two or more defective particles, phage growth may succeed or fail, by hypothesis depending on whether or not at least one copy of each gene is supplied by one or another of the particles. Theoretically, two phage particles, each containing a random segment representing two thirds of a circular

genome, furnish a complete set of genes with a probability of one third. For three particles, the probability is two thirds. These probabilities define a theoretical upper limit to the efficiency with which defective phage particles of the specified number and kind can complement one another to produce phage progeny. In strain B of *Escherichia coli* the measured efficiency is equal within limits of error to the theoretical maximum. This result shows that cooperation among two or three chromosome fragments is highly efficient, that the DNA molecule and the genome are colinear, and that the two-thirds length is cut at random from an effectively circular genome, confirming and extending the conclusions from marker rescue experiments.

In several other bacterial strains, the efficiency of mutual complementation is lower; in some, practically zero. The efficiency measured by this test in different bacterial strains is correlated with the efficiency with which two or more phage particles damaged by ultraviolet light can cooperate to produce phage progeny. Neither efficiency is correlated with the sensitivity of the bacteria themselves to ultraviolet light, or with other known properties of the strains. The frequency of marker rescue also depends, though less strongly, on the bacterial strain. Linkage measured from the frequency of joint rescue of two markers does not.

The bacterial properties affecting cooperation among phage particles are unknown, but suggest a possible means of analyzing the mechanism of cooperation. *Marker rescue* (cooperation between intact and defective chromosomes to produce chromosomes of double parentage) and *mutual complementation* (cooperation between two defective chromosomes to produce intact biparental chromosomes) both call for DNA replication and genetic recombination. In mutual complementation, but not in marker rescue, recombination must precede replication, unless chromosome fragments themselves can replicate. The results show that both rescue and complementation occur with perfect or nearly perfect efficiency in some bacterial strains but not in others. The bacterial defect therefore interferes with some event common to the two phenomena, an event that is especially critical in mutual complementation. The event recognized in this way may be an early genetic recombination, prerequisite for replication of fragmented chromosomes. Comparative experiments with different bacterial strains may serve, with Mosig's material, to determine whether or not chromosome fragments can replicate, and otherwise to analyze the functional competence of the fragments.

The same bacterial defects that interfere with mutual complementation among chromosome fragments also interfere with cooperation among radiation-damaged chromosomes, and the same considerations apply to both phenomena. Radiation damage is more complicated than a missing chromosome segment, because the number and kinds of radiation damages cannot be adequately specified, and because radiation damages are subject to repair as well as complementation. Incidentally, mutual complementation among chromosome fragments confirms the principle by which Luria proposed, in 1947, to explain cooperation among radiation-damaged phage particles: that a complete set of intact genes in two or more phage particles suffices to initiate phage growth. The quantitative failure of his theory for irradiated phage particles should probably be attributed to the complications mentioned.

Another phenomenon from the classical radiobiology of phage should be recalled in this connection. Krieg found that the radiosensitivity of the function of a specific gene, measured from the frequency with which that gene in an irradiated phage particle can complement its mutant allele present in an unirradiated phage particle, is improbably high in view of the small target size of the gene. No doubt the explanation must be sought in the complicated nature of radiation damages. The present experiments show that one must also take into account the efficiency with which two defective chromosomes can cooperate, which depends on properties of the bacterium as well as on the nature of the defects.

During the summer of 1964 Frances Womack of Vanderbilt University joined our group with the intention of studying crosses between defective phage particles carrying multiple genetic markers. The experiments failed because the numerous amber mutations in the phages interfered with mutual reactivation, presumably because the mutant genes function too poorly in the permissive bacterial host. A second try will be made if the difficulty can be circumvented.

The work with phage T4 was critically aided by A. H. Doermann, Robert Edgar, Ruth Hill, Frances Womack, and Michael Yarus, who provided phage and bacterial strains and information about them.

Genetic Recombination With Exceptional Particles of Phage T4

Ehring

Mosig's present work began with her observation that genetic recombination frequencies in phage T4 were correlated with the buoyant densities of the phage particles entering into the crosses. In particular, particles of low density showed high recombination frequencies. The particles of low density were subsequently resolved into two classes (*Year Book 63*, pp. 587, 588): viable particles with slightly curtailed DNA molecules (class 3) and inviable particles with DNA molecules only two thirds the normal length (class 5). Particles of class 5 yield many recombinants for understandable reasons: the phage progeny are necessarily multiparental in origin. Ehring has now made crosses with purified particles of class 3. Yields of recombinants do not differ from those in comparable crosses with particles of average DNA content. Therefore recombination frequencies in crosses with phage particles of the majority class do not depend on the lengths of chromosomal redundancies, or at any rate this test failed to reveal such dependence.

Genetic Linkage in T4 DNA Molecules
Goldberg

Modified bacteria (spheroplasts) can be infected with modified T4 phage particles (particles treated with urea to alter the specificity of attachment). If T4 DNA is present at the time of infection, a few per million of the phage progeny will show genetic markers derived from the DNA. The complicated system (first developed by van de Pol, Veldhuisen, and Cohen) is probably needed to get DNA into the cells, or to get it in without excessive enzymic degradation. The efficiency is inconveniently low with present techniques, but that does not seriously interfere with the experiments. The unusual feature of the system is that fragmented and denatured DNA can still contribute genetic markers. Since DNA fragments are not infective by themselves, even in principle, the genetic contribution by the DNA may be thought of as a rescue of markers by incorporation of specific nucleotide sequences into multiplying chromosomes.

If the DNA is doubly marked, the frequency of joint rescue of both markers provides a measure of genetic linkage. The actual measurement takes the form of a ratio between the number of phage particles showing both markers and the total number showing a specified one of them. This measure is influenced both by the initial rescue and by subsequent genetic recombinations, and is therefore inferior in principle to Mosig's test, which apparently depends only on the presence or absence of markers in the DNA fragments. The advantage of the system introduced by the Dutch workers is that the DNA fragments can be experimentally modified.

Figure 1 shows the effect of varying the size of the fragments on the frequency of marker rescue. It will be seen that rescue of a single marker is little affected by limited enzymic hydrolysis of the DNA. The frequency of joint rescue of two markers, however, decreases progressively with the amount of enzymic degradation. Since, after enzymic treatment, the DNA is denatured, the chief variable is the length of the single polynucleotide chains entering the bacterial cells. This method too, then, must measure the physical distance between markers.

The actual sizes of the fragments are not known and, indeed, precise methods of physical measurement are only now being worked out. The genetic data suggest that fragments of the order of size of single genes can be rescued. It should be possible now to construct a detailed physical map of the chromosome.

Replicating DNA
Smith

Frankel began several years ago to study DNA isolated from bacteria infected with phage T2 (*Year Book 61*, p. 448). He found a component of very high molecular weight but his work was hampered for a long time by the difficulty of extracting DNA from the cells without degradation.

Smith, in Burton's laboratory at Oxford, explained part of the difficulty in work with bacteria infected with phage T5. He found that bacteria treated with detergent and phenol in the usual way yielded two fractions of DNA, one going into solution in the aqueous phase, the other remaining in insoluble form at the interface between phenol and water. Appropriate isotopic labeling showed that the insoluble DNA was precursor to the soluble, and that the soluble fraction contained DNA derived from finished or nearly finished phage particles. The insoluble fraction thus became the center of interest. It turned out that the insoluble DNA was still trapped inside

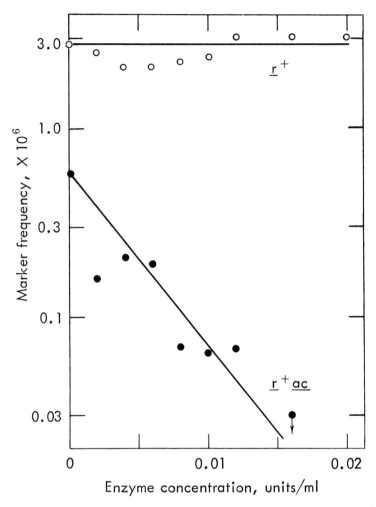

Fig. 1.   Frequency of marker rescue from polynucleotides of varying length. Digestion mixtures contained 1 μg T4 DNA, carrying the marker pair *r73⁺ ac41*, and the indicated amount of pancreatic deoxyribonuclease (Worthington Biochemical Corp.) expressed in units measured by the manufacturer. The enzyme was allowed to act for 5 minutes at room temperature before the mixtures were boiled. Open circles indicate recovery of *r⁺* phage, irrespective of the *ac* allele; filled circles, recovery of *r⁺ ac* phage; both expressed as fractions of the total yield of phage in the test system.

partly dissolved bacteria, from which it could be released, somewhat erratically, by treatment of the cells with lysozyme, an enzyme that hydrolyzes mucopolysaccharides in the bacterial cell wall.

Since joining our group, Smith has studied the properties of the phage-precursor DNA. Its density is the same as that of DNA extracted from phage T5, which shows that it has the expected base composition and is not attached to appreciable amounts of non-DNA material. It sediments as a single, very broad band, most of which moves faster than DNA extracted from phage particles. The fast-sedimenting DNA is sufficiently fragile under shear to suggest a long, threadlike structure. An examination in the electron microscope by MacHattie at The Johns Hopkins University showed, in fact, long threadlike structures without noticeable branching. Fragmentation of the DNA has not yet revealed any preferred subunit length. On denaturation in alkali the DNA sediments more or less like denatured DNA from phage particles, at a rate suggesting that both are composed of relatively short polynucleotide chains.

Frankel (now at the University of Pennsylvania), Smith, Skalka, Mosig, and others are extending work along these lines with several phages. It appears that the formation of very long structures is a common feature of the replication of phage DNAs. The results depend somewhat on the time during the growth cycle at which the DNA is labeled, on the time at which it is extracted, and on the phage species. Under certain circumstances a putative ring form of lambda DNA can be seen, as first found by Young and Sinsheimer. Detailed kinetic studies with several phage species may reveal an intelligible sequence of events.

Genetic Transcription in Bacteria Infected With Phage Lambda

Skalka

It is known from the classical observations of genetics that not all genes function at the same time. According to the operator theory of regulation of gene action, control is exerted at the primary step by which information coded in DNA is transcribed into complementary base sequences in messenger RNA. At any rate it has been shown recently that the specific messenger accumulates only when a given protein is being synthesized.

Transcription of specific genes can be studied only in a few favorable instances. In phage lambda, gross differences in base composition of the two halves of the DNA molecule (*Year Book 63*, pp. 581–583), permit analysis along somewhat different lines. One can ask, for instance: Where in the molecule does transcription start?

An appropriate experiment is performed as follows. Bacteria are infected with phage lambda. At any desired time thereafter, radiophosphate is fed to the culture for a period of two minutes, the culture is promptly chilled to stop synthetic activities, and RNA is extracted from the cells. The significant experimental variable is the time after infection at which the examination is made.

The labeled RNA extracted from the cells consists of several types, among which only that matching lambda DNA in base sequence is of interest in the present context. It is isolated, and its amount judged, by annealing to lambda DNA according to the technique of Bolton and McCarthy.

The labeled RNA isolated by annealing to lambda DNA is analyzed in the following ways. Its base composition is determined by alkaline hydrolysis and electrophoretic separation of the resulting nucleotides. Its purity is verified by annealing to lambda DNA a second time. This time the annealing is 60 to 80 per cent efficient (close to the maximum attainable by the method), and the base composition is not altered by the second annealing. Finally, the RNA is tested for its ability to anneal to the separated right or left halves of lambda DNA.

The yield and composition of lambda-specific RNA at different times after infection are illustrated in Table 1, which shows that lambda messenger RNA comprises only about 5 per cent of the total RNA labeled at early times after infection, but amounts to 30 per cent of the RNA labeled at 30 minutes. The same trend has been reported by Sly and Adler. The guanine-plus-cytosine content of the

TABLE 1.   Lambda-Specific RNA From Infected Bacteria

Interval of Synthesis, minutes	Yield, % of total	Guanine + Cytosine, mole %	Purines, mole %
5–7	6	45.4	52.6
15–17	5	48.5	54.6
30–32	32	53.9	52.6

RNA, not previously measured, is low at early times and high at late times. This trend is reproducible, as are the individual analyses, but the absolute values vary somewhat in different experiments. Finally, the purine content of the labeled RNA is high at all times (in double-stranded DNA it is necessarily 50 per cent), probably because only the pyrimidine-rich strand of the DNA is read, as has been shown directly for phages SP8 and alpha.

One interesting feature of these data is the difference in guanine-plus-cytosine content of the RNA labeled at different times. In principle it could arise from trivial changes in RNA precursor pools in the cells, but that is unlikely because precursor nucleotides have their phosphorus at the 5' end whereas the nucleotides analyzed have their phosphorus at the 3' end, a shuffling that should tend to cancel out a bias of the sort mentioned. Otherwise the data show that different parts of the DNA are transcribed at different times, and suggest that transcription starts in the right half of the molecule, which is low in guanine plus cytosine.

That transcription does start mainly in the right half of the DNA molecule is confirmed by annealing tests. Messenger RNA synthesized at early times anneals almost exclusively to right halves of lambda DNA, whereas messenger synthesized between 30 and 32 minutes after infection anneals with equal efficiency to both halves.

These results can, of course, be interpreted in any of the ways previously considered in connection with problems of gene regulation. The unique feature is that lambda DNA represents a "hybrid" chromosome consisting of two parts differing from each other in base composition just as much as chromosomes of different species usually differ. In general, perhaps, one must postulate mechanisms of control operating at some step in the functioning of individual genes or small groups of related genes. Instead of or in addition to this, there may be in lambda a broader category of control regulating simultaneously all or most genes lying in the same half of the DNA molecule. At any rate, if the composition of DNA reflects the composition of genes, different genes in lambda must use somewhat different coding vocabularies, and perhaps different punctuation marks, depending on their positions in the molecule. These differences may or may not be pertinent to mechanisms controlling transcription. Seemingly we have the opportunity, if we can muster the skill, to explore such possibilities.

A second interesting feature of the data in Table 1 is the increased rate of transcription from both halves of the molecule seen at late times. The rate change implies a second mechanism of control that is neither gene specific nor dependent on base composition. It could reflect simply the increased amount of DNA in the cells at late times, but other explanations are equally plausible.

Nada Ledinko, now at the Salk Institute, started the work described above in our laboratory. We were also assisted by generous advice from Sly and Adler, who independently began work along related lines at the University of Wisconsin.

STRUCTURE AND FUNCTION OF PHAGE DNAs

A. D. Hershey

Edward Goldberg, Gisela Mosig, and Mervyn Smith left our group during the year, and Phyllis Bear and Rudolf Werner joined it. This sort of change in our faces necessarily alters the character of our work, because those who leave tend to take their engagements with them, and those who arrive have to find new things to do. The arrangement ensures variety of interests and at the same time preserves continuity of effort.

Described in general terms, our work this year is typical. It includes a fresh look at old problems, made possible by new methods, and at least one innovation, the experiments of Rudolf Werner.

Ruth Ehring and Phyllis Bear are Carnegie Institution Fellows; Anna Marie Skalka is a Fellow of the American Cancer Society. Other members of the group are Elizabeth Burgi and Laura J. Ingraham. The work with phage is partly supported by a grant, HD01228, from the National Institute of Child Health and Human Development, U.S. Public Health Service.

Internal Structure of Lambda DNA
Hershey and Burgi

Lambda DNA is unusual among known DNAs in that different parts of the molecule differ considerably in base composition (*Year Book 63*, pp. 581–583). We found that the left third of the molecule contains 56% guanine-plus-cytosine (GC) and that most of the remainder contains only about 47%, but we were unable to pursue the analysis owing to lack of suitable methods.

More recently, Nandi, Wang, and Davidson showed that mercury ions combine preferentially with adenine-thymine base pairs in DNA, and that for a mixture of DNAs the buoyant density of the several mercury complexes in cesium sulfate solution depends strongly on base compo-

559

sition. We found, in collaboration with Davern, that separation of DNAs by density-gradient centrifugation could be much improved when tubes were spun in an angle rotor instead of in the usual swinging buckets (*Year Book 64*, p. 520). By combining these two methods, we have succeeded in learning a little more about the internal structure of lambda DNA.

Our procedure is the following. Twenty to 50 μg of radioactive lambda DNA is mixed with Cs_2SO_4 and $HgCl_2$ to make 4 ml of a solution at *p*H 9 that contains 44% by weight of the cesium salt and 3 mercury atoms per 10 nucleotides in the DNA. The solution is overlaid with mineral oil in a centrifuge tube and the tube is spun for 48 hours at 36,000 rpm in a Spinco type 40 rotor at 4°C. Serial fractions of the solution are then collected by drops through a hole pierced in the bottom of the tube, and the band or bands of DNA, which form during centrifugation at positions determined by their buoyant density in the salt-concentration gradient, are located among the fractions by radioassay.

A variant of this method can be used for analysis of trace amounts of radioactive lambda DNA. For this purpose, the bulk of the DNA added to the tube consists of unlabeled T2 DNA, sheared in a French press to reduce its viscosity. It serves to define the combining ratio of mercury to DNA, which is not reproducible at DNA concentrations much below 5 μg/ml. T2 DNA is convenient because it binds mercury more strongly than lambda DNA does and so collects at the bottom of the tube when sufficient mercury is added to bring lambda DNA to the center of the gradient. Under these conditions about 3.6 atoms of mercury per 10 nucleotides is optimum for analysis of lambda DNA. The method is analytical in the sense that one can compare the density of an unknown fraction of lambda DNA, labeled with P^{32}, with the densities of known fractions, labeled with H^3, spun in the same tube.

Native lambda DNA forms a single

narrow band when centrifuged under the conditions described, confirming other evidence that all the molecules are similar in composition. When the molecules are broken in two and then analyzed in the same way, the fragments separate into three bands, as shown in Fig. 1. This result, first seen by Laura Ingraham and Anna Marie Skalka, puzzled us for some time. Properly understood, it reveals practically all we know about the structure of lambda DNA. We give first the explanation, then its proof.

What we call half-length fragments of lambda DNA are products of single breaks per molecule. The fragments actually vary in length between one third and two thirds of the original molecular length. The molecule itself is made up of at least three dissimilar segments. At the left end is a stretch, measuring about 0.42 of the total molecular length, that is uniform in composition and contains 56% GC. Next comes a section of length 0.1 to 0.2, containing only 41% GC. The remainder of the right half contains about 46% GC.

According to this model, all fragments including the left molecular end and not exceeding 0.42 in length have the same

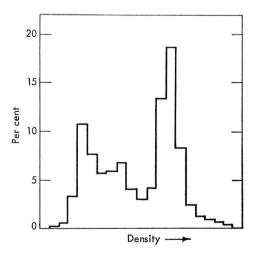

Fig. 1. Density distribution among mercury complexes of half-length fragments of lambda DNA.

density, and form the left band in Fig. 1. Fragments including the left end and exceeding 0.42 in length vary in density depending on length and are increasingly numerous as their length approaches 0.5. They form the middle band in Fig. 1. The composition of right ends does not depend strongly on length, and all lengths appear in the right band of Fig. 1. The observed density distribution can be reconstructed theoretically on the basis of the stated model.

The existence of a segment especially rich in adenine and thymine was proved by analysis of DNA broken into fragments of length 0.11, as shown in Fig. 2. The three components revealed in this way were isolated and their densities were determined in CsCl₂ solution. These measurements indicated major fractions, left to right in Fig. 2, containing 56% and 46% GC, and a minor fraction containing only 41%. The DNA contents of the same three bands measured 0.42, 0.47, and 0.11, expressed as fractions of the total DNA, therefore fractions of the total molecular length. As is reasonable, the fraction of the DNA resolved in the minor band rises when the DNA is broken into somewhat smaller pieces, reaching about 0.18 of the total DNA at length 0.09. That the 41%-GC fragments can be resolved at all among pieces of length 0.11 shows that they originate from one or a very few long sections in the molecule.

The location of a 41%-GC segment at the molecular center was demonstrated as follows. P³²-labeled DNA recovered in in-

dividual fractions of each of the three bands shown in Fig. 1 was broken into pieces of length 0.11 and returned to a Hg-Cs₂SO₄ mixture containing, in addition, H³-labeled marker DNA like that shown in Fig. 2. Centrifugal analysis of the mixtures showed that DNA from the left band of Fig. 1 (short left ends) was composed of 56%-GC fragments only. DNA from the middle band (long left ends) yielded fragments of which 75% contained 56% GC, 12.5% contained 41% GC, and the remainder formed a heterogeneous collection of intermediate composition. DNA from the right band in Fig. 1 (right ends) yielded practically no GC-rich fragments, but contained 41%- and 46%-GC fragments in about the same relative quantities as seen in unfractionated DNA.

The measured lengths of DNA fragments recovered from the bands illustrated in Fig. 1 were also consistent with the interpretation given. The first band on the left contained pieces ranging in length from about 0.43 downward (weight average length 0.35). The middle band contained fragments of rather uniform length, weight average 0.56, maximum 0.65, and no pieces much shorter than the average. The right band contained mostly halves but also longer and shorter pieces.

Inman (University of Adelaide, Australia) examined in the electron microscope lambda DNA that had been partially denatured by heating. He found a major site of incipient denaturation at the center of molecular length, and two minor acentric sites, presumably in the right half of the molecule. Our finding of a large segment of low GC content at the molecular center is consistent with his results. We have not yet looked specifically for AT-rich segments near the right end of the molecule, but it is clear that there are no large ones near the left end.

Our results confirm earlier evidence that the right and left ends of the lambda DNA molecule differ by about 10 percentage units in GC content. What does this tell us about the composition of in-

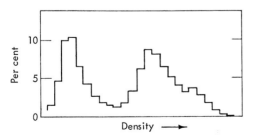

Fig. 2. Density distribution among mercury complexes of 3.5 × 10⁶ dalton fragments of lambda DNA.

dividual genes within each segment? In principle, the question can be answered by analyzing sufficiently small DNA fragments, as in the following experiment.

P32-labeled DNA recovered from central fractions of the left and right bands shown in Fig. 1 (left molecular ends and right molecular halves, respectively) was fragmented to a molecular weight of 7.5×10^5 (about 1000 nucleotide pairs), and the two preparations were separately analyzed by recentrifugation in Hg-Cs2SO4 mixtures containing samples of H3-labeled marker DNA of the kind shown in Fig. 2. Results were plotted so that the common H3-labeled markers in each tube coincided, thus bringing out the relation between the two P32-labeled materials. That relation is displayed in Fig. 3. It shows that left molecular ends contain practically no stretches of 1000 nucleotide pairs with a GC content as low as 46%, and that right molecular halves contain practically no such stretches with a GC content as high as 56%. Thus all or most of the individual genes in each part of the molecule reflect the composition of that part as a whole. This conclusion supports the idea that different genes in a given part of the molecule have unknown functional attributes in common.

For two reasons, this method of analysis underestimates the homogeneity in composition of the small fragments. First, because the fragments are small, their distributions in the salt-concentration

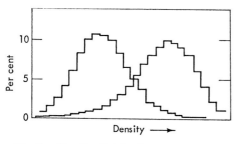

Fig. 3. Density distribution among mercury complexes of 7.5×10^5 dalton fragments of lambda DNA. Left band, fragments from left molecular ends. Right band, fragments from right molecular halves.

gradient are broadened by diffusion. Second, small fragments exhibit a higher average density in the presence of mercury than does the same DNA in the form of larger pieces. This effect may be the result of the exposure of single-stranded ends at breakage points, the ends combining preferentially with mercury. Whatever its explanation, the effect is not likely to be equal in all fragments, and so may broaden the density distribution for a second reason unrelated to base composition.

We intend to pursue the analysis further, partly to see how far the methods will permit us to go, partly in the hope of relating the peculiar structure of lambda DNA to some of its many functions.

Base-Sequence Homologies

Ingraham, Ehring, Hershey

Intramolecular homology. In principle, one might expect base-sequence homologies in DNA to reflect both species-specific and gene-specific characters. Species-specific homology could be recognized if different parts of the DNA of a single species interacted more strongly than DNAs of different, unrelated species.

Ehring has found that right and left end quarters of lambda DNA interact as measured by the homology test of Bolton and McCarthy. The interaction is weak but probably sufficient to exclude purely accidental resemblances. Perhaps it is remarkable that any resemblance at all is found since the two ends of the lambda DNA molecule are quite dissimilar in composition. The example was chosen for technical reasons: left and right ends can be isolated in adequately pure form (ends, by virtue of their ability to join to each other; left and right by virtue of their very different buoyant densities in a mixture containing mercury).

The interaction in this instance could, of course, be a special case, perhaps related to the very efficient interaction between the ends of the native molecules. Comparable tests with other parts of the

DNA molecule will be more difficult technically but now seem worth a try.

Coli-lambda homology. Cowie and McCarthy of the Carnegie Department of Terrestrial Magnetism first demonstrated base-sequence homology between the DNAs of lambda and *Escherichia coli.* Their finding furnished a physical clue to the close relationship between virus and host that is visible at both functional and genetic levels. Cowie and Hershey examined left and right ends of lambda DNA and a segment (b_2^+) defined by a genetic deletion. They found that all three parts of the lambda DNA molecule interacted with *E. coli* DNA. Owing to limitations in the preparative methods then available, they did not stress quantitative aspects of the data, but noticed that all their fractionated materials interacted less strongly with bacterial DNA than unfractionated lambda DNA does.

Results of comparable tests with better materials are shown in Table 1. The results demonstrate again that all parts of the lambda DNA molecule tested can react with *coli* DNA. They also show that the strongest reaction occurs with fragments containing 46% GC, derived from the right half of the molecule. This result,

together with the weak reaction of short right ends observed by Cowie and Hershey, suggests that there may be a long homologous segment just to the right of the molecular center. We have not yet tried to verify the inference.

Table 1 also contains the results of tests with another heterologous DNA, that from the related phage ϕ80. Here it is the left end of the lambda DNA molecule that gives the strong reaction.

One knows from the work of Marmur and Doty and others that the optimum temperature for reactions of this type depends on the composition of the interacting DNAs. We therefore tested hybridization at three different temperatures. Altering the temperature did affect somewhat the extent of hybridization, but it did not change the specificity of the reactions. These results also are recorded in Table 1.

The results show clearly that we are able to detect locally specific structural relationships among DNAs. When adequately defined, such relationships ought to be intelligible in terms of biological function.

Lengths of homologous sequences. Results already presented indicate that

TABLE 1. Hybridization Tests with Lambda DNA Fractions

DNA Agar	41%-GC Fragments	46%-GC Fragments	56%-GC Fragments	Unfractionated DNA
	Annealed at 60°C			
Lambda (21)	86	85	83	...
E. coli (330)	7.9	20	9.2	15
ϕ80 (100)	3.5	13	27	15
	Annealed at 55°C			
E. coli (330)	11	29	...	...
ϕ80 (50)	5.0	12	...	...
	Annealed at 65°C			
E. coli (100)	...	13	4.1	...
ϕ80 (50)	...	9.6	20	...

P^{32}-labeled lambda DNA fractions were isolated from the centers of the three bands shown in Fig. 2. After removal of mercury by dialysis, the materials were sheared by sonication, denatured by heating, and tested according to Bolton and McCarthy. The DNA contents of the agars (μg/g) are shown in parentheses. The other figures in the table are percentages of radioactive DNA eluted by SSC/100 at 75°C after the columns had been washed with 2 $\times$ SSC at the stated annealing temperature.

regions of homology with the bacterial chromosome are dispersed among all large sections of the lambda DNA molecule. How long are these regions?

In experiments of the sort already described, one breaks lambda DNA into fragments about $\frac{1}{80}$ the length of the molecule and asks how many of them can attach to *coli* DNA and how many cannot. The classification is not black and white, because the answer depends somewhat on the conditions of the test, but it is clear that at least 30% can attach specifically.

A fragment that can attach may be able to do so only in one small spot or over its entire length. These extreme alternatives are easily distinguished by breaking the fragment in two: either one or both of its parts should retain the ability to attach.

Laura Ingraham prepared a series of small single-stranded fragments of P^{32}-labeled lambda DNA by limited digestion of the native DNA with varied amounts of pancreatic deoxyribonuclease, followed by phenol extraction to remove the enzyme. She fractionated each product, otherwise very heterogeneous, by zone centrifugation in alkaline cesium chloride solution, recovering only the band center. Finally, to measure molecular weights, she recentrifuged samples of each kind on alkaline sucrose solution together with a molecular-weight reference of sonicated, H^3-labeled DNA (single-strand molecular weight 200,000). The smaller molecular weights measured in this way are only reasonable guesses, but the ranking with respect to molecular weight is unambiguous. A nominal molecular weight of 5000 at the lower end of the series indicates pieces that can no longer form stable complexes in lambda DNA agar, that are close to the limit of action of the enzyme, and that fail to precipitate completely with protein in the presence of trichloroacetic acid. (Homology tests are feasible because membrane filters hold back the fragments in the usual analytical procedure.)

The results are clear. The fragments bind to lambda DNA agar with 100%

efficiency (normalized: observed efficiency 90%) throughout the molecular-weight range from 160,000 to about 12,000. The efficiency of binding then falls steeply to 15% at 5000 daltons. The loss of ability to bind may be a direct effect of the small size, or may signify unknown side effects of the enzyme.

The same fragments bind to *coli* DNA agar with 100% efficiency (normalized from 15%) over the molecular-weight range from 160,000 to nearly 20,000, and still bind with 65% efficiency at 12,000. Thus if fragmentation releases nonhomologous pieces, their average size does not exceed about 10,000 daltons or 30 nucleotides. For purposes of discourse we describe this situation by saying that when a long fragment binds to *coli* DNA it does so because of base-sequence homology that is imperfect but usually extends the full length of the fragment. The results of other, less careful, experiments indicate that this description applies to fragments ranging in size up to at least 4% of the total molecular length. Therefore the 30% homology originally reported by Cowie and McCarthy is distributed over the lambda DNA molecule in not more than eight sections that together span one third of the molecular length. According to the results summarized in Table 1, the homologous sections are not distributed evenly but are nevertheless found in all three parts of the molecule differing in base composition.

One verification of this model is missing. If there are not more than eight homologous segments most of which measure at least 4% of the molecular length, there must be not more than nine nonhomologous segments of average length at least 8%. We ought to be able to isolate representatives of one class or the other, preferably by a method that is independent of the homology test itself.

The interpretation in terms of long stretches of imperfect homology is also supported by measurements of the temperature required to dissociate labeled DNA fragments from their union with unlabeled DNA in agar. In both homologous and heterologous reactions, this temperature is lower the shorter the fragments. For fragments of any given length, the temperature required is lower in the heterologous reaction than in the homologous.

Transcription of Lambda DNA in Phage-Infected Bacteria

Skalka

Analysis of messenger RNA formed at different times after infection of bacteria with phage lambda reveals two distinct phases of transcription (*Year Book 64*, pp. 526–527). During the early phase, transcription is slow, and mainly confined to the AT-rich half of the DNA molecule. During the late phase, transcription is faster, and both halves participate about equally. Skalka reached these conclusions partly by hybridization tests with separated molecular halves of the DNA and partly by analysis of the nucleotide composition of the messenger.

By means of competition experiments, one can determine whether the same or different genes in a given part of the molecule are transcribed at different times. Skalka tested, for example, the ability of unlabeled RNA extracted from cells during the late phase of transcription to compete in the binding of labeled early-phase messenger to the individual molecular halves of lambda DNA. This type of analysis revealed three categories of messenger produced by transcription from the right half of the molecule: one formed only at early times, one formed only at late times, and one formed at both times. RNA produced by transcription from the left half of the molecule could not be clearly subdivided and may represent a single class formed slowly at early times and rapidly at late times.

These results indicate that lambda DNA contains at least four sets of genes whose transcription is subject to independent or semi-independent control. Among these, the one most readily amena-

ble to analysis is the set comprising the GC-rich left end of the molecule, probably genes responsible for directing the synthesis of the structural proteins of the phage particle. It is almost necessary to suppose that these late genes function in response to some signal generated by the functioning of one or more early genes. If they do, one has to identify the signal, the signal generator, and the mechanism of response.

Chloramphenicol is a drug that interferes specifically with protein synthesis, not directly with genetic transcription. Skalka finds that addition of the drug to bacterial cultures before or at the time of infection with phage permits messenger synthesis that continues without acceleration for 35 minutes or more. At both early and late times, the messenger synthesized in the presence of the drug resembles early messenger in that it hybridizes preferentially with right halves of lambda DNA. These results suggest that a particular phage-specific protein gives the signal for the late phase of transcription. The critical protein could act indirectly. For instance, Naono and Gros, in experiments rather similar in intent to Skalka's, found that interference with DNA synthesis by thymine deprivation also interferes with the late phase of genetic transcription. Thus the phasing of transcription could depend on modulation of DNA structure during replication.

Skalka and Harrison Echols (University of Wisconsin) have looked for specific mutational blocks that might identify a gene controlling the late phase of transcription. The only reasonably complete set of mutants, Campbell's *sus* mutants, prove to have rather ambiguous effects on transcription, perhaps because the mutational defects are incomplete. Two other mutants were found that make early-phase but not late-phase messenger. One is a defective, transducing phage in which a large block of genes (A–J) is missing from the left end of the DNA molecule. The other (T11) contains a defective gene of unknown function in the C region of the chromosome. DNA synthesis also is blocked in this mutant, but not synthesis of some early phage proteins.

One generalization emerges. When late functions are blocked in any of a variety of ways, the late phase of transcription also is blocked. Here, as in several other situations, it is not clear whether normal control mechanisms act directly or indirectly to influence the rate of transcription.

Replicating DNA

Smith and Skalka

In bacteria infected with phage lambda, but not, so far, in those infected with T4 or T5, a ring form of DNA can be seen (Young and Sinsheimer; Bode and Kaiser). It apparently consists of two continuous, circular strands forming a supercoiled helix, since it resembles in many respects the structure of that description found in polyoma virus and analyzed by Vinograd. Its formation in lambda-infected bacteria permits a test of the hypothesis that DNA molecules replicate in circular form (*Year Book 63*, p. 592).

Smith and Skalka find that the ring form of lambda DNA is not a major precursor of the DNA eventually incorporated into phage particles. On the contrary, H^3-thymidine fed to cultures while DNA synthesis is in progress scarcely enters rings at all. Instead it promptly appears in structures that resemble in several respects very long threadlike molecules of DNA. The rings, on the other hand, seem to be formed chiefly from the DNA of the infecting phage particles, and to persist during the course of the infection. This finding is consistent with the results of Bode and Kaiser, who found that DNA from infecting lambda phage particles is converted into rings in "immune" bacteria in which DNA synthesis does not occur.

Evidently the ring form of lambda DNA does not replicate as such. It could play a direct role in initiation of DNA synthesis, or could, as suggested by Bode and Kaiser, be a form of DNA to which replication is expressly forbidden.

The formation of large DNA-containing structures is common to the replication of T4, T5, and lambda DNAs (*Year Book 64*, pp. 524–526, and Frankel's work at the University of Pennsylvania). It must be admitted that the precise nature of these structures is uncertain. Possibly they are very long DNA molecules, that is, polynucleotide structures held together by phosphodiester linkages and base-pairing forces only. One support for this hypothesis is the extreme fragility of the structures under shear, but an exact description of their breakage characteristics is still wanting. Another support comes from the fact that the T4 and lambda structures sediment in alkaline solutions as if they contained single DNA strands some of which are longer than those in the DNA of the respective phage particles. Unfortunately, sedimentation rates do not distinguish between long strands and shorter strands that are cross-linked in some way.

The T5 example is somewhat reassuring in this respect. Here the finished DNA from phage particles contains only short polynucleotide strands, and so does the rapidly sedimenting DNA extracted from infected bacteria. One can suppose that, in both forms of T5 DNA, long DNA molecules are constructed of shorter polynucleotides by an arrangement that staggers the chain breaks in antiparallel strands. John Abelson at Johns Hopkins University has shown that the length distribution of strands in the DNA of T5 phage particles is compatible with that hypothesis.

Rightly or wrongly, the rapidly sedimenting DNA structures found in phage-infected bacteria shift the center of interest from circular structures to linear polymers as characteristic replicating forms of DNA. Phage ϕX174 now seems to be the only uncontested example, a rather special one, in which DNA replicates as rings. Rings and linear polymers are functionally equivalent in certain respects. The properties of the cohesive ends of lambda DNA show that a single structure can spontaneously generate either form (*Year Book 62*, pp. 482–483). And, as Streisinger and his colleagues pointed out in advance of the physical facts, circular genetic maps and related complexities can originate either way.

In one respect, though, the circular and linear models are not equivalent, and the genetic facts support the linear model. Streisinger *et al.* (unpublished experiments) find that genetic deletions increase the length of the terminal redundancy in T4 DNA molecules. In effect, this means that in T4 the characteristic molecular length of DNA is maintained independently of the length of its unique base sequence. This conclusion is understandable if the DNA content of a phage particle is derived by excision of a measured length from an indefinitely long polymer. It is not understandable if the DNA content of a phage particle has to be derived by modification of a closed double helix, whose contour length would be the length of its unique base sequence.

Note that this argument does not apply to the DNA in phage lambda, the length of which depends on its information content as modified by deletions and substitutions. Phage T4 measures lengths of DNA; phage lambda recognizes punctuation marks.

Replicating Points in T4 DNA

Werner

A bacterium is supposed to contain a single DNA molecule that replicates at one or a very few growing points to generate Y-shaped structures containing the parental duplex ahead of the fork and two daughter duplexes (one strand parental, one strand newly synthesized) behind the fork. According to ideas of Jacob, Brenner, and Cuzin, this seemingly inefficient way to duplicate a very long DNA molecule is the efficient way to ensure the proper segregation of daughter molecules to daughter cells.

A bacterium infected with phage T4 contains about 17 molecular equivalents of DNA that could be replicating. (The figure cited was measured by Koch and Hershey in 1959 for the related phage T2.) In the past, I think, most people assumed that all 17 molecules were replicating, probably independently of one another. The results of Smith and Skalka just presented change the issues somewhat. If several molecular equivalents of DNA are replicating as a single tandem structure there is a segregation problem of sorts, though it places no obvious constraint on the mechanics of replication. However, one can imagine at least three alternatives concerning the number of growing points: one per molecular equivalent of DNA, one per larger hypothetical unit of replication, or one per bacterium. John Cairns pointed out a year or two ago that there was no very compelling reason for ignoring the third alternative, and so he is partly responsible for Werner's current activities. Werner also benefits from the advice of Cedric Davern, who was a pioneer in the development of the applicable experimental techniques.

The specific experiment is basically that designed by Bonhoeffer and Gierer some years ago to count the number of growing points in the DNA of a single bacterial cell. They fed C^{14}-labeled 5-bromouracil to growing bacterial cultures for a fraction of a generation time. Under the conditions of their experiment, 5-bromouracil enters DNA in place of thymine. Thus short sections in each branch of the Y immediately behind a growing point should be "hybrid" in structure: one strand containing radioactive 5-bromouracil, the other thymine. All parts of the DNA ahead of or distant from a growing point should contain thymine only. According to these expectations, DNA extracted from the cells and fragmented to sufficiently small size should contain all its radioactivity in hybrid structures. Hybrid structures can be recognized because their buoyant density in cesium chloride is increased in a decided and characteristic manner by the substitution of bromine for methyl groups.

Two measurements are possible. The

size of fragments to which the DNA must be reduced in order to liberate radioactive pieces of hybrid density measures the lengths of individual hybrid sections. The total amount of radioactivity incorporated into DNA measures the sum of the lengths of all hybrid sections. This sum, expressed per bacterium and divided by twice the individual length, gives the number of growing points per bacterium.

Needless to say, there are numerous difficulties in this type of experiment, and still more when it is applied to phage-infected bacteria. Werner believes that he has encountered and circumvented most of them. His experiments so far completed show that there are more than six growing points per bacterium infected with phage T4, and about one per molecular equivalent of phage-precursor DNA in the cells. Apparently the unit of replication is the T4 chromosome as defined by the DNA content of a finished phage particle, and the infected cell contains a number of such units that replicate continuously.

STRUCTURE AND FUNCTION OF PHAGE DNA'S

A. D. Hershey

The following report derives from the work of Phyllis Bear, Elizabeth Burgi, Laura Ingraham, Shraga Makover, Anna Marie Skalka, Rudolf Werner, and myself. Bear, Makover, and Skalka are Carnegie Institution Fellows. Our work as a whole is partly supported by a research grant, HD01228, from the National Institute of Child Health and Human Development, U.S. Public Health Service.

Nucleotide Distribution in λ DNA

Burgi, Skalka, and Hershey

Last year we described methods for analyzing the distribution of guanine and cytosine along the length of the λ DNA molecule. These methods depend chiefly on breaking the DNA into fragments of known average length and then sorting the fragments with respect to nucleotide composition. The resolution that can be achieved in this way depends on the length of the fragments. Pieces of fractional length 0.12, for instance, fall into three discrete classes (*Year Book 65*, pp. 559–562). Pieces of fractional length 0.06 fall into four classes, as illustrated in Fig. 1. These and other results show that the λ DNA molecule contains three large segments that differ in composition. From left to right these measure 0.44, 0.10, and 0.46 in fractional length, and 57, 37, and 46 mole per cent guanine plus cytosine (GC) in composition. The two central components visible in Fig. 1 come from the 46%-GC segment, which is therefore made up of two subclasses measuring 43%

and 48.5% in GC content. The composition of right-terminal fragments of various lengths shows that the 43%-GC DNA is more abundant toward the molecular center than toward the molecular end. However, a short stretch poor in guanine and cytosine at the right end of the molecule (see below) complicates analysis of terminal fragments.

The molecular ends provide a special opportunity in that terminal fragments can be isolated individually, owing to the specific left-to-right joining of terminal cohesive sites (*Year Book 63*, pp. 581–585). Burgi has found that left and right molecular ends of fractional length 0.14 do not differ appreciably in composition from the larger terminal segments from which they come. When reduced to fractional length 0.012, however, left ends contain only 48% GC, and

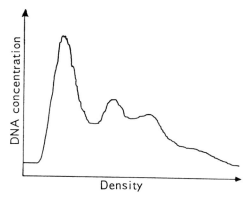

Fig. 1. Density distribution of mercury complexes of λ DNA fragments. Fractional length of fragments, 0.06. Hg/nucleotide ratio, 0.22. Cs_2SO_4 concentration, 42.8%. Centrifugation, 40 hours at 44,770 rpm in the Spinco analytical centrifuge.

right ends only 42%. These results support the anticipated conclusion that the shorter the fragment, the less it is obliged to resemble its neighbors in composition. More interestingly, the two molecular ends show a common tendency toward diminishing GC content. Since the two terminal genes in the genetic map function late during the phage growth cycle, and late functions are generally associated with DNA of high GC content (*Year Book 64*, pp. 526–529), it may be desirable to locate the terminal genes with respect to the changing base composition near the molecular ends.

Campbell's Model for Prophage Insertion

Escherichia coli and its phages T4 and λ are in many ways the best-known biological species in existence, each having presented first-rate biological problems and favorable opportunities for their investigation. Lambda is important mainly because it can recombine genetically with its host, mingling cellular and viral inheritance in ways that are fascinating to contemplate and, very likely, of practical importance to humans. In short, λ is one of a small class of biological elements to which Jacob and Wollman gave the name episome, in expectation, I presume, that many and varied examples remained to be discovered.

Phage λ was found by Esther Lederberg (1951) as a prophage residing in the K12 strain of *E. coli*, with which J. Lederberg and Tatum had first demonstrated sexual recombination in bacteria 5 years earlier. To understand λ, one must keep in mind that it is a creature with three potentialities, each of which can be studied separately in other, less versatile, elements.

First, like other phages, λ can in-

fect and lyse bacterial cells, producing numerous phage progeny in the process. During the lytic cycle, genetic recombination occurs between suitably marked phage chromosomes. Wollman and Jacob (1954) and Kaiser (1957) constructed the genetic map summarizing the results of such crosses. I shall call their map, based on ordinary phage crosses, the ordinary genetic map of λ. About 20 genes have been ordered so far, leaving a considerable gap in the center of the map where functions not essential to the lytic cycle remain to be identified. The length of DNA in λ, about 44,000 nucleotide pairs, could accommodate 30 or 40 genes.

Second, like some but not all other phages, λ can take up residence in the bacterial chromosome, giving rise to viable cell lines of modified inheritance. Such cell lines are said to carry prophage and to be lysogenic. (To keep matters of classification straight, I note that among phage species able to produce typical lysogens, some do and some do not occupy identifiable sites in the bacterial chromosome, and among those that do, some, like λ, occupy a unique site and some may occupy any of several sites.) Lwoff (1953) defined the main attributes of lysogenic cells. They are immune to the lytic development of the prophage they carry and to superinfection by phage particles of the same species. On the other hand, lysogenic cells can spontaneously regenerate phage particles when their immunity breaks down. Initiation of the lytic cycle in this way is called induction and can in many instances be brought about by irradiation with ultraviolet light.

The characteristic structure and properties of the chromosome of *E. coli* lysogenic for λ is revealed by the following facts.

1. Crosses between lysogenic and nonlysogenic bacteria bring to light

a determinant of lysogeny that is linked to a cluster of galactose-fermentation (*gal*) genes (Lederberg and Lederberg, 1953; Wollman, 1953).

2. When the bacterial chromosome is transferred from a lysogenic donor to a nonimmune recipient during bacterial mating, the recipient cell lyses (Jacob and Wollman, 1956). Also, in crosses between lysogens carrying marked prophages, the prophage markers show linkage to *gal* (Wollman and Jacob, 1954). The determinant of lysogeny identified by bacterial crosses is therefore the prophage itself, which is inserted into the bacterial chromosome between *gal* and a biotin gene called *bio* (Rothman, 1965).

3. Genetic analysis of lysogens also shows that the order of genes in the prophage is a cyclic permutation of the order in the ordinary map of λ (Calef and Licciardello, 1960; Rothman, 1965).

4. When a lysogenic bacterium reverts to nonlysogeny, a phenomenon conveniently observed if the prophage is λ*dg*, the "cured" cell originates a line that can, as a rule, be lysogenized again by reinfection with phage. Thus lysogenization is reversible with respect to both phage and bacterial chromosomes.

In its third manifestation, λ and a few related phages, unlike many other phages, can form transducing phage lines of a special kind. The best-known examples are called "*gal* transducing" because they carry genes concerned with galactose fermentation from the cell in which they originate to cells they infect. This propensity of λ is called specialized transduction, to distinguish it from the rather different phenomenon of generalized transduction discovered in phage P22 by Zinder and Lederberg (1952).

When a culture of *E. coli* lysogenic for λ is induced to produce phage, the lysate contains, with rare exceptions, λ of the genetic type with which the culture was originally infected. The exceptions prove instructive.

Morse (1954) found that lysates obtained by induction of λ lysogens, used as a source of phage to infect galactose-negative bacteria, yielded about one bacterial clone able to ferment galactose per million infecting phage particles. He had discovered specialized transduction, characterized as follows by the work of Morse, Lederberg, and Lederberg (1956).

1. The genetic modification of the recipient bacteria is brought about by phage particles, and is allele specific for *gal* markers present in the lysogenic donor.

2. Only markers adjacent to the prophage site in the bacterial chromosome, namely, *gal* and (as found later) *bio*, are transduced by λ.

3. Transducing phage particles originate only by induction of lysogens, not by infection with ordinary λ.

4. Cells genetically modified by transduction are typically immune to λ, either because they are lysogenic or because they carry a defective prophage.

5. The genetically modified cells revert to galactose-negative with rather high frequency. The revertants typically show the *gal* genotype of the recipient; that is to say, reversion occurs by loss of genes brought in by the phage.

6. When a culture made galactose-positive by lysogenization is induced in turn, and the lysate is used to infect galactose-negative bacteria, about one galactose-fermenting clone per 10 phage particles is obtained. These secondary lysates are called "high frequency transducing," as opposed to the "low frequency transducing" lysates obtained by induction of an ordinary λ lysogen.

Item 1 above is the criterion of transduction of all kinds. All the other features of transduction by λ are peculiar to specialized transduction. Note particularly item 5, which shows that *gal* transduction is usually brought about by addition of genes to the bacterium, not by substitution of gene for gene. Note also that comparison of the properties of nontransducing, low frequency transducing, and high frequency transducing lysates shows that transducing phage lines originate by some rare event associated with multiplication or induction of prophage. However, the nature of this event could not be investigated until an important source of confusion had been cleared up.

Clarification came from the simultaneous work of Arber, Kellenberger, and Weigle (1957) and Campbell (1957), work that depended for its success on the simple expedient of making sure that bacterial cells in the experimental cultures were infected with single phage particles. Experiments performed with that precaution quickly showed that transducing lysates contain a mixture of normal phage particles and transducing phage particles, and that transduction clones of the sort previously studied usually arise by double infection with particles of both kinds. It is worth noting that Morse and the Lederbergs (1956) and Weigle (1957) had performed experiments that narrowly missed bringing out these facts.

The conclusions reached by Arber, Kellenberger, and Weigle (1957) and Campbell (1957) may be summarized as follows.

1. High frequency transducing lysates contain about equal numbers of transducing particles and wild-type particles, which can be separately enumerated because both kinds of particles kill bacteria but only wild-type particles form plaques. By inference, low frequency transducing lysates contain about one transducing particle per million wild-type particles.

2. Transducing particles adsorb to bacteria and lyse them but produce no offspring. Transducing particles are found only in mixed yields coming from bacteria doubly infected with transducing and wild-type particles. Situations of this kind are well known, and are understandable, since gene function defective in one chromosome can often be supplied by the corresponding gene in another chromosome present in the same cell. Arber named the transducing particles λ*dg*, signifying "defective" and "*gal* transducing."

3. "Transductants," that is, galactose-fermenting clones of bacteria originating by transduction, are of two sorts: defective lysogenic, produced rarely by infection with a single particle of λ*dg*, and actively lysogenic, produced more frequently after simultaneous infection with both λ*dg* and wild-type particles. Actively lysogenic transductants also arise with low frequency when single particles of λ*dg* infect bacteria already lysogenic for λ.

4. Defective lysogenic transductants yield, on induction, sterile lysates. Actively lysogenic transductants yield, on induction, the high frequency transducing lysates already defined.

5. Defective lysogenic transductants revert spontaneously to galactose-negative clones that are nonlysogenic. Actively lysogenic transductants revert to galactose-negative clones that are typical stable lysogens or, sometimes, nonlysogenic (Campbell, 1963).

These facts permit a number of important inferences. First, the correlation between defectiveness and transducing power in λ*dg* shows that the inclusion of bacterial genes in a phage particle entails the loss of phage genes. Second, the instability of lysogens carrying λdg shows that λdg is defective as a prophage, as well as defective in functions expressed after induction, though the nature of the prophage defect is not yet clear. Finally, phage genes and bacterial genes must be combined in a single structure, the λdg chromosome, for two reasons. Both categories of gene can multiply during the lytic cycle of phage growth to produce high frequency transducing lysates (Weigle, 1957). Both categories of gene are lost simultaneously when a defective lysogenic transductant reverts to galactose-negative (Campbell, 1957). Thus prophage insertion and the origin of λdg are different manifestations of genetic recombination between phage and bacterial chromosomes.

Arber (1958) made crosses between λdg and λ that revealed two remarkable features of the structure of the λdg chromosome. First, the missing phage genes, notably a host-range marker, correspond to a subterminal section of the ordinary genetic map. Second, in different lines of λdg, the right ends of the deleted segments fall at the same point in the map, just to the right of the host-range marker.

In the meantime, Campbell (*Year Book 57*, pp. 386–389) had discovered in λ what are now known as suppressor-sensitive (amber) mutations. These mutations are important not only because of their role in the elucidation of gene function but also because they can be found in any gene that has an essential function. Campbell (*op. cit.* and subsequent papers) exploited them, by what is known as the overlapping deletion method, to analyze the structure of λdg in great detail. He found, first of all, that λdg's of independent origin tend to differ from one another, whereas the λdg particles in a high frequency transducing lysate derived

from the same transduction clone are identical. This conclusion meant that λdg's, like wild-type phage, are stable during replication in both the prophage condition and the lytic cycle (verified by Campbell, 1960). It also meant that the events by which λdg's originate, doubly exceptional because they give rise to λdg's of various kinds, occur at the moment of induction, not during replication of λ prophage (verified by Campbell, 1963).

The common feature of all λdg's, besides the inclusion of *gal* genes for which they are selected, is the absence of a single block of phage genes. The deleted region is variable in length, but always includes genes I, J, K, and part of L on the right. The left end of the deletion may fall at any of 25 or more places in any gene A through L. (The genes identified by complementation tests with suppressor-sensitive mutants, and relevant here, were named as follows, starting from the left end of the ordinary genetic map: A through H in alphabetical order, then M, L, K, I, J. Thus the right-hand ends of all the deletions, placed to the right of the host-range marker by Arber, lie to the right of J, which may be the same gene.)

Weigle, Meselson, and Paigen (1959) analyzed λdg's by physical methods. They found that lines of λdg of independent origin differ in density, each line maintaining its characteristic density during growth. They ascribed the differences in density to differences in DNA content per particle, correctly as it turned out. Thus crossovers between phage and bacterial DNA's giving rise to λdg's are nonequational in the sense that the lengths of phage DNA deleted and of bacterial DNA inserted are unequal.

The facts recapitulated above impose severe restrictions on possible mechanisms of recombination between λ and *E. coli.* Normal prophage insertion and excision, conserving both chromosomes intact, already impose restrictions. Campbell (1962) proposed what seems to be the only acceptable model. It is presented in Fig. 2, which is intended to be self-explanatory.

Note what the model accomplishes. The ring configuration makes prophage insertion and excision credible as consequences of single, reciprocal crossovers. The ring structure also permits, though it does not demand, map permutation. Given the map permutation, the model explains the origin of λ*dg* as a terminal prophage deletion associated with substitution of bacterial DNA (Kayajanian and Campbell, 1966). Their terminal posi-

tion in the prophage map explains in turn the fixed right end and variable length of different deletions. In an analogous manner, *bio* transducing particles can arise concomitantly with deletions at the left end of the prophage. The *bio* transducing particles, unlike *gal* transducing particles, need not be defective (Wollman, 1963), presumably because the distance in the prophage map between N and *bio* is shorter than that between *gal* and J (see Fig. 2).

Note too that the model predicts permutations that are cyclic, defined symbolically as ABC ⇌ BCA (Webster's New International Dictionary, second edition, 1934). Cyclic permutation should be clearly distinguished from circular permutation,

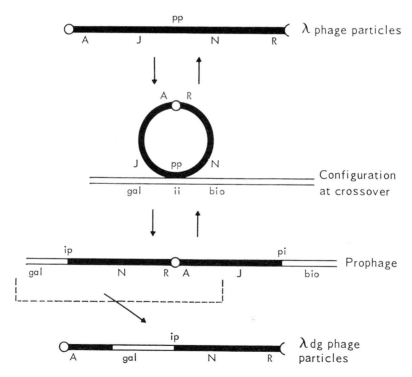

Fig. 2. Structural relations of λ, λ prophage, and λ*dg* according to Campbell's model. Genetic maps are indicated by the phage genes A, J, N, R and the bacterial genes *gal* and *bio*. Phage and bacterial components ore indicated by shading, terminal cohesive sites by arbitrary symbols. The symbol pp stands for the locus of permutation points and ii for the locus of prophage insertion sites, loci that become crossover regions ip and pi in the prophage map.

defined by Streisinger, Edgar, and Denhardt (1964) as a process giving rise to sequences among which all cyclic permutations are equally frequent.

Kayajanian and Campbell (1966) clarified certain features of the recombinations giving rise to λdg's. They selected lines of λdg containing some but not all of the gal genes, thus fixing within narrow limits one terminus of the substituted bacterial DNA. If λdg's arise as terminal prophage deletions, the selected lines should contain a fixed amount of bacterial DNA and their density should be inversely related to the size of the deletion. Kayajanian and Campbell found a strict correlation. Equally important, they found that the lengths of the deletions varied over the same range as deletions in unselected λdg's. Thus restriction of the crossover point to a specified region in the bacterial DNA does not markedly restrict its location in the phage DNA. If the crossovers depend on local similarity of base sequences, matching sequences are not ordered in the same way in the two DNA's. This is a very clear demonstration of illegitimate crossing over.

The selected λdg's varied in DNA content from −14.2% (deletion ending in A) to +3.4% (deletion ending in M), expressed in relation to the DNA content of wild-type phage particles. The range of variation signifies that the genetic distance between genes A and M corresponds to about 18% of the length of the λ DNA molecule. If no genes in this region remain to be found, the measurement requires 1100 nucleotide pairs per gene, which is a reasonable number.

The results of Weigle, Meselson, and Paigen (1959) as well as those of Kayajanian and Campbell show that λdg, like wild-type λ, enters and emerges from the prophage state without gain or loss of DNA. There-

fore both types of phage, though differing in the frequency with which they enter and leave the bacterial chromosome, must do so by crossing over at points separated by a fixed distance, either a unique pair of points or, as originally suggested by Campbell, points determined by legitimate crossing over in homologous regions of appreciable length.

A proper appreciation of Campbell's model, which, together with the facts that go with it, is one of the important contributions of microbiology to genetic thought, calls for an assessment of its historical origins. (Its actual origin, in the inventive mind of its creator, is another matter.) Note first of all that the structures invoked, with the exception of a properly situated region of homology between phage and bacterial DNA's, have by now been demonstrated. Only the origins of the structures are subject to hypothesis. Note, too, that prophage excision and insertion are strictly analogous to deletion and insertion in general. The formal models for genesis of all chromosome rearrangements are basically the same as Campbell's. (See, for example, Fig. 46 in Sturtevant and Beadle's *Introduction to Genetics*, 1939, where the general scheme is attributed to Serebrovsky.) Campbell's perceptiveness lay, perhaps, in recognizing early that the prophage has to be inserted, in spite of some experimental results that seemed to point in another direction.

Sturtevant and Beadle pointed out in their book that crossing over is a process concerning which only the results are known, and defined illegitimate crossing over as crossing over between nonhomologous chromosomes or between different parts of the same chromosome. Therefore no hypothesis is implied in the use of these terms except, perhaps, that normal prophage excision and the origin of λ*dg* share common mechanisms. It is doubtful too whether use of the word "homology" implies a hypothesis, and whether it should be used at all in the present connection. In Campbell's model, "region of homology" could be translated as "locus of crossover points." The meaning of homology is clear when it refers to chromosomes identical except for experimentally introduced genetic markers. Otherwise, in molecular contexts at least, it is better to speak of lengths and distributions of common base sequences, on the one hand, and crossover frequencies on the other, the relation between the two being unknown. The solid fact coming from the analysis of λ*dg*'s is that illegitimate crossing over, if it depends on common base sequences in this instance, can occur between DNA segments in which the common sequences are short compared with lengths of genes, and are, moreover, erratically distributed.

The basic question here concerns the mechanism of crossing over in general, but one can ask more modest questions, too. These have to do with similarities or differences between the illegitimate crossing over that gives rise to λ*dg*'s and the crossing over that is responsible for normal prophage excision. Do the two processes depend on the same or different enzymes? Current work may answer this question (Signer and Beckwith, 1967; Zissler, 1967). Are both examples of crossing over reciprocal? Prophage insertion is reciprocal in the sense that, in the overall process, four strands are cut and rejoined in new combinations. Prophage excision is probably reciprocal in the sense that a single lysogenic bacterium can give rise to daughter cells that lyse and daughter cells that are nonlysogenic (Weisberg and Gallant, 1967). These processes are also reciprocal in the sense that they are reversible.

Questions that can be asked about the origin of λ*dg* are severely limited because rare events are involved. All we know is that, in effect, a single piece of DNA is cut out of the lysogenic bacterial chromosome, the ends of the piece are joined, and additional cuts are made to create ends resembling the ends of normal λ DNA molecules. And, of course, that bacteria can sometimes survive deletions. These facts are plausibly accounted for if deletions in general result from single, reciprocal crossovers, as in the models of Serebrovsky and Campbell. Experiments designed to detect bacterial deletions in the vicinity of the prophage, arising concomitantly with prophage induction, might serve to indicate whether or not the illegitimate crossovers that generate transducing phage are reciprocal events. Such experiments should be feasible with phage φ80 if not with λ (Franklin, Dove, and Yanofsky, 1965).

Campbell's model beautifully organizes facts and clarifies questions. The same cannot be said of other models that might be experimentally equivalent to it.

Deletions in λ DNA

Skalka and Burgi

A mutant of phage λ known as λ*b*₂ contains a DNA molecule 15% to 20% shorter than that of wild-type λ (Kellenberger, Zichichi, and Weigle, 1961; Burgi, *Year Book 62*, p. 482). It has suffered a deletion near the chromosomal center to the right of the host-range marker (Jordan, 1964). The mutant grows normally except that it does not produce stable lysogens. Indirect evidence suggests that it has lost the crossover locus within which genetic recombination with the bacterium occurs during lysogenization (Campbell, 1965). Burgi and Skalka have found that

b_2 DNA lacks all or most of the 37%-GC segment present in the DNA of wild-type λ. Therefore b_2^+ function resides in or adjacent to that segment, and the segment does not contain genes whose functions are essential during the lytic cycle of phage growth.

Skalka has also analyzed DNA from a defective strain of λ known as λdg(A–J), which is a typical defective, gal-transducing phage whose deletion spans genes A through J in the left third of the genetic map.

Her results reveal two features of the DNA of λdg(A–J) that are best seen by analysis under somewhat different conditions (see legends, Figs. 3 and 4). As shown in Fig. 3, a large fraction of the DNA present in wild-type phage, including all the left-terminal 57%-GC section, has been replaced in λdg(A–J) by DNA having the GC content characteristic of the DNA of *E. coli*. As shown in Fig. 4, the 37%-GC section also has

been deleted. The deleted DNA therefore represents a continuous stretch measuring at least 54% of the wild-type molecular length taken from the left arm of the DNA molecule (but not including the extreme tip, because cohesive function remains). This physical structure, like the genetic results, is best interpreted as the consequence of a terminal prophage deletion according to the model shown in Fig. 2. Given that interpretation, four conclusions follow.

1. The 37%-GC segment must lie near the right end of the prophage map.

2. The locus of permutation points in the λ DNA molecule lies near or to the right of the right end of the 37%-GC section, at a position distant at least 54% of the molecular length from the left molecular end.

3. The 37%-GC segment cannot lie in a region of exact homology between λ and *E. coli*. If it did, loss of the λ homologue would be compen-

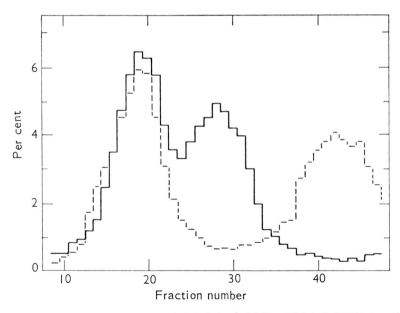

Fig. 3. Substitution of bacterial for viral DNA in λdg. Solid line, H^3-labeled DNA from λdg(A–J). Broken line, P^{32}-labeled DNA from wild-type λ. Size of fragments, 4 × 10^6 daltons. Hg/nucleotide ratio, 0.3. Cs$_2$SO$_4$ concentration, 45.5%.

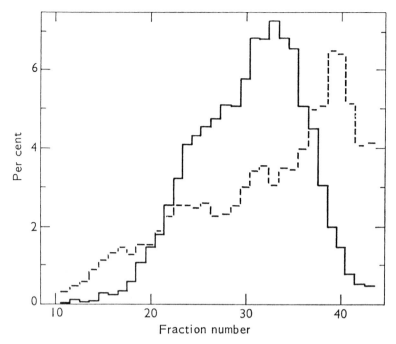

Fig. 4. Absence of 37%-GC section in DNA of λ*dg*. Solid line, H³-labeled DNA from λ*dg*(A–J). Broken line, P³²-labeled DNA from wild-type λ. Size of fragments, 1.9 × 10⁶ daltons. Effective Hg/nucleotide ratio, 0.20. Cs₂SO₄ concentration, 42%. The two DNA's were mixed before shearing and centrifuged after the addition of unlabeled T2 DNA fragments.

sated by gain of the bacterial homologue in the origin of λ*dg*.

4. All λ*dg*'s should lack the 37%-GC section plus varying fractions of the 57%-GC section depending on the lengths of their deletions, that of λ*dg* (A–J) being one of the longest.

Independently of models, comparison of the DNA's of λ*b₂* and λ*dg* (A–J) shows that the genes A through J, present in *b₂* but not in λ*dg*, must lie in the 57%-GC segment of the molecule.

The physical data show that λ*b₂* has suffered a 10% deletion of DNA near the right prophage terminus and an additional 5%–10% deletion of DNA of unknown location. We suggest as a plausible hypothesis that λ*b₂* arose by an illegitimate crossover deleting some of the DNA corresponding to both prophage ends.

Base-Sequence Similarities between λ and E. coli DNA's

Ingraham and Hershey

From previous work, we concluded that base-sequence similarities between the DNA's of phage λ and *E. coli* are generally though not uniformly distributed throughout the length of λ DNA (*Year Book 65*, pp. 562–565). As far as it goes, this conclusion is consistent with the hypothesis that the illegitimate crossovers giving rise to diverse lines of transducing phage depend on irregularly distributed base-sequence similarities.

The preferred crossover locus, pp in Fig. 2, responsible for normal prophage insertion and excision, must be something different. To account for the constancy in DNA content of phage λ through its lysogenization cycle, that locus must contain either a unique crossover point or a region of matching base sequences that is co-linear in the DNA's of λ and coli. Skalka concluded that the λ*dg* she analyzed had arisen from a prophage inserted by crossing over near the right end of the 37%-GC section of the DNA molecule. To examine base-sequence similarities in this region, we broke λ DNA into fragments of fractional length 0.32, removed re-joined ends (*Year Book 63*, pp. 581–582), and then selected mercury com-plexes of the remaining central sec-tions that were rich in adenine and thymine. Such fragments should con-tain most of the 37%-GC section and an adjacent part of the 46%-GC sec-tion, and should be shorter than un-selected fragments. When the selected fragments were reduced in size to about 0.1 of the original molecular length and were again analyzed by Hg-Cs$_2$SO$_4$ fractionation, two distinct components, with densities corre-sponding to 37% and 46% GC, in the approximate ratio 1/1.5, were recov-ered. The effectiveness of the selec-tion was also indicated by complete absence of a 57%-GC component. Both components recovered from the molecular centers proved to bind poorly to the DNA of *E. coli*, as shown in Table 1.

TABLE 1. Hybridization Tests with λ DNA Fractions

λ DNA Fraction	*E. coli* DNA Filters, 100 μg	λ DNA Filters, 10 μg
37%-GC centers	6, 6	83
46%-GC centers	7, 8	86
Right-end thirds	17, 19	87
Left-end thirds	11, 13	83
Unfractionated	13, 13	82

The numbers express percentages of soni-cated, P^{32}-labeled DNA fractions bound to unlabeled DNA attached to membrane filters. Two numbers signify duplicate measurements. Procedure according to Denhardt (1966).

We also examined short right and left molecular ends (fractional length 0.14), donated by Elizabeth Burgi. They were indistinguishable from the corresponding terminal thirds in tests like those shown in Table 1.

We conclude that the measurable base-sequence similarities between λ and coli DNA's are strongest (in λ DNA) near the right molecular end, weakest in the 37%-GC and 46%-GC sections near the molecular center, and intermediate near the left molec-ular end (see also *Year Book 65*, pp. 562–565).

These results do not of course ex-clude the possibility of a critical re-gion of exactly matching base se-quences near *gal* in *E. coli* and near the molecular center in λ DNA, be-cause genetic considerations suggest that the crossing over responsible for prophage insertion ought to be rather precisely defined, and that a short matching sequence in a region of poor matching would be all to the good. The concentration of matching se-quences in other parts of the mole-cule, however, argues for a recogni-tion device that does not depend on homology alone. Signer and Beckwith (1967) and Zissler (1967) propose, in fact, that phage λ employs a spe-cial enzyme that somehow directs the normal insertion and excision of the prophage. In principle, such an en-zyme could act by recognition of one matching base sequence among many, or even different, specified sequences in phage and bacterial DNA's.

Genetic Transcription in Bacteria Infected with Phage λ

Skalka

Skalka and Harrison Echols (Uni-versity of Wisconsin) have studied the effects of mutational defects in λ on production of messenger RNA during phage growth. Their results, which are being published in detail elsewhere, identify two genes whose primary function may be control of transcription.

Mutations in gene N block production of all messenger excepting a small amount similar to that formed when protein synthesis is inhibited by chloramphenicol. The simple inference, also suggested by the work of R. Thomas, is that a product of gene N directly initiates messenger synthesis characteristic of the lytic cycle of phage growth.

Mutations in gene Q, which do not block DNA synthesis or production of early-phase messenger, selectively depress transcription of genes responsible for late functions. The simple inference is that a product of gene Q specifically facilitates transcription of those genes. Needless to say, the facts are more reliable than the inferences at this time.

DNA Replication in Bacteria Infected with Phage T4

Werner

Last year Werner reported that a T4-infected cell contains a number of sites of DNA replication, about one for each molecular equivalent of phage DNA. His measurements were made at 45 minutes after infection of cultures growing at 25°C. Such cultures are entering a steady state of phage growth in which a constant rate of DNA synthesis is matched by an equal rate of phage particle formation to maintain an intrabacterial pool of replicating DNA of constant size.

When bacteria are infected with phage T4, DNA synthesis starts about 10 minutes later and attains a rapid rate very quickly, a rate that is clearly not proportional to the amount of phage-precursor DNA present in the cells. The approach to the steady state therefore calls for regulation of DNA synthesis. The in-fected cell could control its rate of DNA synthesis in either of two ways: by varying the number of growing points per unit length of DNA, or by varying the rate of synthesis at individual growing points. Werner has examined this question, and has found that the variable factor is the number of growing points. The rate of DNA synthesis at individual growing points remains constant.

Werner performed experiments of three types, described here in terms of specific examples.

Experiment 1. Infect thymine-requiring bacteria (*E. coli* B3) with thymine-requiring phage (T4*td*8) and allow growth to proceed in medium supplemented with 2 μg/ml of thymidine. At minute 40, add 20 μg/ml of H[3]-labeled 5-bromouracil. At minute 42, add a large excess of unlabeled thymidine and permit growing points to move away from the labeled sections in the DNA. Extract DNA from the cells at minute 45, break samples into fragments of various sizes, and measure both the 5-bromouracil content of the DNA and the size to which the DNA must be broken to liberate pieces, identifiable by their density, that contain one heavy and one light strand. The critical size turns out to be 0.10 of the length of a T4 DNA molecule. The 5-bromouracil content of the DNA corresponds to 6.0 DNA molecules per cell. Therefore the individual cells contain an average of 6.0/0.1 or 60 growing points at 40 minutes after infection. In this experiment, the total amount of DNA per cell is not measured.

Experiment 2. Start the infection in medium supplemented with C[14]-thymidine, then switch to H[3]-labeled 5-bromouracil by centrifugation and washing. The procedure is the same as in experiment 1 except for the additional measurement of the amount of DNA synthesized after infection. Because some cells lyse during centrifugation, the only reliable estimates are L, the average length of DNA segments of hybrid density, and F, the fraction of the recovered DNA containing 5-bromouracil in place of thymine. The ratio L/F gives the amount of DNA per growing point. This ratio measures 1.5 molecular equivalents at 45 minutes after infection.

Experiment 3. Start growth in thymidine-containing medium and add an excess of H^3-labeled 5-bromouracil at time t. After an additional interval, Δt, extract DNA, reduce it to small fragments by sonication, and measure the ratio between H^3 counts in heavy DNA (both strands labeled) and in hybrid DNA (one strand labeled). This ratio increases in proportion to Δt, and results can be interpolated to find Δt corresponding to the ratio heavy/hybrid equal to 0.5. The interpolated Δt is roughly the interval during which two growing points move over an average segment of replicating DNA. Since the rate of movement of growing points is known, the length of DNA between growing points can be calculated. The distance between growing points, D, depends both on the time t at which 5-bromouracil is added and on the thymidine concentration of the medium. At 2 $\mu g/ml$ thymidine, D = 0.17 of a T4 length when t = 20, and 0.29 of a T4 length when t = 40. At higher concentrations, D is greater at t = 40, about 0.73 of a T4 length.

Results of these three types of experiment permit the following conclusions.

1. Growing points first appear at 10 minutes after infection and increase in number at the rate of 3 per bacterium until they number 60 at 30 minutes, after which the number remains constant (experiments of type 1). The number of growing points found during a short pulse with 5-bromouracil does not depend on the amount of DNA in the cultures as influenced by the prevailing thymidine concentration.

2. Individual growing points move at the rate of 5% of a T4 length per minute in the presence of 5-bromouracil. This rate does not depend on the time after infection at which the measurement is made (experiments of types 1 and 2).

3. The distance between growing points in replicating structures varies from about 0.2 of a T4 length at early times to nearly 1 at late times (experiments of type 3). Replicating DNA can therefore take the form of a multiply branched structure.

4. During the steady state of phage growth, cultures maintained in the presence of excess thymidine produce 4.5 phage particles per bacterium per minute and synthesize DNA at the equivalent rate. If there are 60 growing points per cell, they are moving at the rate of 0.075 of a molecular length per minute. The local rate of DNA synthesis measured in the presence of 5-bromouracil is 0.05. Brief labeling with H^3-thymidine or with C^{14}-5-bromouracil shows that growing points move 1.5 to 2.0 times faster in the presence of thymidine than in the presence of 5-bromouracil. Thus each result checks fairly well with two independent measurements. The same is true at earlier times when the rate of DNA synthesis per cell is increasing.

Werner concludes that the rate of DNA synthesis in the presence of thymidine is controlled by the number of growing points, not by their rate of movement, and that there is no severe limitation to the number of growing points per length of DNA or per cell. A similar conclusion was suggested by Sueoka and his colleagues concerning replication of bacterial DNA in *Bacillus subtilis*.

According to Werner's results, growing points accumulate rapidly but move rather slowly and tend to remain clustered in the templates on which they originate. This conclusion suggests an unanticipated role for genetic recombination: to distribute growing points over the newly synthesized DNA. Eckhart found that genetic markers introduced into a T4-infected cell by a superinfecting phage replicate mainly after recombination with markers contributed by the primary infection. His finding can perhaps be explained, wholly or in part, by the clustering effect mentioned above.

Sedimentation Rates of Polynucleotides

Ingraham

Molecular weights of the polynucleotide chains released by denaturation of DNA can be measured from their rates of sedimentation in alkaline solutions according to an equation of the type $D_2/D_1 = (M_2/M_1)^\alpha$, where the D's and M's refer to distances sedimented and molecular weights of two DNA's spun in the same tube. The exponent α was estimated at 0.40 by Studier and at 0.38 by Abelson and Thomas, the difference possibly reflecting the use of different reference DNA's and different solvents. In any case, it remains uncertain whether or not the exponent α is really constant over a wide range of molecular weights. With the assistance of Dr. Gobind Khorana, Ingraham has made a partial check.

Khorana supplied two samples of C^{14}-labeled thymine deoxyoligonucleotides, one containing the heptanucleotide, the other containing mixed nucleotides of somewhat greater length. Since the mixed sample was larger, it was calibrated against the heptanucleotide and then used as a reference in other measurements.

In principle the check is simple. The molecular weight of single strands of λ DNA is 15.5 million, that of the heptanucleotide 2320. Measurement of the relative distances sedimented permits an estimate of α. Owing to the very great difference in sedimentation rates, the comparison has to be made in several steps. Ingraham used a sample of DNA broken by stirring and a sample of enzymically hydrolyzed DNA as intermediate references of unknown molecular weight. Other than the thymine oligonucleotides, the materials were prepared from P^{32}-labeled λ DNA.

Adjacent pairs in the molecular weight series were spun in concentration gradients containing 5%–20% sucrose, 10^{-3} M ethylenediaminetetraacetate, 0.3 M NaOH, and 0.7 M NaCl. Time and speed of centrifugation were chosen to bring the faster-sedimenting member of the pair well down the tube.

Results for two or three trials of each kind are given in Table 2. They show that the molecular weight ratio $15.5 \times 10^6/2320$ corresponds to a distance ratio falling between 28.0 and 31.4. Substitution of these limits in the equation gives estimates of α ranging from 0.378 to 0.392. Ingraham also verified that the exponent 0.38 serves in the range of molecular weights between 15.5 million and 320,000 under the conditions specified. The provisional molecular

TABLE 2. Relative Sedimentation Rates of Polynucleotides

DNA Pair	Distance Ratio
Intact λ–sheared λ	2.78, 2.81, 2.82
Sheared λ–enzyme digest	3.15, 3.17, 3.23
Enzyme digest– oligonucleotide pool	2.73, 2.81, 2.88
Oligonucleotide pool– heptanucleotide	1.17, 1.20
Intact λ–heptanucleotide, calculated	28.0–31.4

weights she cited last year are therefore about right (*Year Book 65,* pp. 564–565).

The check is incomplete in at least one respect. Nobody knows how strongly the sedimentation rates of oligonucleotides may depend on their composition.

STRUCTURE AND FUNCTION OF PHAGE DNA'S

Segmental Distribution of Nucleotides
in λ DNA

Skalka, Burgi, and Hershey

For some years it has been clear that the DNA molecule in particles of phage λ contains segments of unlike composition. This is conveniently demonstrated by breaking the molecules, which are initially uniform in density, into fragments. The fragments exhibit a density that depends on average nucleotide composition and varies with place of origin in the molecules. As shown by Nandi, Wang, and Davidson, one can accentuate the density differences by combining the DNA with mercury. We described in previous reports our principal methods of

---

*Stent cites Schrödinger and Bohr as prophets of "other laws of physics" presumed to reside in the hereditary substance. The curious reader should be able to find a textbook of physiological chemistry, probably out of date when I read it around 1930, whose author expressed in his preface the opinion that atoms composing living matter would turn out to possess unique attributes.

Fig. 1. Nucleotide distribution in λ DNA. The compositions of segments, expressed in mole per cent guanine plus cytosine, are given above the line; fractional molecular lengths below. The orientation places gene A on the left.

analysis (*Year Book 65*, pp. 559–562) and preliminary results (*Year Book 66*, pp. 650–651; reprint, pp. 6–7). This year we pursued the analysis about as far as the methods permit, with rather greater success than we had anticipated. We identified six intramolecular segments. Their arrangement in the molecule is illustrated in Fig. 1. A full account of the work has been published in the *Journal of Molecular Biology*. There we discuss the significance, actually obscure, of the proposed structure.

The Molecular Site of the Repressor Gene

Bear and Skalka

In bacteria lysogenic for phage λ, the functioning of most phage genes is in-

hibited by a protein, called repressor, that is the product of a gene called c_I situated in the right half of the genetic map of the phage. As far as is known, only c_I and an adjacent gene called *rex* function under these conditions, and other genes are not transcribed.

Phyllis Bear took advantage of the circumstance described above to locate the genes c_I and *rex* in our physical map of the λ DNA molecule. Though she encountered many difficulties, her experiment is simple in principle. She broke λ DNA molecules into small fragments, sorted them into fractions of unlike nucleotide composition by density analysis, and passed equal volumes of each fraction, heated to separate the DNA strands, through membrane filters. This gave her a series of filters containing

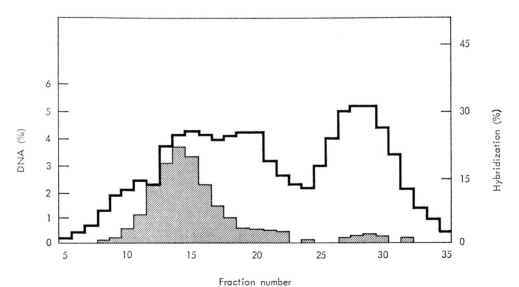

Fig. 2. Hybridization of messenger RNA with λ DNA fractions. Outer histogram: λ DNA fragments of fractional molecular length 0.05 separated according to density in 42.5% Cs_2SO_4 containing 0.22 mole $HgCl_2$ per mole of nucleotides. The four bands, left to right, contain DNA of 37%, 43%, 48.5%, and 57% GC. Shaded histogram: hybridization of λ messenger RNA, isolated from uninduced lysogenic bacteria, with DNA from each fraction.

various amounts of DNA but representing, in effect, equal numbers of copies of individual genes. She then isolated labeled RNA from lysogenic bacteria, fractionated it by hybridization with λ DNA to enrich for phage-specific messenger, and tested the messenger for hybridization with the DNA fractions affixed to filters. Her results, presented in Fig. 2, show that the RNA binds specifically to DNA of 43% GC content. Reference to Fig. 1 shows that this DNA originates mainly from the segment of fractional molecular length 0.17 whose center lies at fractional distance 0.38 from the right molecular end. Therefore c_I and *rex* are two of perhaps five genes that could be accommodated in this segment.

Dove (*Ann. Rev. Genet.*, 1968), citing data from several sources, puts c_I at fractional distance 0.25 from the right molecular end. His estimate is consistent with Bear's, even though neither may be very exact.

The results verify in a direct way that repression of gene function specifically suppresses the transcription of the repressed genes. Bear's experiments also demonstrate a useful means for determining the GC content of RNA.

Transducing Phages

Yamagishi and Skalka

Phage λ has contributed in many ways toward an understanding of genetic recombination. To appreciate this contribution, one must clearly distinguish three kinds of recombination seen in λ.

The first kind, observed in ordinary phage crosses involving homologous chromosomes, is the molecular equivalent of the process described in textbooks of genetics. Meselson showed that the parallels include reciprocal exchanges and, probably, four-strand exchanges. Meselson, Weigle, and others brought to light recombination by literal breakage and reunion of DNA molecules. The same authors, and, with particular clarity, Kaiser and Hogness, demonstrated the congruence of the genetic map with molecular structure in DNA. The now commonplace fact that genetic crosses yield information about topology of DNA molecules is less than ten years old.

Ordinary genetic recombination can occur at any internucleotide bond (for example, within anticodons, as shown by Henning and Yanofsky), and is equational in the sense that the parental numbers of nucleotides per gene are conserved in recombinants (as shown by Crick, Barnett, Brenner, and Watts-Tobin in the drastic effect of exceptions). These properties place recombination in the same category as replication: processes dependent on pairing between complementary nucleotide sequences. In λ, ordinary recombination is catalyzed by the function of a phage gene, *red*, and by the function of an equivalent bacterial gene, *rec* (Signer and Weil; Echols and Gingery).

The second kind of genetic recombination seen in λ is responsible for the reversible insertion of prophage into the bacterial chromosome, a process for which Campbell invented the proper model (*Year Book 66*, pp. 651–657; reprint, pp. 7–13). Prophage insertion and excision differ from the exchanges described above in three respects: matching nucleotide sequences cannot be confidently invoked (*Year Book 66*, pp. 657–660; reprint, pp. 13–16), the exchanges occur at a unique site in each DNA, and they depend on the function of a phage gene *int* (Zissler; Gingery and Echols; Gottesman and Yarmolinsky). The function of *int* is itself site-specific, but is otherwise equivalent to that of *red*. Thus *int* function can substitute for *red* function with respect to ordinary recombination at the appropriate site, but *red* function does not measurably assist in prophage insertion. Other differences, implied by the words insertion and recombination, are topological, not mechanistic.

The third type of recombination seen in λ brings about excision of prophage in such a manner as to give rise to structures in which both phage and bacterial genes are represented. The recoverable structures, usually chromosomes of transducing phage particles, consist of a continuous segment of prophage DNA joined to a continuous, adjacent segment of bacterial DNA (Arber, Kellenberger, and Weigle, 1957; Campbell, 1957). Kayajanian and Campbell (1966) identified this type of recombination as illegitimate crossing over, in the technical sense that it does not depend on linear homology between phage and bacterial chromosomes.

The generation of transducing phage lines by illegitimate crossing over of the sort described is a rare event largely inaccessible to direct analysis. It is not even clear as yet whether or not the process depends on *rec, red,* or *int* functions, because when these are not provided, transducing phage particles of an atypical kind make their appearance (Gottesman and Yarmolinsky; Gingery and Echols). This result, which may signify a fourth category of recombination catalyzed by the hypothetical enzyme that cuts the molecular ends of λ DNA, illustrates a point made in the Introduction: that in biology one process often interferes with the analysis of another.

The origin of typical transducing lines of λ can perhaps be accounted for by a process that resembles recombination of the first kind except for one known difference: the interacting DNA's are not homologous. According to this view, recombination of the third kind would depend on a nonlinear pattern of matching nucleotide sequences in the DNA's of λ and its host, *Escherichia coli.* Then the only resemblance to normal prophage excision (recombination of the second kind) would be topological.

Yamagishi and Skalka undertook to determine whether or not matching nucleotide sequences in phage and host DNA's are distributed in the same manner as the observed exchanges that generate transducing phage lines. If they

are, bacterial DNA recovered from *gal*-transducing phages must resemble the left half of λ DNA, and bacterial DNA recovered from *bio*-transducing phages must resemble the right half of λ DNA. This prediction follows from the model of Kayajanian and Campbell for the origin of transducing phage lines (*Year Book 66,* pp. 651–657; reprint, pp. 7–13).

Yamagishi and Skalka retrieved bacterial DNA from the λ*dg*(A–J) of Adler and Templeton, and from the λ*dbio* M55-3 of Kayajanian. Skalka's previous work with λ*dg* illustrates the principal method (*Year Book 66,* pp. 657–659; reprint, pp. 13–15). She found that the segment of bacterial DNA lying between *gal* and the prophage has the uniform composition 51% GC, the same as that of the bacterial DNA as a whole. The segment found in λ*dbio,* on the contrary, proves to contain subsections ranging in GC content from 37% to 50%. Thus a certain resemblance is evident between the *gal* region of E. coli DNA and the left half of λ DNA (Fig. 1), and between the *bio* region of E. coli DNA and the right half of λ DNA.

The distribution of matching nucleotide sequences, expressed in terms of results of the appropriate DNA-DNA hybridization tests, shows the following features. (1) Bacterial DNA recovered from either transducing phage binds more efficiently to λ DNA than does unselected bacterial DNA. (2) Bacterial DNA from λ*dgal* binds equally to right and left halves of λ DNA. (3) Bacterial DNA from λ*dbio* binds preferentially to right halves of λ DNA. Thus the distribution of matching nucleotide sequences is such as to permit and even encourage the kinds of illegitimate crossing over that generate transducing phages.

The results suggest a resemblance between the left half of λ DNA and the *gal* region of E. coli DNA, and another between the right half of λ DNA and the *bio* region of E. coli DNA. The resemblances are weak but are perceptible both in local composition and in sequence of nucleotides, a correlation that is obligatory if matching sequences are numerous and clustered (*Year Book 65*, pp. 562–565). Given the observed types of λ transducing phages, and the hypothesis of their origin by crossing over at match points, one sees the germ of a functional explanation for the observed distributions of nucleotides in phage and host DNA's in the constraints imposed by the pattern of their habitual interactions. Differentiated segments in two interacting DNA's should resist the manifest selective influences that otherwise favor uniformity. More important, one might expect positive selection for differences in composition between elements whose interaction could be profitably discouraged. These are not idle thoughts. They lead to the prediction that the sort of structure found in λ DNA should be characteristic of specialized transducing phages, not of phages like P1 that disseminate bacterial genes by another mechanism.

Related experiments performed by Laura Ingraham some years ago were less edifying, though not, in fact, inconsistent with the ideas just presented. Ingraham found, in collaboration with Naomi Franklin, that the DNA of bacterial strains, deleted in the vicinity of the prophage insertion site of φ80, did not show diminished ability to hybridize with the DNA of that phage. Such experiments should be repeated with appropriately defined bacterial deletion mutants.

We suggest as a general inference that all recombination processes represent variations on a single theme. Those giving rise to transducing phages are inefficient by design. Because illegitimate crossing over is a rare event, it can afford to sacrifice precision for the sake of adventure.

Phage φ80

Skalka

The phage φ80 is one of several species related in many ways to λ. Comparison of structure of the two DNA's might be expected to clarify relatedness, on the one hand, and to provide clues to interdependence of structure and function on the other, notwithstanding the perennial difficulty that these two aspects of comparative structure tend to defeat each other.

Skalka has analyzed structure and transcription of the DNA of φ80 by methods previously applied to λ. She finds, in molecules of φ80 DNA, a GC-rich half that, by analogy with λ DNA (Fig. 1), can be called the left half. Then it turns out that the two DNA's terminate in identical cohesive sites at their left ends, as shown by the fact that left molecular halves of either DNA can join only to right molecular halves of either. Both DNA's contain long segments of high GC content (55% in φ80) in their left halves, and both terminate on the left in short segments of lower GC content. Hybridization tests show a moderate cross-reaction (20% for the unfractionated DNA's), which is strongest in the left half of each. In both DNA's the late-functioning genes reside in left molecular halves, as determined by hybridization tests with messenger RNA.

Right molecular halves of the two DNA's terminate in identical cohesive sites, contain early-functioning genes, and show similar discontinuities in nucleotide composition. They differ appreciably, however, in average composition (45% GC in λ, 51% in φ80), and show relatively weak cross-reactions in hybridization tests. The central segment of low GC content found in λ has no counterpart in φ80 detectable either by density analysis or by hybridization tests.

Messenger RNA synthesis in bacteria infected with the respective phages is entirely similar, characterized by slow transcription of right molecular halves at early times and rapid transcription of both halves, but predominantly the left, at late times.

The relatedness of the two phages is therefore clearly evident as linear homology in their DNA's. The differences are instructive, however. Our work with λ was prompted in part by our biological prejudice that structural singularities necessarily serve some useful purpose. Some of Skalka's results with λ were compatible, for instance, with the hypothesis that the switch from early to late phases of transcription depended directly on some device sensitive to local nucleotide composition (*Year Book 64*, pp. 526–527). If that were true, one should expect early and late messengers to differ markedly in composition even in a phage like φ80, where the average compositions of large DNA sections do not differ greatly. Since Skalka's results with φ80 fail to verify the prediction, she concludes that patterns of transcription are not directly dependent on nucleotide composition.

Skalka plans to extend comparative analysis along the lines indicated to a few additional phage species.

Origin of λ DNA Synthesis

Makover

If a replicating DNA molecule contains a number of growing points that started from a single origin and are moving in the same direction, a nucleotide sequence lying ahead of the first growing point should be represented only once in the replicating structure, whereas a sequence lying between the nth and $n+1^{th}$ growing points should be represented $n+1$ times. Sueoka and his colleagues exploited this principle to determine the sequence of replication of genetic markers in *Bacillus subtilis*. Makover is using it to determine the sequence of replication of the segments of λ DNA of dissimilar nucleotide composition (Fig. 1).

Makover finds, first of all, that the appropriate experiments are feasible. He treats bacteria with mitomycin to destroy their capacity for synthesis of their own DNA, infects them with phage, and feeds tritiated thymidine to the culture for just three minutes to label replicating phage DNA. He then extracts DNA from the cells, mixes it with differently labeled DNA from phage particles, breaks the DNA in the mixture to small fragments, and fractionates these according to nucleotide composition as illustrated in Fig. 2. He finds that the replicating and nonreplicating DNA's in the mixture form identical distributions except for a quantitative bias that he attributes to sequential replication of the several molecular segments. His results show that the 48.5%-GC segment is one of the first to be replicated. He hopes that further experiments will define both origin and direction of replication. What is more important, he may be able to verify the general model of replication in a single direction from a unique starting point.

Makover's results are consistent with other facts. A line of λdg from which the 57%-GC section of DNA (Fig. 1) has been deleted is able to synthesize DNA (Brooks; Echols, Skalka, and others; also analogous experiments by Dove and Franklin). A chromosomal locus (ri), situated near the right end of the genetic map, is a plausible candidate for the starting point of DNA synthesis (Dove, Hargrove, and Haugli).

R. L. Sinsheimer, in whose laboratory the mitomycin treatment was first applied in experiments with λ, furnished the requisite bacterial strain.

Requirements for Replication of T4 DNA

Mosig and Ehring

Gisela Mosig, now at Vanderbilt University, is continuing her physical and genetic study of T4 phage particles which, by some sort of biological accident, end up with a DNA fragment of two thirds the normal molecular length (*Year Book 64*, pp. 521–524). Such particles contain a random segment of the T4 chromosome, as shown, among other ways, by the fact that two or more par-

562   CARNEGIE INSTITUTION

ticles attaching to a single bacterium can usually initiate infection, whereas one particle cannot.

In the summer of 1966, Mosig returned to our laboratory to collaborate with Ruth Ehring in an effort to determine whether or not a single DNA fragment can replicate. Mosig and Ehring infected bacteria with the incomplete phage particles, added radiophosphorus to the culture, extracted DNA from it some time later, and measured the amount of radioactivity in T4 DNA by hybridization tests. Unfortunately, there is an irreducible limit to the sensitivity of such experiments, because some bacteria necessarily get infected with two or more phage particles. However, Mosig and Ehring found that little or no DNA synthesis ensued in singly infected bacteria.

In the genetic map of T4, known in great detail from the work of Epstein, Edgar, and colleagues, there are a number of genes essential for DNA synthesis. Seldom could all these genes be represented in a DNA segment of two thirds the normal length. This consideration sufficiently explains the observed result. However, as Mosig and Ehring point out in their paper, DNA fragments lack terminally repetitious nucleotide sequences, which may also be essential for DNA synthesis. Mosig is investigating this possibility.

The T4–λ Universe

Hershey

I discuss here certain aspects of DNA structure which, though not especially new, are still sufficiently novel to provoke thought. The serious reader may find it helpful to have at hand my previous discussions of related matters (*Year Book 63*, pp. 589–592; *Year Book 66*, pp. 647–657; reprint, pp. 3–13).

The main historical facts are worth recalling for their own sake. Streisinger, Edgar, and Denhardt, in the first of a series of three papers written between 1961 and 1963, and published between 1964 and 1967, predicted faithfully on the basis of genetic experiments the structure of T4 DNA verified by Thomas and Rubenstein (1964) and MacHattie, Ritchie, Thomas, and Richardson (1967). So much for the complementarity principle according to which genetics and chemistry were to provide immiscible information.

Campbell in 1962, likewise on the basis of genetic experiments, predicted a rather different structure for λ DNA. That structure was verified by Hershey, Burgi, and Ingraham (1963), Ris and Chandler (1963), and Wu and Kaiser (1968).

My universe of discourse is presented in Fig. 3, which shows information dia-

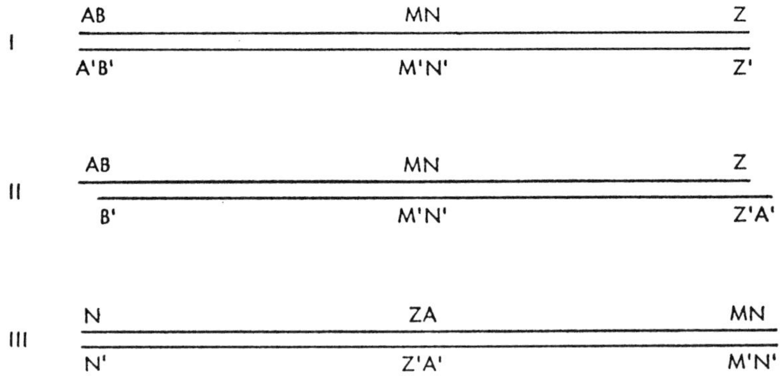

Fig. 3. Three bihelical DNA molecules represented as information diagrams. Each letter signifies a nucleotide sequence of arbitrary length; each primed letter, the corresponding complementary sequence.

grams of three bihelical DNA molecules. Each letter in the figure stands for a nucleotide sequence of arbitrary length, and each primed letter for the corresponding complementary sequence.

Structure I, though it conforms to the classic model of 1953, is not known to exist. It may be characterized as an open sequence possessing a twofold redundancy of language in a code of pairwise complementarity.

Structure II could be derived from structure I by Maxwell's demon without the performance of work. But he would have to violate the principles of information theory that dominate DNA synthesis, because structure II contains the sequence Z′A′ missing in I. Structure II is slightly less than twofold redundant, and contains slightly more information per nucleotide than structure I. Note that joining the complementary ends of structure II, which does not alter information content because scientists can do it reversibly, results in a closed sequence with exactly twofold redundancy. Thus the extra information in structure II is the potentiality of closure; messages I and II contain the same number of sentences. Structure II, characterized by a non-repetitive terminal redundancy, is that of λ DNA.

Structure III could be derived from structure II by an energetic but mindless demon or, in fact, by Kornberg's DNA polymerase. The two structures have the same information content but differ in mass. The increased mass is associated with the terminal repetition, which makes structure III appreciably more than twofold redundant. Structure III also possesses the potentiality of closure, albeit in a slightly hidden form. The ends of structure III can be joined in the laboratory with the aid of nonspecific enzymes to form a closed sequence with exactly twofold redundancy.

Structure III as I have drawn it also differs from structure II in sentence order. In fact, the sentence alphabet permits 26 cyclic permutations, which are

equivalent since the cyclic structure is common to all. Structure III and its cyclic permutations represent T4 DNA, characterized by terminal repetition and circular permutation. T4 DNA molecules exist as isomeric systems representing, perhaps, 200,000 permuted structures.

Phage λ also permutes its nucleotide sequences to form prophage, but only the one cyclic permutation is known, and DNA molecules isolated from λ phage particles are all alike (*Year Book 66,* pp. 650–657; reprint, pp. 6–13).

I conclude that DNA language is not human language and that information theory of the message variety is not serviceable for both. Consider a message A through Z in which sentence Z is a paraphrase of A. Then, for purposes of human communication, we have A–Z = A–YA = A–Y, and Z serves only to minimize errors of transmission. In the language of DNA chemistry, A–Y and A–YA are not equivalent, and open sequences of the type A–Y may well be meaningless. Furthermore, we can write A–YA = N–YA–N for DNA language, a rule implying a degree of independence in the meaning of individual sentences not common in human messages. This second difference is only one of degree, and DNA sentences must surely remain intact in the long run. Lambda sees to this by cutting between sentences. T4 sees to it by cutting terminal repetitions of generous length. In short, DNA language is a chemical language obeying special topological rules, rules that are enforced by the special kind of twofold redundancy usually found in its messages.

If the redundancy of DNA language plays a special role in reclaiming permuted messages, it also serves, of course, for the ordinary purpose of minimizing accidental losses of information. Single strands of DNA are vulnerable physically because of mechanical fragility and because they are susceptible to a variety of nucleases to which bihelical molecules are refractory. They are vulnerable as messages because every internucleotide bond is essential to the meaning of the whole. It cannot be accidental that unpaired DNA messages of appreciable length have been found only in encapsulated form in phage particles of several species, and it seems likely, too, that production and encapsulation of single-strand DNA are more or less simultaneous events during phage particle formation. This reasoning also applies to the unpaired ends of λ DNA, which, in spite of their complementary structure, are vulnerable because they form together a joining sequence without which the whole message becomes illegible. Seen in this light, the T4 DNA molecule is no more redundant than its nonrepetitive circular equivalent, and the phrase "terminal redundancy," which derives from the theory of open messages, becomes inappropriate to DNA language theory. I therefore substitute hereafter the phrase "joining sequence."

In 1963 a number of people suggested, mainly on the basis of genetic evidence, that DNA replication starts at a unique nucleotide sequence and proceeds in one direction (Cairns; Jacob, Brenner, and Cuzin; Nagata; Sueoka and Yoshikawa). The model will probably survive as a prototype if not as a universal rule. It introduced the starting sequence as a genetic element of novel kind on which replication depends. It also contained the implication, not obvious in 1963, that joining sequences constitute a second obligatory element. (The model was actually proposed for circular structures.)

Let us notice at the outset that there are at least two difficulties in this model. One of the reasons for proposing a unique starting point in the first place was to account for the fact that not just any piece of DNA can replicate. If joining sequences are also obligatory, that particular argument evaporates. The second difficulty is more pervasive. Doermann and Boehner showed in beautiful experiments that certain heterozygotes in phage

T4 can be characterized as structures with overlapping ends that replicate as such. According to Streisinger and Stahl and their colleagues, this means that, to a limited extent, T4 replicates its terminally repetitious nucleotide sequences without joining them. Readers of the following paragraph will note that the results of Doermann and Boehner contradict all aspects of the 1963 model for DNA replication. In discussing that model, I am in effect assuming that T4 replicates its DNA in two ways: one to account for the results just mentioned, one to account for other facts. For similar reasons, I ignore evidence recently gained by Huberman and Riggs that DNA sometimes replicates in two directions from a common starting point. As it happens, the model proposed by Huberman and Riggs nicely accounts for the results of Doermann and Boehner.

If DNA replicates from a unique starting sequence, it is unlikely in general, and is clearly impossible among the permuted molecules of T4, that the starting point is situated at a molecular end. If, in addition, replication proceeds in a single direction, end-to-end joining becomes necessary for completion of the replication cycle, and DNA language is seen to have a mechanical origin. Joining of either of the two sorts seen in λ DNA can serve: ring formation or chain formation. (Incidentally, catenate and concatenate are misused verbs and adjectives. Perhaps catemer is a preferable noun.) Thus end joining, not circularity as such, is the characteristic feature of DNA messages. Both rings and chains are needed: rings to simplify replication of single molecules and to serve other genetic purposes (*Year Book 66*, pp. 651–657; reprint, pp. 7–13); chains to permit length variation in the joining sequences of T4 (Streisinger, Emrich, and Stahl, 1967). In T4, owing to the permuted sequences, chains cannot be formed by literal end-to-end joining but must be synthesized directly from a circular template or arise by recombinative processes.

In λ, the possibility that molecular chains are formed by joining is structurally plausible and cannot be excluded on kinetic grounds. In view of the vulnerable nature of open ends, few replicating DNA molecules are likely to possess them. However, in vitro λ DNA molecules with complementary ends separated by an average distance of at least 13.5 μm form rings at appreciable rates. In a bacterium, a single pair of complementary ends, whether belonging to one molecule or two, cannot be distant from each other by more than about one micrometer. Thus, other things being equal, single DNA molecules in bacterial cells should form rings at least $(13.5)^3$ times faster than they do in bulk solution. Moreover, in cells the competition between chain formation and ring formation becomes a numerical problem indifferent to DNA concentration as such. For instance, if one could introduce several λ DNA molecules into a bacterium simultaneously, formation of a unimolecular ring would be an exceptional event.

In contrast to the continuously variable joining sequences in T4, phage λ has evolved two specific kinds: the molecular ends already discussed, and the ends of the cyclically permuted prophage. For each a specific cutting mechanism must be postulated. The molecular ends are better known than the prophage ends, but a cutting mechanism has been put in evidence only for the prophage ends. The two situations reinforce the common model for λ, on the one hand, and the more general model including T4 on the other. Since I shall have few occasions to speak of the prophage ends of λ, my phrase "joining sequences" will refer to those at the molecular ends.

The nature of the T4–λ DNA message suggests that DNA replication depends on at least four categories of genetic elements: one or more starting sequences; one or more enzymes (and perhaps antienzymes) directly responsible for DNA synthesis, including something that rec-

ognizes starting sequences; joining sequences; and a mechanism for the regeneration of joining sequences from joined sequences. In λ, the last function is supplied by a hypothetical gene *ter* that ensures production of a hypothetical cutting enzyme able to open joined ends. In T4, the corresponding function is supplied by a hypothetical device for measuring lengths of DNA. The fourth element of either kind is not essential for DNA synthesis as such, but serves as a primitive segregation mechanism for the preservation of molecular structure (Frankel, Weissbach).

The known joining sequences provide a model for a mechanism of genetic recombination catalyzed by at least one sort of cutting enzyme. In fact, Signer and Weil have shown that the hypothetical cutting enzyme that opens λ prophage ends specifically biases recombination frequencies measured in ordinary crosses: a nice demonstration of the abstract power of genetic analysis.

In λ the specific joining sequences and the hypothetical starting sequence differ in operational respects from genes. Presumably they function without translation, and their structural role could hardly be subject to complementation. They belong to a category of DNA structures called recognition sites by Dove. The operator serves as the classic example and the joining sequences at the molecular ends in λ DNA provide the least ambiguous example. Note, however, that in the molecular ends of T4, and for purposes of genetic recombination between homologous chromosomes in general, all sequences become recognition sites.

A unique starting sequence for DNA replication is in principle identifiable by deletion analysis, which for λ has been carried out in a remarkably complete way by Kayajanian and Campbell. I state here only limited aspects of their results.

Recall that joining sequences in λ DNA lie at the center of the permuted map of the prophage, that deletions of the right prophage end can be selected by recovery of *gal*-transducing phage lines, and that deletions of the left prophage end can be selected by recovery of *bio*-transducing phage lines (*Year Book 66*, 651–657; reprint, pp. 7–13). Deletions of either class span various lengths of DNA up to, but never including, the joining sequences. Left-end deletions, the longest of which include all the known genes essential for DNA synthesis, result in defective phage particles able to replicate DNA when helped by normal λ present in the same bacterium. Right-end deletions result in defective phage particles able to replicate DNA on their own.

From these results, Kayajanian and Campbell drew conclusions of the following sort.

1. The joining sequences are essential for encapsulation of DNA in phage particles. That they are also essential for DNA synthesis is suggested by the experiments of N. Franklin and Dove with bacteria containing partially deleted prophages.

2. A unique starting sequence for DNA synthesis cannot lie in the right half of the prophage, that is, in subterminal parts of the left half of the DNA molecule shown in Fig. 1.

3. Rather unexpectedly, the results do not put in evidence any starting sequence recognizable as a structure with non-complementable function essential for DNA synthesis. Simply interpreted, this would mean that there is no unique starting sequence, or that the starting sequence is situated in or near the joining sequence at either molecular end. It could also mean something else, and that alternative I present in the arbitrary statements that follow (also suggested, in part, by Kayajanian).

(a) Any piece of DNA terminated by λ joining sequences can be replicated in the λ system because absence of a starting sequence in one member of a tandem complex can be made good by the starting sequence in a neighboring member.

According to this hypothesis, the starting point for λ DNA synthesis could lie anywhere in the right molecular half.

(b) From (a) it follows that unimolecular rings cannot be strongly preferred templates for synthesis of DNA incorporated into λ phage particles.

(c) Any piece of DNA of suitable length, bounded by λ joining sequences, can be encapsulated by the λ packaging system.

I should add that these statements are not literally true. Phages λ and φ80 possess, as far as we know, DNA's with identical joining sequences. Yet Dove (*Ann. Rev. Genet.*, 1968) finds, in contradiction to my assertion (a), that certain mutations in the DNA of one of these phages prevent its replication in the presence of the other. Dove also states that λ and φ80 possess distinct DNA-encapsulating systems, in contradiction to my assertion (c). As he points out, the latter contradiction disappears if one imagines that a species-specific cutting enzyme recognizes a few nucleotide pairs adjacent to the joining sequences proper. If that is so, assertion (a) may also be true with trivial qualifications. My point here is that a proper answer to any of these questions should include the answer to all—especially the one concerning the starting point for DNA synthesis.

Diverse evidence from the work of Dove, Weigle, Weissbach, Siminovitch, and others shows that a defect in any of several DNA encapsulation genes of λ interferes with the opening of the joining sequence in newly replicated DNA. This suggests that the encapsulation genes, lying at the left molecular end, the hypothetical gene *ter* (cutting enzyme), and the joining sequences, form a functional unit ensuring that the joining sequences will open only at the proper moment, a provision otherwise desirable in view of the vulnerable character of the open molecular ends. Synchronization could be achieved by control of transcription; localization in space could be achieved by suitable protein-protein interactions. These considerations, together with the principle of the clustering of genes of related function, place the hypothetical gene *ter* at the left molecular end.

In λ, clustering of genes of related function is not a principle but a fact. The principle can be debated. R. Thomas pointed out a few years ago the significance of linkage between the structural gene and target site for λ repressor. According to some of my colleagues, the evolutionary significance of clustering was anticipated by R. A. Fisher in 1930. Actual clustering was first clearly demonstrated by Demerec among genes of *E. coli*.

In λ, the joining sequences at the molecular ends form the most obligatory sort of functional unit but are scarcely clustered in the genetic map. This exception to the rule of clustering is a necessary one, but is no less instructive for that. It doesn't help to say that molecular ends are joined in the prophage map, because there the prophage ends come apart. It doesn't help either to say that some historical λ possessed a circular map, because we are looking at λ now. In fact, it appears that clustering is a useful device but one seldom exploited, perhaps, in chromosomes at large. Clustering is a radical principle in relation to genetic recombination; its opposite plays a conservative role. For instance, odd numbers of crossovers (with respect to the ordinary genetic map) between λ and a phage with dissimilar joining sequences should be lethal. In view of Dove's analysis, it seems likely that the same restriction applies to crosses between λ and φ80. Otherwise, λ is the radical creature par excellence: as seen in the transducing lines; as seen in the DNA of φ80, which is λ at its conservative ends and, according to Skalka, a sort of ghost of λ in between. I suggest that Fisher's radical clustering principle is opposed by a conservative unclustering principle, and that history is inscrutable to us precisely because, as Pareto noticed some years back, it mixes incompatible principles. That conclusion didn't daunt Pareto's thinking and doesn't mine.

Why has λ chosen to place its cluster of genes responsible for prophage insertion at the center of the DNA molecule, which automatically places the joined molecular ends at the center of the prophage map? The second consideration seems to be the crucial one. Since the molecular joint is the sole prophage structure indispensable to the formation of a transducing phage, the effect of the existing arrangement is to give λ an equal reach in both directions along the bacterial chromosome for the acquisition of transducible genes. Thus what might be mistaken for an abstract principle of symmetry actually ensures a maximum number of genes transducible in small pieces of DNA. A similar optimizing principle in message theory accounts for the fact that the DNA universe contains equimolar amounts of adenine and guanine.

I conclude this essay by turning for a moment to another phage known as P22, mainly to show that we can do so without leaving the T4–λ universe. Consider just two facts. P22 is a generalized transducing phage; it transports, in rare defective particles, pieces of bacterial DNA selected more or less at random. Its own DNA comes in terminally repetitious and circularly permuted form, like that of T4. As pointed out by Rhoades, MacHattie, and Thomas, who analyzed the DNA, the two facts are not unrelated. Both suggest that P22, like T4, encapsulates DNA by the length-measuring principle: in this case, any available DNA. Thus *E. coli*, in persuading λ to center its joining sequence in the prophage map, just made the best of a second-rate opportunity.

DNA Phenotypes

Many years ago methylcytosine was found as a minor constituent in wheat germ DNA. However, no clue to the significance of unusual bases in DNA appeared until the discovery of glucosylated hydroxymethylcytosine in phages T2, T4, and T6. In these phages, the replacement of cytosine by its hydroxymethyl derivative was found to be complete (Wyatt and Cohen, 1953). The glucosylation also proved to be massive, but showed a distinctive pattern in each of the three species (Volkin, 1954; Sinsheimer, 1956; Lehman and Pratt, 1960; Kornberg, Zimmerman, and Kornberg, 1961). Since the three phage species were

known to be very similar in function, the pattern of glucosylation could be recognized at once as part of the phenotype. A more general argument was clear too: since diverse phage species can multiply in cells of a single bacterial species, many phages must use a common genetic language. Therefore hydroxymethylcytosine is equivalent to cytosine, and uracil (found in some phage DNA's as well as in RNA) is equivalent to thymine, in the genetic dictionary. In fact, experiments with Kornberg's DNA polymerase later showed that a dozen or more bases, including artificial ones, are equivalent to one or another of the four kinds of which DNA is typically composed. Thus the genetic message is a specified sequence of four nonequivalent units. Equivalent units are those expected and found to be interchangeable in the base pairing rules of Watson and Crick. The choice among equivalent units generates optional phenotypes, optional sometimes at the discretion of the experimenter.

The notion that DNA, the bearer of the genetic message, itself exhibits diverse phenotypes occasioned some surprise, though the biological rationale was clear enough. Speciation may be regarded as the acquisition of devices by which living things compete (and sometimes cooperate) to preserve and disseminate their genes. One might have anticipated modification of DNA structure as a particularly direct means to this end. Indeed, Seymour Cohen suggested that hydroxymethylcytosine in T2 DNA might serve to protect against the action of degradative enzymes. His suggestion has proved correct for the glucosylated DNA.

The role of glucosylation as a species marker in the DNA of phage T2 is particularly dramatic. When its DNA contains glucose, this phage multiplies in bacterial cells and destroys the nonglucosylated DNA of the host. When the DNA of the phage does not contain glucose, it is rejected by the host, though it can function normally without gluco-sylation under special conditions (Arber, *Annual Reviews of Microbiology*, 1965).

The example of phages T2, T4, and T6 is rather special since few DNA's contain glucose. However, similar purposes are accomplished by more subtle chemical means in other species. In *Escherichia coli* and many of its phages, methylation serves as a strain-specific marker. Here the common features of several systems are "modification" of DNA at a few spe-cific sites by a bacterial methylating enzyme, and "restriction" by a nuclease that can cleave the DNA at the same sites provided they have not been meth-ylated previously. A number of such ge-netic systems are known, each character-istic of a different bacterial strain. Thus in *E. coli* strain A, the DNA of phage λ is methylated (or cleaved) at just one critical site lying between genes c_{II} and O. In *E. coli* strain B, λ DNA is meth-ylated or cleaved at two or more sites not including the A-specific site. The terms "modification" and "restriction" refer to the biological consequences of methyla-tion and cleavage: in general, phage par-ticles cannot infect a given strain of *E. coli* with high frequency unless they con-tain DNA previously marked by the methylating system of that strain. The bacterial DNA is subject to the same modifications and restrictions, which therefore give to DNA itself a number of alternative mating types (Arber and Linn, *Annual Reviews of Biochemistry*, 1969).

It should be added that the DNA of *E. coli* contains numerous methylated adenine and cytosine residues that do not play any known role as compatibility factors, although their distribution in the DNA is strain specific. Their significance is unknown.

Diverse phenotypes are seen also in gross structure of DNA. Thus the DNA of phage φX comes in single-strand rings, T2 DNA as circularly permuted rods, T5 DNA with characteristic single-strand cuts, several phage DNA's with

terminal repetitions, others with terminal cohesive sites. The significance of these variations is obscure but they probably reflect modalities of DNA replication on the one hand and, on the other, alterna-tive means of getting the proper length of DNA into phage particles. The varia-tions repeat a common theme: exploita-tion of the structural principle of com-plementary base sequences to permit cleavage and rejoining of DNA mole-cules without loss of message content. As expected according to this principle, one cut in the single-strand ring of φX DNA appears to be biologically irreparable. Idiosyncrasies of DNA structure have been discussed in annual reports from this laboratory for several years. Impor-tant examples are reviewed by C. A. Thomas, Jr. (*Journal of Cellular Physi-ology*, Supplement 1, 1967).

The amount of DNA per cell is an-other complex variable with phenotypic aspects. The primary component of the variation is the species-specific number of genes per set, which varies from three or four to a few hundred just among the viruses. Since evolutionary specialization often calls for new genes without making old ones obsolete ("ontogeny recapitu-lates phylogeny"), the maximum number must be very large. The question of non-genic DNA remains open and may prove unanswerable since some genes probably function only under special conditions, during embryonic life for instance. Sev-eral genes in T4 are dispensable or not, depending on functions provided by the host. Perhaps the proper way to phrase the question about nongenic DNA is to ask what functions of DNA remain to be discovered.

Repetitious DNA is fairly common (Britten and Kohne, *Science*, August 9, 1968). It is of course an obligatory fea-ture of DNA replication, especially dur-ing phage growth where it probably plays a physiological role in terms of gene dosage. A more interesting example has been analyzed by Brown and Dawid

GENETICS RESEARCH UNIT   663

(*Science,* April 19, 1968). Oocytes of the African toad contain large amounts of DNA that consists mainly of sequences matching those present in ribosomal RNA. Apparently the repetitious DNA is used for rapid synthesis of ribosomal RNA during oogenesis.

The extreme case of variation in amount of DNA is of quite another sort. RNA viruses dispense with the DNA phase of genetic determination entirely, having invented one or two genes permitting direct replication of RNA. What once seemed a major historical puzzle turns out to be a typical biological quirk.

Britten and Kohne give evidence for the existence of repeating sequences within single gene sets. They detect such sequences only in vertebrate species, but their methods may not be applicable to species with fewer genes. The striking feature of the data is multiple repetition of a few sequences, which cannot represent simply production of supernumerary gene copies because the repeated sequences are not perfectly identical. Britten and Kohne interpret their data in historical rather than functional terms, but the possibility of special function should be considered too. Current ideas about the genetic origin of antibodies are a case in point.

Adams, Jeppesen, Barrell, and Sanger (Cold Spring Harbor Symposium, 1969) have detected a complementary sequence in the RNA of phage R17 (in DNA, the equivalent structure would be called an inverted repetition). These authors directly determined the sequence of 57 ribonucleotides found in a particular fragment of the viral nucleic acid. The sequence can be written in the form of a hairpin cross linked by 19 out of 25 possible base pairs. Evidently such a structure could be accounted for by a series of historical accidents. More interesting is the likelihood that specified sequences affect secondary structure in RNA to permit control of replication or translation or both. The general implica-

tion seems to be that nucleotide sequences are subject to evolutionary constraints that have nothing to do with the genetic message proper—an inevitable correlate, perhaps, of the redundancy of genetic language.

The arrangement of genes within DNA molecules has subtle phenotypic consequences (Stahl, *Journal of Cellular Physiology,* Supplement 1, 1967). These are well illustrated by a single example. In λ prophage, nearly all phage functions have to be repressed, and the function of a single gene called c_I serves this purpose. Ptashne and Hopkins showed that the c_I product is a protein that attaches specifically at two binding sites in the DNA bracketing the genes c_I and *rex*. Szybalski and Taylor showed that in the absence of the c_I product, transcription starting in the vicinity of c_I proceeds outward in both directions. Evidently the repressor interferes with transcription at two starting points to meet the needs of the prophage in a remarkably direct way. This scheme of control depends on the arrangement of two genes and two repressor binding sites and, owing to the polarity of the genetic message, on the orientation and control mechanisms of outlying genes as well. Thus a highly specified chromosomal arrangement that serves functional needs also links together several genetic elements whose shuffling by genetic recombination has to be discouraged. The example conforms nicely to the operon model of Jacob and Monod, with special features attributable to the fact that lysogeny compresses the entire phage genome into just two mutually exclusive functions.

The λ repressor system also illustrates an evolutionary principle that is too often ignored: biological adaptation always means coadaptation, ultimately involving entire genomes, organisms, and populations. This principle accounts in part for the paradox that evolution utilizes mutations that are individually deleterious. The same principle suggests that attempts to distinguish between adaptive variation and "non-Darwinian evolution" through neutral mutation (King and Jukes, *Science*, May 16, 1969) are doomed to failure. Lest my remarks be construed as a defense of Darwinism, I offer the following propositions. Only strong theories generate alternatives. Darwinian theory is characteristically weak.

Perhaps the most puzzling aspect of DNA phenotypes has to do with the distribution of nucleotides within the molecules. Since recent discussions of this subject (Skalka, Burgi, and Hershey, *Journal of Molecular Biology*, 1968; *Year Book 67*, pp. 558–560) are already out of date, I recapitulate here the main historical facts before presenting some recent results obtained by Yamagishi and Skalka.

Perhaps the best way to state the problem is to describe the methods of study. Owing to the base-paired structure of DNA, the average composition of a molecule or fragment can be expressed by a single number, the molar fraction of guanine plus cytosine (G+C), which is equivalent to the fraction of guanine-cytosine pairs. The remaining fraction, if we neglect exceptional bases, represents adenine-thymine pairs.

The distribution of nucleotides within molecules can be determined by breaking them into fragments of known size, separating the fragments into classes of diverse composition, and measuring the G+C content in each class. Since the distribution is necessarily dependent on size of fragments, the analysis has to be repeated with fragments of various sizes. This sort of analysis has now been carried out for a few phage and bacterial species.

The nature of the problem could be seen only dimly in 1953, when interest was first focused on base sequence as the clue to the genetic message. The DNA species known at that time contained about 44% G+C, which seemed reasonable in a way, since an efficient language

would use all letters with similar frequency. This thought was short lived because Lee, Wahl, and Barbu (1956) and Belozersky and Spirin (1958) reported a number of bacterial DNA species whose G+C contents ranged from 26% to 74%. Thus it appeared that DNA language, like human language, was not designed primarily for efficient communication.

The discovery that the buoyant density of DNA is strongly dependent on composition (Rolfe and Meselson, 1959; Sueoka, Marmur, and Doty, 1959) yielded the first results concerning nucleotide distribution in DNA. For instance, Rolfe and Meselson found that the standard deviation of G+C content among fragments of *E. coli* DNA (fragment length probably about 10^4 nucleotide pairs) was less than $\pm 3\%$, to be compared with the 48% range covered by variations among species. Thus diverse bacterial species, surely possessing many functions in common, do not contain many DNA segments of similar composition. Rolfe and Meselson concluded that the compositions of protein and DNA could not be directly related to each other by a universal code.

Sueoka (1961) studied directly the relation between composition of DNA and composition of protein by analyzing the whole cellular protein of a number of microbial species. He found that the frequencies of the amino acids leucine, valine, and threonine showed no correlation with the G+C content of DNA. However, glycine, alanine, and arginine showed a weak positive correlation, and lysine, glutamic acid, and isoleucine showed a weak negative correlation. Sueoka's results can now be interpreted in terms of the degeneracy of the genetic code, in which 61 codons specify one or another of just 20 amino acids (Crick, *Cold Spring Harbor Symposia on Quantitative Biology*, 1966). Thus there are four valine triplets each containing either one or two guanine or cytosine residues, and the abundance of valine could not be favored by either extreme DNA composition. Alanine triplets contain two or three guanine or cytosine residues, and lysine triplets zero or one, in agreement with Sueoka's results. Sueoka's main conclusion, that compositions of DNA and protein are not strongly correlated, is also consistent with the coding dictionary, which allows a stretch of DNA specifying one each of 15 frequently occurring amino acids to vary in G+C content between 29% and 67%. Furthermore, both mutational study of individual proteins and comparative analysis of homologous proteins from different species show that functional requirements do not impose severe restrictions on the composition of protein. Therefore the observed variations in composition of DNA cannot signify diverse requirements with respect to the composition or function of proteins.

Having reached the conclusion just stated, Sueoka (1962) and Freese (1962) proposed that the composition of DNA was determined mainly by the genetically determined rates of mutational interconversion between guanine-cytosine pairs and adenine-thymine pairs. These authors also assumed that DNA composition as such had no functional significance and therefore could not respond to selective pressures. The latter assumption was perhaps superfluous to their main proposal because, however divergence in composition of DNA among different species may arise, one might expect it to be accompanied by coadaptive variation in mutational habit.

If DNA composition in a given species were determined primarily by mutational habit, guanine-cytosine pairs should be distributed at random among DNA fragments of gene size or larger. Recent analyses of several phage and bacterial DNA's by Yamagishi and Skalka show that the distributions are never random (see below). One must conclude either that DNA composition does reflect specialized functional adaptations or that interspecific genetic re-

combination is frequent with respect to the evolutionary time scale. Perhaps both possibilities should be considered likely. In any case, the hypothesis of domination by mutational equilibria loses its force.

Last year Yamagishi and Skalka proposed that an asymmetric distribution of G+C in bacterial DNA in the vicinity of λ prophage might be designed to favor the types of genetic recombination that give rise to transducing phage (*Year Book 67*, pp. 558–560). They are no longer enthusiastic about this hypothesis for two reasons. First, it now appears that the *bio*-transducing phage analyzed last year is atypical, having picked up bacterial DNA not proper to the *bio* region of *E. coli*. Its structure may not be relevant to the hypothesis under test. Second, the recognition that unselected fragments of *E. coli* DNA are rather dissimilar in composition neutralizes the significance of departures from the average composition in the vicinity of prophage insertion sites.

Miyazawa and Thomas (1965) first demonstrated that the DNA of *E. coli* contains segments of dissimilar composition. Yamagishi has carried the analysis further, and some of his results are presented in Fig. 1. The upper part of the figure confirms previous work in showing that large fragments of the DNA are uniform in composition. The lower part of the figure shows that fragments of the order of size of individual genes range in G+C content from 39% to 56%. The distribution is asymmetrical, with an average at 51%. The distribution is nevertheless rather compact: its standard deviation is ±3.8 percentage units in G+C content, as compared with ±6.7 units for λ DNA (Skalka, Burgi, and Hershey, 1968).

Yamagishi also examined *E. coli* DNA fragments of other lengths. His results show that stretches of the extreme composition 39% G+C range in length up to about 35,000 nucleotide pairs and comprise 3% of the total DNA. The asym-

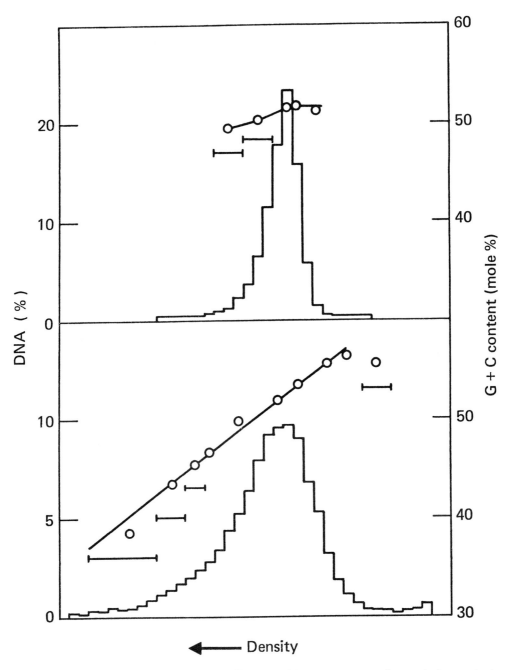

Fig. 1. Distribution of fragments of *E. coli* DNA with respect to guanine + cytosine content. Upper part: fragments of molecular weight 70 million (about 10^5 nucleotide pairs). Lower part: fragments of molecular weight 1.3 million (2000 nucleotide pairs). In both parts, distributions of DNA with respect to buoyant density in Hg-Cs$_2$SO$_4$ are shown by histograms, and the G + C content of fractions by the curves. Single fractions, or pooled fractions indicated by horizontal bars, were analyzed directly to get the points on the curves. For methods, see Skalka *et al.*, *Journal of Molecular Biology, 34,* 1–16, 1968.

metry of the distribution shown in Fig. 1 is characteristic, and signifies that long stretches of low G+C content are more numerous than long stretches of high G+C content.

The DNA of *Bacillus subtilis* is generally similar to that of *E. coli* except that its fragments range from 35% to 50% in G+C content, with an average of 44%. In collaboration with I. Takahashi of McMaster University, Yamagishi could show by genetic tests that regions of exceptional G+C content in *B. subtilis* include typical bacterial genes. Therefore local variations in composition do not reflect merely temporary residents in the bacterial chromosome such as prophages.

Yamagishi also analyzed several specific segments of *E. coli* DNA recovered from various $\phi80$ transducing phage lines. Here the content of bacterial genes can be identified by genetic tests, and the corresponding DNA can be recognized by fractionation with respect to density combined with hybridization tests to distinguish between components of phage and bacterial origin. A segment containing the tryptophan operon consists of DNA ranging in G+C content from 45% to 57%. A segment containing lactose genes is more homogeneous, with an average G+C content of 54%. Among the various segments examined, only the *gal* region contains DNA corresponding to the average for the entire chromosome, 51% G+C (Yamagishi and Skalka, *Year Book 67*, p. 559).

Skalka has examined a number of phage DNA species by density analysis of molecular halves and smaller fragments (about 2000 nucleotide pairs). By this method λ DNA molecules are readily shown to consist of dissimilar halves and to be made up of four or more distinct segments containing 37%, 43%, 48.5%,

and 57% G+C (Skalka, Burgi, and Hershey, 1968). The closely related phages 434, 82, and 21 are very similar to λ except that the 37% G+C section is absent in phage 21. Phage $\phi80$, related to λ, and the unrelated phage 186 resemble each other in containing only two distinct segments, the molecular halves, measuring approximately 50% and 55% G+C, respectively. The DNA of phage P2 also contains dissimilar halves, and resolves into three widely dissimilar segments. The DNA of phage P22 contains at least two dissimilar segments. Molecular halves of this DNA have the same composition, presumably because the molecules come with circularly permuted nucleotide sequences. Unlike the others, phages T5, T7, and P1 contain DNA's that are strikingly uniform in composition, though not absolutely so because small fragments exhibit asymmetrical density distributions. Phage P1 contains 5% of DNA of only 37% G+C.

Two conclusions emerge. First, all DNA's so far examined contain relatively long segments that differ in composition. Second, the phage DNA's so far examined fall into two classes. DNA molecules from phages λ, 186, P2, and probably P22 are composed of a few long segments of dissimilar composition. Since the effect is to produce dissimilar halves, these may be called asymmetric DNA's. By contrast, phages T5, T7, and P1 contain DNA's of relatively uniform composition. The grouping suggests that phage λ may be taken as representative of a class. If so, asymmetry of DNA structure, clustering of genes of related function in the genetic map, and propensity toward interspecific genetic recombination form a seemingly harmonious set of class characteristics.

BIBLIOGRAPHY

Bear, P. D., and A. Skalka, The molecular origin of lambda prophage mRNA. *Proc. Natl. Acad. Sci. U.S., 62,* 385–388, 1969.

Makover, S., A preferred origin for the replication of lambda DNA. *Cold Spring Harbor Symp. Quant. Biol., 33,* 621–622, 1968.

Skalka, A., Nucleotide distribution and functional orientation in the deoxyribonucleic acid of phage ϕ80. *J. Virology, 3,* 150–156, 1969.

Skalka, A., *see also* Bear, P. D.

Yamagishi, H., Single strand interruptions in PBS 1 bacteriophage DNA molecule. *J. Mol. Biol., 35,* 623–633, 1968.

The last three research reports from Al's lab are reflections on aspects of phage chromosome structure and function. Even today they appear truly wise.

PERSISTENT HETEROZYGOTES IN PHAGE T4

Parma and Ingraham are studying the genetic structure of phage lines that carry chromosomal duplications. The point of interest in their work, as in much of phage biology, lies in the possibility of understanding at the molecular level phenomena known to have their counterparts in higher organisms. Before summarizing the experimental results, it is necessary to recall some peculiarities of phage T4. They are interesting in their own right.

The rII gene of T4 is known from the work of Benzer (1955), who used it in experiments that bridged the gap between the structure of DNA and the structure of genes. The rII gene consists of two adjacent DNA segments, A and B, that together comprise about one percent of the DNA of the phage. Except for their cooperative function, A and B behave like separate genes, each subject to mutation from the active $(+)$ to the inactive $(-)$ form. Thus Benzer recognized four genotypic classes: A^+B^+, A^+B^-, A^-B^+, and A^-B^-. Only A^+B^+ can perform rII function. The three inactive classes are distinguishable both by breeding analysis and by the fact that only the chromosome pair A^+B^- and A^-B^+ can cooperate, when present in the same bacterial cell, to carry out rII function. This fact, together with evidence that A^- and B^- mutations occur in adjacent, nonoverlapping segments of the DNA, reveals the bipartite structure of the gene.

The function of the gene is superfluous to phage growth in ordinary bacterial strains but is essential for growth in bacterial strains lysogenic for phage λ. The critical function in phage λ has been traced to a gene called rex. Neither rII nor rex function is known in chemical terms. The technical importance of the rII gene resides in its bipartite structure and in the ease of recognition of its partial functions.

If two phages of genotypes A^+B^- and A^-B^+ are crossed, one gets among the progeny a few percent of A^+B^+ recombinants, and about one percent of A^+B^-/ A^-B^+ heterozygotes. Among the heterozygotes some are "internal hets" (two DNA strands of different parental origin) and some are "terminal hets" (repetitions at the ends of the DNA molecule marked by genes of different parental origin). If the particular mutants $rJ101$ and $r1589$ are crossed, one gets no internal hets (because the mutants contain mismatching deletions) and no A^+B^+ recombinants (because the two deletions correspond to overlapping nucleotide sequences). Nevertheless, owing to their exceptional structure, the respective mutants are functionally A^+B^- and A^-B^+, and the cross does generate A^+B^-/A^-B^+ terminal hets. The mutants $rJ101$ and $r1589$ are especially useful in genetic experiments, including those described below, that require detection of exceptional events. In principle, though, any functionally equivalent mutants would serve, and for clarity I shall stick to the more descriptive generalized notation.

F. W. Stahl and his colleagues (*Proceedings of the National Academy of Sciences*, 1965) first prepared A^+B^-/A^-B^+ terminal hets and proved their structure. Among other things they found that such phage particles could carry out rII function in a lysogenic host cell, but yielded progeny that were exclusively of the original mutant types. The prompt segregation identified the particles as heterozygotes rather than recombinants, and the functional complementation was consistent with the expected structure of terminal hets, but not with that of internal hets. (Internal hets would contain part of their genetic information in the unreadable DNA strand, information that could be expressed only after replication.)

717

At the time of Stahl's work, interest in T4 genetics was focused on terminal repetition and circular permutation of nucleotide sequences, and much effort was devoted to the search for mutants lacking one or both of these features. No such mutants were found, but the purposes of the search are being accomplished in other ways.

Mosig (*Year Book 64*, p. 521) described exceptional particles of phage T4 that contain single DNA fragments of only two-thirds the normal molecular length. Such DNA fragments represent a random, continuous segment of the genome. They lack terminal repetitions but also lack about one-third of the normal set of genes. The doubly defective particles are of course noninfective. Mosig (now at Vanderbilt University) and Werner (now at the University of Miami School of Medicine) sought to determine what sort of lesion the absence of a terminal repetition entails.

Mosig and Werner (*Proceedings of the National Academy of Sciences*, 1969) obtained physical evidence that many of the DNA fragments start to reproduce, but on the average duplicate only half their length. Mosig and her students subsequently showed that replication starts at a point between genes 41 and 42, proceeds in the direction of gene 43, and presumably stops when it reaches the end of the fragment. Terminal repetitions in DNA molecules therefore represent "joining sequences" that are essential for ring formation, on which replication ultimately depends (*Year Book 67*, p. 562). Botstein has more direct evidence that in phage P22 replication depends on ring closure and that ring closure is brought about by intramolecular crossing over between repetitious molecular ends. In view of these facts, mutants of T4 lacking repetitions could hardly exist. Independently of replication mechanisms, the only model we have for packaging of DNA into T4 phage particles leads to the same inference. It too requires terminal repetition and generates

circular permutation (Streisinger, Emrich, and M. M. Stahl, *Proceedings of the National Academy of Sciences*, 1967). Nevertheless, the search for phage lines with a nonpermuted gene sequence has proved fruitful in unexpected ways.

Weil, Terzaghi, and Crasemann (*Genetics*, 1965) reported that rII crosses of the type $A^+B^- \times A^-B^+$ yielded about one phage particle in 10 million of a special type that could form a plaque on plates seeded with the lysogenic bacterial host. The resulting clones contained physiologically A^+B^+ phage, but proved to segregate with relatively high frequency particles of the ancestral genotypes A^+B^- and A^-B^+. Such particles could be recognized as heterozygotes of the genotype A^+B^-/A^-B^+, and Weil *et al.* concluded that they were stabilized terminal hets, or else carried duplications of another sort.

Duplications of another sort have in fact been revealed by the work of Weil and Terzaghi and of Parma and Ingraham. The experimental results can be appreciated best if the applicable genetic principles are first brought to mind.

The local event that gives rise to recombinant chromosomes is called crossing over. Typically, crossing over is observed between homologous partners and produces no persistent local lesions: parental and recombinant chromosomes differ only with respect to content of genetic markers. The overall process therefore consists, in effect, of interchange of exactly corresponding segments of homologous chromatids. Since crossing over can occur at practically any chromosomal site, one is forced to think of a matching principle based on complementary nucleotide sequences.

In T4, a single crossover involves pairing restricted to the appropriate short segment of DNA (Drake, *Proceedings of the National Academy of Sciences*, 1967). It is therefore plausible to imagine occasional local mispairing accompanied by "illegitimate" crossing over. The hypothetical products of single illegitimate crossovers between homologous chromatids are deletions and tandem duplications, as illustrated in Fig. 1. Note that the tandem duplication (II) is not a cyclic permutation or a biological equivalent of the corresponding terminal repetition (IV), because IV does not contain the sequence ···mk··· and cannot generate it except by something equivalent to illegitimate crossing over. Thus a tandem duplication can be exactly defined as one product of a single illegitimate crossover. In terms of DNA structure it is characterized by a single repeat in a nucleotide sequence containing just one joint of novel origin.

Illegitimate crossing over, like typical crossing over, is known by its products. Neither process is understood in mechanical or chemical terms.

Illegitimate crossing over is unequal in the sense that the reciprocal products are not homologous with respect to each other or their parents. Nevertheless, in chromosomes containing tandem duplications the ancestral genome is not lost and is, in fact, spontaneously regenerated with rather high frequency. The reasons for this were made clear by Sturtevant

in experiments with the Bar eye mutant of *Drosophila*. His experiments also provided the criteria for recognizing tandem duplications.

Figure 2 shows the results of unequal crossing over between homologous chromosomes containing a duplication. Here the parental type represents Bar and the products are normal eye and "double Bar." The crossover is "legitimate" in that it involves local homology just as ordinary crossing over does. The main point is that the reduced product of unequal crossing over will be homologous with the ancestral type only if the original duplications are tandem. The genetic results obtained by Sturtevant were further clarified by Bridges, who ascertained by examination of salivary gland chromosomes that the Bar mutant carries a tandem duplication of a set of bands that occurs once in the normal segregant and three times in the double Bar segregant. The interpretation in terms of a duplication, originating by illegitimate crossing over and erased by unequal crossing over according to Figs. 1 and 2, was thus verified in the most satisfactory manner possible.

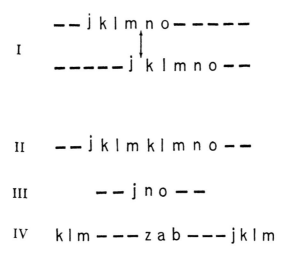

Fig. 1. Illegitimate crossing over. Nucleotide sequences corresponding to parts of the genome are represented by alphabetical sequences in which each letter stands for many nucleotides. In I, two molecules are mispaired. An illegitimate crossover at the double arrow generates a tandem duplication (II) and the corresponding deletion (III). Structure IV is the terminal repetition analogous to II.

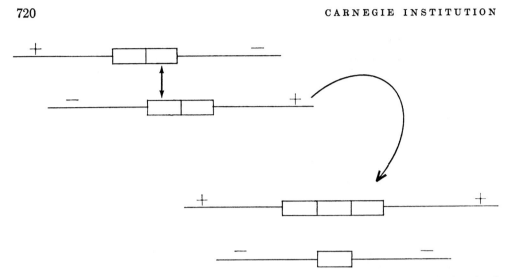

Fig. 2. Unequal crossing over between DNA molecules containing a duplication. Each unit of the repeated sequence is indicated by a rectangle. The plus and minus signs signify genetic markers bracketing the repeats.

The foregoing line of thought carries over from *Drosophila* to T4 with just one complication. In genetic experiments with T4, recombinant DNA molecules are detected only if they can be put into phage particles, a requirement that prohibits lengthening. Thus a DNA molecule containing a duplicated DNA segment must usually suffer deletion of another, dispensable, DNA segment if it is to be recovered in viable form. Matvienko (*Molekulyarnaya Biologiya*, 1969) verified by density analysis that persistent hets of one line contained only the normal length of DNA per particle.

According to the ideas outlined above, Weil's persistent hets might be expected to contain tandem duplications of a DNA segment including the rII gene, and to have suffered deletions of nonessential DNA elsewhere in the chromosome. This is the hypothesis that Parma and Ingraham have investigated. Their conclusions are summarized below.

1. Persistent hets produce three types of progeny besides particles like themselves: A^+B^- segregants, A^-B^+ segregants, and inviable particles. The segregants comprise 15% to 45% of the viable particles, their proportion being char-

acteristic for each line of hets, depending, most likely, on the length of the particular duplication. The number of inviable particles is about equal to the number of segregants. The inviable particles therefore can be interpreted as triplications generating a gene set too large to go into a phage particle without loss. Then the segregants and triplications are just reciprocal products of unequal crossing over according to Fig. 2.

2. The r segregants from persistent hets marked by revertible rII mutants are themselves revertible to r^+, showing that the revertants are haploid rather than homozygous diploid in structure (Weil *et al.*, 1965). The same inference follows from the fact that crosses between pairs of segregants do not regenerate persistent hets at very high frequency (*cf.* item 4 below). These facts support the interpretation in terms of tandem duplications as opposed to duplications of other kinds.

3. Crosses of the type $A^+B^- \times A^-B^+$, in which the phages are segregants from persistent hets, generate terminal (nonpersistent) hets at higher than normal frequency, the actual frequency being characteristic of the particular line of

persistent hets from which the segregants were derived. This Parma and Ingraham interpret as evidence that the duplications are often accompanied by deletions, the net effect being to lengthen terminal repetitions. Weil and Terzaghi have evidence for deletions of known genes in some of their persistent hets.

4. In crosses of the type last mentioned, segregants from persistent hets also regenerate new persistent hets with appreciably higher frequency than do the corresponding ancestral lines. This too may be an indication that the segregants contain deletions, since lengthened terminal repetitions might be expected to favor viability of newly formed duplications.

5. Parma and Ingraham prepared their persistent hets from stocks containing multiple genetic markers. The distribution of these markers in the hets clearly shows that the duplications arise by a single crossover in the vicinity of the rII gene, and that the resulting duplication is usually not more than a dozen or so recombination units in length.

6. The markers recovered in the persistent hets, as well as other genetic tests, show that the duplications usually have the genetic structure $\cdots B^+A^-B^-A^+\cdots$ rather than $\cdots B^-A^+B^+A^-\cdots$. This result is expected if short duplications are more likely to survive than long ones. Persistent hets can be recognized only if they contain intact A^+ and B^+ segments. In the first structure this requirement can be met without duplicating the entire rII gene, whereas in the second it cannot.

These results are evidently consistent with the hypothesis that persistent hets contain tandem duplications accompanied by compensating deletions. The alternative hypotheses that seem to be excluded are: stabilized terminal repetitions, inverted duplications, and nontandem duplications in general. None of the alternatives can account for the fact that persistent hets sport haploid segregants and inviable particles at similar high frequencies.

The origin of the tandem duplications seems to be accounted for satisfactorily by the hypothesis of illegitimate crossing over. It is not so easy to account for the accompanying deletions. If the two aberrations arise independently, crosses between appropriately marked stocks already carrying deletions should generate persistent hets with much higher frequency than they do. The alternative, that generation of a repeat during phage growth favors a concomitant deletion in the same chromosome, is not an intelligible hypothesis at present.

Apart from questions about detailed genetic mechanisms, what do these findings signify? The mere fact that species exist tells us that genetic replication and genetic recombination are conservative processes. The mere fact that species evolve should tell us that both processes are likely to exhibit limited fidelity. Replication errors (mutations) are better known than recombination errors (illegitimate crossing over), but apparently only because errors of the first kind are easier to detect. Heretofore, tandem duplications were known chiefly from Sturtevant's analysis of Bar in *Drosophila*. The work of Weil and Parma and their colleagues translates Sturtevant's findings into molecular terms and goes further in showing that illegitimate crossing over occurs with surprisingly high frequency. No numerical frequency can be given except to say that duplications of a specific region in T4 DNA reveal themselves in one out of 10 million progeny of the appropriate cross in spite of the fact that the great majority of similar aberrations must go undetected unless accompanied by a viable deletion. In short, gross aberrations, both additions and subtractions, are now part of the molecular chemistry made visible by genetic analysis.

BIBLIOGRAPHY

Hershey, A. D., Genes and hereditary characteristics, *Nature, 226,* 697–700, 1970.

Hershey, A. D., Idiosyncrasies of DNA structure, Nobel Lecture published in *Science, 168,* 1425–1427, 1970.

Yamagishi, Hideo, Nucleotide distribution in the DNA of *Escherichia coli, J. Mol. Biol., 49,* 603–608, 1970.

COMPARATIVE MOLECULAR STRUCTURE AMONG RELATED PHAGE DNA'S

A. D. Hershey

Much of the work at the Genetics Research Unit in recent years has been devoted to the structure of DNA of phage λ. The main facts brought to light may be summarized as follows. The DNA extracted from each phage particle is a linear double helix of molecular weight 31 million (Hershey and Burgi, 1963). The molecules are uniform and terminate in cohesive sites that permit formation of rings and chains (Hershey, Burgi, and Ingraham, 1963). Strack and Kaiser (1965) and Wu and Kaiser (1967) showed that the cohesive ends are short, complementary single strands. The right cohesive end consists of the 12-nucleotide sequence reading, from left to right, CCCGCCGCTGGA (Wu and Taylor, 1971). The left cohesive end is the antiparallel complement; that is, when the ends of a molecule pair, the result is a simple double-helical ring without single-strand gaps. The molecules consist of several segments differing in nucleotide composition (Skalka, Burgi, and Hershey, 1968). Phage races able to form hybrid recombinants with λ have DNA's that show both resemblances to and differences from λ DNA in nucleotide distribution (Skalka, 1971). All possess cohesive ends of the λ type (Yamagishi *et al.*, 1965; Baldwin *et al.*, 1966). Such phages form a natural group that may be called the λ family.

Phages belonging to the λ family can multiply in bacterial cells either as viruses, or as prophages occupying a characteristic site in the bacterial chromosome. The viral mode of growth requires autonomous replication of phage DNA, packaging of the DNA into phage particles, and lysis of the cells to release the particles. The prophage mode of growth requires insertion of phage DNA into the bacterial chromosome and con-tinued cell division, in the course of which the prophage is replicated passively. The two modes of replication are mutually exclusive. Both depend on ring formation. The nature of the dependence is particularly clear in the transition from phage to prophage. The transition is brought about by a single reciprocal recombination between phage DNA in its ring form and the bacterial chromosome. The recombination occurs at specific sites in both DNA's and thus has two effects: it generates a particular cyclic permutation of gene order in the phage DNA, and inserts the prophage at a characteristic site in the bacterial chromosome. Figure 1 shows the linear molecular map of λ DNA. To form the prophage, the molecular ends are joined, and the DNA is cut at the point labeled "prophage ends" in the figure.

Besides possession of like cohesive ends and ability to form hybrid recombinants, phages of the λ family show other evidences that their DNA molecules are functionally homologous. Nevertheless, their genes are not individually interchangeable. Two examples will illustrate this relationship.

First, phages λ, φ80, and 21, whose prophages occupy different sites in the bacterial chromosome, differ from each other in two other respects: DNA structure in the vicinity of the prophage ends, and the structure of a gene (*int*) whose product catalyzes prophage insertion. (The gene *int* lies just to the right of the prophage ends.) On the other hand, phages λ, 434, and 82 occupy the same bacterial site and are alike in the other respects mentioned.

Second, phages belonging to the λ group owe their stability as prophages to a protein, called repressor, that prevents independent viral growth. In λ, the re-

3

pressor gene and the sites ("operators") at which the repressor acts lie in the longer "regulation" segment of the map of Fig. 1. Other members of the λ family also make repressors, but their repressors often differ in specificity of action from that of λ. Phages whose repressors differ must differ at least in their repressor genes and the corresponding operators. All the phages listed in Fig. 1 differ in the DNA segment containing these elements (and called the immunity region). Distinct phage races with identical immunity regions can be found too, but they haven't been studied much (Parkinson, 1971).

Davidson and Szybalski and their colleagues (citations in the legend of Fig. 1)

have developed methods that enable them to map genes with new precision and at the same time delineate nucleotide-sequence homology among related phage DNA's. Because of its general interest, some of their work will be summarized here. More complete information can be found in *The Bacteriophage Lambda*, published by the Cold Spring Harbor Laboratory in 1971.

In the procedure mentioned, the experimenter forms "heteroduplex" DNA structures by allowing DNA strands of one kind to pair with potentially complementary strands of a second kind, and examines the product in an electron microscope. Under suitable conditions, heteroduplexes show double-helical seg-

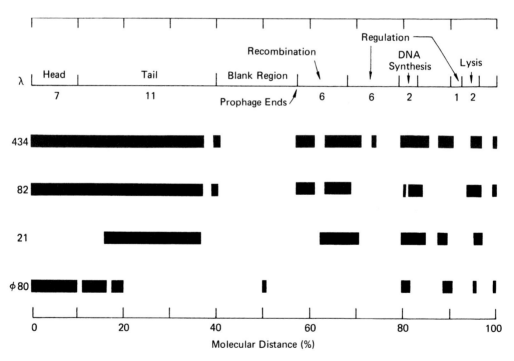

Fig. 1. Homology between λ DNA and the DNA's of several related phages. The upper map shows the distribution of genes in λ DNA molecules. "Head" genes are involved in packaging DNA into phage particles. "Tail" genes are necessary to make the particles infective. At least three "blank" regions, containing no genes with essential functions, are visible. One of them is labeled because it is known to be dispensable in deletion mutants. The "prophage ends" are explained in the text. Numbers of genes in each segment are indicated below the line. The shaded rectangles in the four lower lines delineate segments in λ DNA that match with respect to nucleotide sequence corresponding segments in DNA's of the four phages identified at the left. (Data from papers by Davidson and Szybalski; Simon, Davis, and Davidson; and Fiandt, Hradecna, Lozeron, and Szybalski; in *The Bacteriophage Lambda*.)

ments where the two DNA strands are complementary in nucleotide sequence, and single-stranded loops where they are not. On a statistical basis, measurements of the lengths and positions of complementary segments prove remarkably accurate.

The scale of the genetic map shown in Fig. 1 was worked out in this way chiefly by use of heteroduplexes involving deletion mutants. The other results shown in the figure refer to heteroduplexes formed from pairs of different bacteriophage races.

The figure shows that the five races chosen for study possess few DNA segments common to the group (except for the cohesive ends, which require other methods of detection). On the other hand, all pairs of DNA's show common segments. Moreover (though this is not evident in the figure) the common segments lie at corresponding positions in both members of the pair. Only regions of perfect duplex formation are shown in the figure. In addition, a few segments, notably the left end of DNA from phage 21, form imperfect helices with λ DNA, presumably owing to imperfectly matched nucleotide sequences. Finally, segments of two phages that differ from λ may resemble each other. For example, phages 434 and 82 are identical in their

"blank" regions as well as in their left halves.

A number of correlations between nucleotide sequence and function can be made. For example, all the DNA's mentioned differ from λ (and presumably from one another) in their immunity regions, as expected since all the phages have different repressors. Also, DNA's of phages 434 and 82, but not 21 or φ80, resemble λ in their prophage ends. Here the functional correlate is site of insertion of prophage: identical for λ, 434, and 82; unique for 21 and φ80.

One conclusion seems clear: that the several races were derived from still other races by reassortment of unlike segments. What is not so clear is the evolutionary mechanism by which DNA molecules are permitted to diversify locally while preserving their overall functional homology. Presumably both the diversity and the capacity to form hybrid recombinants have served a joint evolutionary purpose. DNA segments brought in from outside the λ family by illegitimate recombination may have played a dominant role in the family history (N. Franklin, see book cited). That hypothesis may be needed anyway to explain the differing average nucleotide composition of the various DNA segments (Skalka, Burgi, and Hershey, 1968).

In His Own Words

Words

Int, etc. I guess everybody knows I am trying to suppress these symbols in our book, except when used in lower case italics as names of genes. I won't repeat arguments here. I suggest that mutineers form a committee of three to present counterarguments. (I admit that there are special cases, like Fec^+.)

map (verb). Genes don't, geneticists do.

map (name). I am plumping for two kinds only: prophage and molecular. Further qualifications depend on methods, for instance, a map of recombination frequencies. The term genetic map no longer says anything about methods. It gives the positions of genes. The term physical map means nothing at all: all maps are physical.

molecule. There are only two kinds of λ DNA molecule, rods and rings, both with the molecular weight of 31×10^6. They are never seen replicating, sedimenting fast, immature, or broken. Other things are DNA, fragments, or structures. Hence "molecular map."

lysogenic. Means "generating lysis," practically the same as "lytic." A temperate phage is not lysogenic, it is lysogengenic. I think the first misuse of this word occurred on page 205 of "Molecular Biology of the Gene," 1965 edition. "Lysogenic" has since become quite virulent.

dilysogenic?

lysogenic excision?

dilysogenic excision??

These words have a fine ring until you ask yourself what they mean.

system. I began to feel uneasy about this word several years ago. Now I know why. It is pretentious, meaningless, and overworked. I recently read a paper whose author used "system" to mean a biological function, an experimental set-up, and a fractionation procedure, practically in adjacent paragraphs. Signer seems to be dropping "system" in favor of "pathway," as in "the *rec*-promoted pathway." This is by analogy with "the histidine pathway." If you don't like the analogy you might prefer "mechanism" or "process." Anyway, let's preserve "system" to go with "Copernican."

λ-*related.* Unpleasant and ambiguous. It could mean "a transient derived from λ" or "a species belonging to the λ family."

genome. To my mind this word signifies the complete set of genetic determinants or nucleotide sequences that defines a species. This is an informational concept, not a name for a thing. The word should not be used (Paris notwithstanding) as a synonym for DNA or chromosome. Chromosomes multiply by replication, genomes multiply by speciation. Maybe this is not just my prejudice. Who wants to say "a solution containing 10 μg/ml of λ genomes?"

repressor. Echols and Ptashne debated usage here. They agreed that the terms "immunity repressor" and "*cI* repressor" should be given up because confusing. The term "λ repressor" or just "repressor" can stand in many contexts. When ambiguity threatens, Echols favors "*cI* protein." Franklin surmounted her problem by something like "dual repression of *trp* function by tryptophan and by *cI* product." The old term "immunity substance" may still be serviceable too.

Clear but nasty: replication inhibition, lambdoid, transcription initiation, and too many others to list.

Nasty but interesting: "heteroduplex mapping." Here heteroduplex could be an adjectival noun, which is merely bad. It could also be an adjective and drive you crazy.

Unclear, nasty, and dull: "immediate early."

I once observed to Chargaff that scientists don't have time to read each other's papers anymore. Speaking as an editor he said, "They don't have time to read their own papers."

I have lots of time.

Hershey

To all authors of general chapters
November 20, 1970

Funeral Sentences

*A*s its title suggests, my speech has to do with death. It was planned before death came to George Streisinger. After some debate with myself, I decided to proceed as if he were here to listen.

I used to expect that as I grew older, my mind would turn to eternal questions. That didn't happen. I still find myself pushed by the same compulsions that pre-empt the faculties of the young. Eventually it dawned on me that there is no rational escape from the business of the day short of lunacy. Here Isaac Newton is our object lesson. The difficulty, of course, lies in the nature of rational thought itself. Not only is wisdom beyond our reach, even the concept of wisdom has to be given up.

Peter Medawar called science the art of the soluble. It isn't hard to see that his phrase prescribes the limits to all common sense. His phrase also neatly explains how science differs from common sense, namely in the proportion of art that goes into it. Let's admit that a mind attuned to common sense, no matter how artful, cannot break out of the confines of the soluble.

The scientist's professional way of getting around this difficulty is to say that the insoluble problem doesn't exist. In real life this won't do. The word crisis in its many uses refers precisely to the dilemma of the insoluble problem, that is, to occasions where wisdom fails.

Growing old and facing death are two of the universal human crises, for which rational thought is by definition useless. Nevertheless they can dominate the imagination, as expressed in all arts and religions. Art and religion can therefore be viewed as contrivances serving to override the limitations of rational thought.

There is, then, an art of the insoluble; several arts, which are at least as venerable as science. The two I shall talk about are poetry and music. My thesis is that poetry, and especially music, derive in part from the human capacity to experience crises in sensual terms; even, in a not too radical deviation, to confuse pleasure and pain. Where can we explore this thought if not at the scene of death?

Let me redefine somewhat arbitrarily my subject matter. Poetry in the sense needed here is pure allusion and suggestion. It expresses the inexpressible or nothing. It is an art of the insoluble. Music, meaning programless music, is quintessential poetry. All other language is prose, metered or not. Thus most of Shakespeare becomes prose, either narrative or philosophical cliché, made cliché by history if not by the author. On the other hand the story of Judas Maccabaeus or of Israel in Egypt is poetry, made poetry by history if not by their authors. Bible stories are poetry of a unique kind because of their ancestral relation to western culture. In short, I require that poetry and music engage identical organs once they get past eyes and ears. There they evoke what cannot be conveyed.

Each turning point in human affairs has its appropriate music, by no means always composed for the occasion. Because of the abstract character of music, its sentiment must be established by custom, or be left to the intuition of the listener. Thus the theme of mortality can be found in Bach's Goldberg Variations, in Beethoven's quartets, or even in Mozart's C-minor Serenade for Wind Instruments, as well as in masses for the dead. But all major composers wrote music expressly commemorating death, often music representing their finest work. Even Haydn, whose music can be obnoxiously cheerful, so admired his own "Seven Last Words of Christ" that he published it in three arrangements. He left no requiem mass. For his memorial service in Vienna, music was borrowed from Mozart.

The funeral music of Henry Purcell is a particularly concise and eloquent example of the genre. His "Funeral Sentences" are settings of three biblical texts quoted in the graveside service of the Book of Common Prayer. At some risk, the poetry can be taken as a proxy for the music. The first sentence reads: Man that is born of a woman has but a short time to live and is full of misery; the second reads: In the midst of life we are in death; the third: Thou knowest, Lord, the secrets of our hearts. The music was composed as three separate anthems before 1683, when Purcell was not yet 24, and it is unlikely that they were then meant for funerals. But Purcell used them with minor changes in his music for the funeral of the second Queen Mary in 1694. The following year the same music was played at his own funeral. In some recordings, the anthems are embellished with fanfares of remarkable solemnity and strength, equally suitable for a queen and the chiefest composer of her realm.[1]

How are we to account for the appeal of music of this kind? Remember that funeral music is not intended for the dying, but for survivors. In fact, it is mostly heard in concert halls or from recordings, not at funerals. It exploits the sensuality of the idea of death, of the imagined experience of death. That sensuality is there can scarcely be denied. How else explain the universal proneness to thoughts of voluntary death? In biological terms, death is a presence that had to be domesticated. The precise biological means may be impossible to pin down, but it must go back to the first human tribe or beyond. Surely it predates anything as sophisticated as a death wish.

At any rate, funeral music is rich in sensuality and is perhaps unmatched in this respect by any other form of art. Listen, for example, to Bach's 47th aria in the St. Matthew Passion: *erbarme dich*, have mercy, *mein Gott*. Or consider a typical funeral verse which might run: Grieve not, but envy the departed. These constructions are to be savored as music and poetry. They do not call for belief.

Thoughts of death do not characteristically belong to old age. On the contrary, preoccupation with death seems to be especially prevalent in the young. I remember a popular song of the 1930s called "Gloomy Sunday" that touched off a world-wide

---

[1] The fanfares mentioned belong with the funeral anthems only as a plausible example of music Purcell may have supplied for the Queen's funeral procession. They are derived in part from music for Shadwell's play "The Libertine" (about 1692). Possibly also from a version for slide trumpets of "Thou Knowest, Lord..." (Z 58C, 1695), music that I haven't heard.

spate of suicides among young people. Recall too that Purcell went back to his early anthems when he needed music for the Queen's funeral.

Cherubini, a French composer born in Italy, wrote in 1816 a requiem for the anniversary of the death of Louis XVI. The music was extravagantly praised by Beethoven, Schumann, Brahms, and even Berlioz, who otherwise disliked Cherubini's music. After another performance in 1834, the mass was banned by the Bishop of Paris because of its use of women's voices. So Cherubini composed at the age of 76 a second requiem to be performed at his own funeral. The first has been described (by Basil Green in *The New Grove Dictionary*) as "universal, grandiose, conciliatory"; the second "individual, tense, pessimistic".

So much for anecdote. The notable fact is that the contemplation of death, and the poetry and music of death, pervade human lives. Truly in the midst of life we are in death, and we have learned to celebrate both.

Why do solemn occasions demand poetry or music? In part because prose, which serves common needs well enough, turns banal when introverted. Joyce taught us that if the lesson were needed, poetry, and especially music, succeed by skirting intelligence. It seems that what we know best cannot be made explicit.

One ingredient of artistic sensibility can be put into words. Men and women have a short time to live, but their works live after them. This familiar principle accounts in human terms for our strongest drives: to procreate, to pursue tangible ends, to leave a notice that we were here. The same principle enters into what we listen to in poetry and music. The artistic experience consists, in part, in vicarious participation in lasting works. (This was made by a creature like me!) Such is the only rational basis for the capacity of abstract art to speak to us.

Science too is abstract to the non-specialist. For him, the content of science must be apprehended as a kind of poetry. In the long run, that is the only way in which the spirit of science, as opposed to its works, can gain public recognition.

Let me close these remarks by introducing some music that will probably be unknown to you. In fact, I consider it my private discovery. It is Edward Elgar's "Dream of Gerontius," an oratorio based on the poem by John Henry (Cardinal) Newman. It represents the struggle and reconciliation to his fate of a dying man (or his dream?) Both subject and manner depart radically from the traditional oratorio made familiar by Bach and Handel. The text is a parody of Catholic doctrine, as befits the mind of an ordinary man of the 19th century. In my view, it escaped banality mainly by touching Elgar's imagination. The music is intense and subjective, so much so as to be painful on first hearing. But its novelty and strength outlast the discomfort.

The first performance of Gerontius in 1900 was a disaster, and the work was not often performed thereafter. It has been recorded only four times to my knowledge, and the best of these, under Barbirolli, is long out of print. But you can hear it. To satisfy my missionary spirit I will send a copy on tape to anyone seriously interested.

Name Index

Subject Index